ENVIRONMENTAL HEALTH

DADE W. MOELLER

ENVIRONMENTAL HEALTH

Fourth Edition

HARVARD UNIVERSITY PRESS
Cambridge, Massachusetts
London, England
2011

Library of Congress Cataloging-in-Publication Data

Moeller, D. W. (Dade W.)
Environmental health / Dade W. Moeller. — 4th ed.
  p. cm.
Includes bibliographical references and index.
ISBN 978-0-674-04740-2 (alk. paper)
1. Environmental health.   I. Title.
RA565.M64 2011
362.1—dc22   2010047294

To Betty Jean, who for more than fifty years—until her death in 1998—was, and continues to be, the joy of my life, and to Rad, Mark, Kehne, Matt, and Anne, their spouses, and our sixteen grandchildren, who have never ceased to make us proud

CONTENTS

PREFACE TO THE FOURTH EDITION

In preparing this new edition, the goal has been to review the complex nature of the environment, how humans interact with it, and the accompanying impacts of one on the other. Nonetheless, it is important to recognize that this will require describing a complex of unlimited numbers of interacting systems, and they must be discussed independently if the task is to be presented in an understandable manner. In seeking to accomplish this goal, the discussions that follow have been divided into 18 segments/chapters. Whatever your religious beliefs may be, you will realize that the interactions of humans and the environment involve a host of interacting forces that only a power, far superior to ours, could have developed and made to work as amazingly as they do.

The goals in preparing this fourth edition were to incorporate new developments in the field and to increase the depth of coverage of the more important subject areas. This involved reviewing almost 1,000 scientific papers and incorporating their salient features into the appropriate chapters. Although, in most cases, the key new information and associated conclusions were straightforward (e.g., new scientific findings, recommendations of the advisory bodies, and regulations of federal agencies), others, such as the status of and goals for addressing climate change, were not. In fact, as good as science has become, there are still controversies in even the most important challenges that are being faced. In the long run, the ensuing discussions will lead to an improved final product.

As part of this effort, this edition has been enriched with new tables and figures, supported by the updating of those that have been retained. As in the past, a concerted effort has been made to write a book that provides comprehensive coverage of the field and its challenges, and to do so in terms of both their local and global implications, their short- and long-range impacts, their importance to people in both the developing and the developed nations, and the roles that individuals can play in helping to resolve them. In all cases, every effort has been made to ensure that the information being presented is based on "good science," and that the ensuing discussion offers a balanced assessment of current conditions. These include topics such as the protection of endangered species; application of the so-called Precautionary Principle; uncertainties in extrapolating data obtained from laboratory studies of toxic materials in small animals for estimating their potential health effects in humans; the effects on humans of the inhalation of extremely small airborne particulates; how best to address the problem of drunk drivers as a major contributor to deaths in vehicular accidents; the proper approach for estimating the risks of toxic materials that have a documented threshold for health effects; how to address our insatiable demand for energy, especially petroleum; and the ongoing threat of terrorism.

Care has also been exercised to ensure that readers understand the limitations associated with techniques, such as epidemiology and risk assessment, that are routinely applied in evaluating the impacts of various environmental stresses. At the same time, the potential for advancing the evidence that can be derived and the conclusions that can be reached through applications of newer techniques, such as molecular toxicology and epidemiology, is clearly enunciated. The effort to ensure that readers understand the differences among clinical medicine, public health, and environmental health, introduced in the third edition, has been continued. Explaining these relationships is important because of the widespread use of the book for teaching environmental health to Master of Public Health (MPH) students, whose primary education is largely in the field of medicine.

As in the past, the goal has been to provide in-depth coverage of each topic, based not only on a review of the latest publications in the field but also on the knowledge and experience of people who are deeply involved in the "real world" aspects of each subject. Another goal has been to emphasize the importance of applying a "systems approach" in assessing the potential impacts of new projects prior to their construction and operation. Far too often, problems arise because planners overlooked a key impact of a project that surfaced, often as long as a decade later. This emphasizes that it is not only important to manage and control various pollutants within individual components of the environment (air, water, soil, and food), but that it is also imperative to account for their interrelationships. These include the necessity that both human and natural resources be protected. This is exemplified by the establishment of primary and secondary standards for airborne and waterborne contaminants—primary to protect the health of humans; secondary, to protect the environment.

During my 60 years of professional work in the field of environmental health, I have had the privilege of working in essentially all facets of the subject. Nonetheless, readers will note that I frequently use examples related to ionizing radiation in discussing many topics. This is in recognition that the leaders in this field, both nationally and internationally, have continually led in the application of innovative new approaches and in solving environmentally related problems. This began with the establishment of the International Commission on Radiological Protection (ICRP) as a link to the International Congress of Radiology. Within a few years, its field of work was widened from protection in medical radiology to all aspects of radiation protection. During subsequent years, the ICRP has provided a forum through which leading scientists can participate on a global basis in the development of radiation protection recommendations. By maintaining an impeccable reputation for scientific integrity, the outcome of these efforts has provided recommendations that are globally accepted and applied. This has been exemplified in many ways, one example being that they have never hesitated to change long-established concepts when the scientific evidence warrants. Another was their pioneering efforts in basing their dose-limit recommendations on the total detriment to health, rather than on that of cancer alone.

As anyone who has written a book of this magnitude recognizes, its preparation requires the unselfish support of multitudes of people willing to share their talents and expertise, both by providing information on specific topics and by reviewing what has been written. In this case, these include former colleagues at the Harvard School of Public Health, a host of coworkers at Dade Moeller & Associates, Inc., and many others who possess expertise in specialized areas of environmental and public health. The more prominent of these, especially those who read and provided detailed critiques of one or more chapters, include Ray Berube, Steve Bump, Bill Craig, Joseph Delfino, John Fomous, Mickey Hunacek, Bill Kennedy, Judson Kenoyer, Tracy Ikenberry, Ed Maher, Steve Merwin, Matthew Moeller, Richard Monson, Gene Rollins, and Steve Sohinki. Offering outstanding support throughout this effort was Casper Sun, a longtime colleague.

Others who played lesser but still very important roles were Joseph Brain, Theodore Daniell, Douglas Dockery, Eric Krouse, John Little, Carrie Moeller, and Walter Willett. To each of these, I extend my gratitude and heartfelt appreciation.

A special note of appreciation is also extended to Robin Klein, artist, Dade Moeller & Associates, Las Vegas, Nevada, who worked tirelessly, including nights and weekends, to ensure that new figures were prepared and old ones updated so that the deadline for submission of the manuscript to the publisher could be met. In a similar manner, a large measure of my success was due to the onsite support provided by David McKeon in helping me manipulate the figures and tables and counter the all-too-frequent contrary behavior of my computer. Finally, a special expression of gratitude is due to Edward Wade for his editorial suggestions, and to Michael Fisher and Anne Zarrella of Harvard University Press, who provided guidance throughout the planning, preparation, and completion of this new edition.

And God pronounced a blessing upon Noah and his sons and said to them, be fruitful and multiply and fill the Earth. And the fear of you and dread and terror of you shall be upon every beast of the land, every bird of the air, all that creeps upon the ground, and upon all the fishes of the sea. Into your hands they are delivered.

Genesis 9: 1–2

Hurt not the Earth, neither the sea, nor the trees . . .

Revelation 7:3

ABBREVIATIONS

ACGIH	American Conference of Governmental Industrial Hygienists
ACRS	Advisory Committee on Reactor Safeguards
ACSH	American Council on Science and Health
AEA	Atomic Energy Act
AEC	Atomic Energy Commission
AIDS	Acquired immune deficiency syndrome
AIHA	American Industrial Hygiene Association
$Al_2(SO_4)_3$	Alum
ALARA	As low as reasonably achievable
AMA	American Medical Association
APHA	American Public Health Association
AQI	Air-quality index
ASHRAE	American Society for Heating, Refrigeration, and Air-Conditioning Engineers
ASTM	American Society for Testing and Materials
ATSDR	Agency for Toxic Substances and Disease Registry, U.S. Department of Health and Human Services
AWWA	American Water Works Association
BACT	Best available control technology
BART	Best available retrofit technology
BAT	Best available technology
BCCT	Best conventional control technology
BCE	Before the common era
BEIs	Biological exposure indices
BEIR	Committee on the Biological Effects of Ionizing Radiation, National Research Council
BOD	Biochemical oxygen demand
BPA	Bisphenol-A
BSE	*Bovine spongiform encephalopathy* (mad-cow disease)

Bt	*Bacillus thuringiensis*
Bti	*Bacillus thuringiensis israeliensis*
Btk	*Bacillus thuringiensis kurstaki*
BWR	Boiling-water reactor
C & C	Command and control
CAFE	Corporate average fuel economy
CAFO	Confined Animal Feeding Operations
CAT	Computed assisted tomography
CCA	Chromated copper arsenate
CDC	Centers for Disease Control and Prevention, U.S. Department of Health and Human Services
CEQ	Council on Environmental Quality, Executive Office of the President
CERCLA	Comprehensive Environmental Response, Compensation, and Liability Act (Superfund Act)
CFC	Chlorofluorocarbon
CFR	Code of Federal Regulations
CIIT	Chemical Industry Institute of Toxicology
CO	Carbon Monoxide
CO_2	Carbon Dioxide
COD	Chemical Oxygen Demand
CRS	Congressional Research Service
CRT	Cathode-ray tube
DDD	1,1-dichloro-2,2-bis(p-chlorophenyl)ethane
DDE	1,1-dichloro-2,2-bis(p-chlorophenyl)ethylene
DDT	1,1-trichloro-2,2-bis(p-chlorophenyl)ethane
DEET	Diethyltoluamide
DHS	U.S. Department of Homeland Security
DMC	Dimethyl carbonate
DNA	Deoxyribonucleic acid
DO	Dissolved oxygen
DOE	U.S. Department of Energy
EEOICPA	Energy Employees Occupational Illness Compensation Program Act
EIA	Environmental Impact Assessment
EIS	Environmental Impact Statement
EMAP	Environmental Monitoring and Assessment Program
EPA	Environmental Protection Agency
EPCRA	Emergency Planning and Community Right-to-Know Act
EPRI	Electric Power Research Institute
EU	European Union
FDA	Food and Drug Administration, U.S. Department of Health and Human Services
FIFRA	Federal Insecticide, Fungicide, and Rodenticide Act
GHPs	Geothermal heat pumps
GI tract	Gastrointestinal tract
GM	Genetically modified

GRAS	Generally recognized as safe
HACCP	Hazard Analysis and Critical Control Points
HCs	Hydrocarbons
HHA	Health Hazard Assessment
HHS	U.S. Department of Health and Human Services
HVAC	Heating, ventilating, and air-conditioning
Hz	Hertz (cycles per second)
IAEA	International Atomic Energy Agency
IARC	International Agency for Research on Cancer
ICNIRP	International Commission on Non-Ionizing Radiation Protection
ICRP	International Commission on Radiological Protection
IFT	Institute of Food Protection
IIHS	Insurance Institute for Highway Safety
INPO	Institute of Nuclear Power Operations
IPCC	UN International Panel on Climate Change
ISM	Integrated Safety Management
IVHS	Intelligent vehicle highway systems
JAMA	Journal of the American Medical Association
LD_{50}	Lethal dose for 50 percent of the exposed population
LEDs	Light-emitting diodes
LLRW	Low-level radioactive waste
LLRWPAA	Low-Level Radioactive Waste Policy Amendments Act
LNT	Linear Non-Threshold Hypothesis
MADD	Mothers against Drunk Driving
MARLAP	Multi-Agency Radiological Laboratory Analytical Protocols Manual
MCL	Maximum contaminant level
MPH	Master of Public Health
MSC	Multiple chemical sensitivity
MSW	Municipal solid waste
MTBE	Methyl tertiary-butyl ether
NAAQS	National ambient air-quality standards
NADP	National Atmospheric Deposition Program
NAE	National Academy of Engineering
NAFTA	North American Free Trade Agreement
NASA	National Aeronautics and Space Administration
NCEA	National Center for Environmental Assessment
NCRP	National Council on Radiation Protection and Measurements
NEI	Nuclear Energy Institute
NEPA	National Environmental Policy Act
NEWWA	New England Water Works Association
NHEXAS	National Human Exposure Assessment Survey
NIEHS	National Institute for Environmental Health Sciences. U.S. Department of Health and Human Services
NIOSH	National Institute for Occupational Safety and Health. U.S. Department of Health and Human Services

NIST	National Institute of Standards and Technology, U.S. Department of Commerce
NLVs	Norwalk-like viruses
NO_2	Nitrogen dioxide
NO_x	Nitrogen oxides
NPDES	National Pollution Discharge Elimination System
NPL	National Priorities List
NRC	National Research Council
NSC	National Safety Council
NSR	New Source Review
O_3	Ozone
OECD	Organization for Economic Cooperation and Development
OSHA	Occupational Health and Safety Administration, U.S. Department of Labor
PAHO	Pan American Health Organization
PCBs	Polychlorinated biphenyls
PET/CT	Positron emission tomography/computed tomography
$PM_{2.5}$	Particulate matter, 2.5 micrometers or smaller in size
PM_{10}	Particulate matter, 10 micrometers or smaller in size
PNNL	Pacific Northwest National Laboratory, U.S. Department of Energy
PPA	Pollution Prevention Act of 1990
ppb	Parts per billion
ppm	Parts per million
PRA	Probabilistic risk analysis
PWR	Pressurized-water reactor
RCRA	Resource Conservation and Recovery Act
REMAP	Regional Environmental Monitoring and Assessment Program
RETS-REMP	Radioactive Effluent Technical Specifications—Radiological Environmental Monitoring Programs
RFF	Resources for the Future, Washington, DC
S & H	Safety and Health
SARS	Severe acute respiratory syndrome
SO_2	Sulfur dioxide
TLVs	Threshold limit values
TSCA	Toxic Substances Control Act
UHF	Ultrahigh frequency
UN	United Nations
UNSCEAR	United Nations Scientific Committee on the Effects of Atomic Radiation
USDA	U.S. Department of Agriculture
USNRC	U.S. Nuclear Regulatory Commission
UVR	Ultraviolet radiation
VOCs	Volatile organic compounds
VRE	Vancomycin-resistant enterococci
WANO	World Association of Nuclear Operators

WEF	Water Environment Federation
WHO	World Health Organization
WNV	West Nile virus
w_R	Radiation weighting factor
w_T	Tissue weighting factor

1

THE SCOPE

MANY ASPECTS of human well-being are influenced by the environment, and many diseases can be initiated, promoted, sustained, or stimulated by environmental factors. For this reason, the interactions of people with their environment are an important component of public health. Although this is common knowledge today, it required a long time to be recognized and gain public support. Achieving this status is largely due to two pioneers in this field who not only recognized the need but also took action to have it addressed. Interestingly, one of them was a woman and one was a man. They were:

- Rachel Carson who, with the publication of her now-classic *Silent Spring* (Carson, 1962), brought to the attention of the world that with the surge in the development of artificial pesticides beginning in the 1940s, multiple chemical compounds were being marketed and applied for the control of insects. Although such applications continued for years without much concern about their health implications, her book stimulated a dramatic change. She sounded the alarm to the potential toxicity of pesticides to nontarget species, including many wild animals, as well as humans.

- Gaylord Nelson, U.S. Senator from Wisconsin, whose wisdom and foresight led to the establishment in 1970 of what is now known worldwide as Earth Day and is celebrated annually on April 22nd.

- The activities of the first of these two pioneers were accompanied by the passage of the National Environmental Policy Act of 1969 (U.S. Congress, 1970) and the creation of the Environmental Protection

Agency (EPA), by President Richard M. Nixon, through an executive order.

Defining the Environment

To accomplish their goals effectively, environmental health professionals must keep in mind that there are many components of the environment. Some of the more prominent are described here. Addressing the related problems requires a wide range of professional disciplines.

THE INNER VERSUS OUTER ENVIRONMENT

From the standpoint of the human body, there are two environments: the one within the body and the one outside it. Separating them are three principal protective barriers: the skin, which protects the body from contaminants outside the body; the gastrointestinal (GI) tract, which protects the inner body from contaminants that have been ingested; and the membranes within the lungs, which protect the inner body from contaminants that have been inhaled (Figure 1.1; Table 1.1).

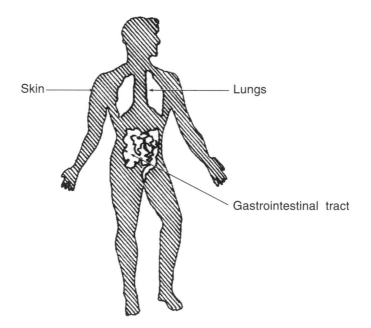

Skin — Lungs

Gastrointestinal tract

Figure 1.1   Barriers between the inner and outer environments

Table 1.1 Characteristics of the principal barriers between the outer and inner body

Barrier	Area		Thickness		Weight		Daily exposure	
	m²	ft²	μm	in	kg	lb	kg	lb
Skin	2	21	100	4×10^{-3}	12–16	30	Variable	
GI tract	200	2,150	10–12	4×10^{-4}	7	15	2–3	4–6
Lungs	140	1,500	0.2–0.4	1×10^{-5}	0.8–0.9	2	24	50

Although they may provide protection, each of these barriers is vulnerable under certain conditions. Contaminants can penetrate to the inner body through the skin by dissolving the layer of wax generated by the sebaceous glands. The GI tract, which has by far the largest surface area of any of the three barriers, is particularly vulnerable to compounds that are soluble and can be readily absorbed and taken into the body cells. Fortunately, the body has mechanisms that can protect the GI tract: unwanted material can be vomited via the mouth or rapidly excreted through the bowels (as in the case of diarrhea). Airborne materials in the respirable size range may be deposited in the lungs and, if they are soluble, may be absorbed. Mechanisms for protecting the lungs range from simple coughing to cleansing by macrophages that engulf and promote the removal of foreign materials. Unless an environmental contaminant penetrates one of the three barriers, it will not gain access to the inner body, and even if a contaminant is successful in gaining access, the body still has mechanisms for controlling and/or removing it. For example, materials entering the circulatory system can be detoxified in the liver or excreted through the kidneys.

Although an average adult ingests about 1.5 kilograms (3.3 pounds) of food and 2 kilograms (2.2 quarts) of water every day, he or she breathes roughly 20 cubic meters (215 cubic feet) of air per day. This amount of air weighs more than 24 kilograms (53 pounds). Because people usually cannot be selective about what air is available, the lungs are the most important pathway for the intake of environmental contaminants into the body. The lungs are also by far the most fragile and susceptible of the three principal barriers.

THE PERSONAL VERSUS AMBIENT ENVIRONMENT

In another definition, people's "personal" environment, the one over which they have control, is contrasted with the working (Chapter 4) or ambient

Table 1.2   Relative importance of various causes of cancer—United States

Risk factor	Estimated percentage of total cancer deaths attributable to this factor
Tobacco	30
Adult diet/obesity	30
Sedentary lifestyle	5
Occupational factors	5
Family history of cancer	5
Viruses/other biologic agents	5
Perinatal factors/growth	5
Reproductive factors	3
Alcohol	3
Socioeconomic status	3
Environmental pollution	2
Ionizing/ultraviolet radiation	2
Prescription drugs/medical procedures	1
Salt/other food additives & contaminants	1

(outdoor) environment, over which they may have essentially no control. Although people commonly think of the working or outdoor environment as posing the higher threat, environmental health experts (as noted in the discussions that follow) estimate that the personal environment, influenced by hygiene, diet, sexual practices, exercise, use of tobacco, drugs, and alcohol, and frequency of medical checkups, often has much more, if not a dominating, influence on human well-being. The estimated contributions of these factors to cancer deaths in the United States are summarized in Table 1.2. As may be noted, the personal environment and the lifestyles followed by individuals account for about 70 percent or more of such deaths. For this reason, the influence of the personal environment on cancer will be discussed in more detail in one of the sections that follow.

THE GASEOUS, LIQUID, AND SOLID ENVIRONMENTS
The environment can also be considered as existing in one of three forms—gaseous, liquid, or solid. Each of these is subject to pollution, and people interact with all of them (Figure 1.2). Particulates and gases are often released into the atmosphere, sewage and liquid wastes are discharged into

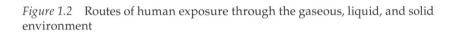

Figure 1.2   Routes of human exposure through the gaseous, liquid, and solid environment

water (Chapter 8), and solid wastes, particularly plastics and toxic chemicals, are disposed on or in the land (Chapter 9).

THE CHEMICAL, BIOLOGICAL, PHYSICAL, AND
SOCIOECONOMIC ENVIRONMENTS

Another perspective considers the environment in terms of the four avenues or mechanisms by which various factors affect people's health. Addressing the related issues requires knowledge of toxicology (Chapter 2), epidemiology (Chapter 3), as well as the management and disposal of liquid and solid waste (Chapters 8 and 9).

1. *Chemical* constituents and contaminants include toxic wastes and pesticides in the general environment, chemicals used in the home and in industrial operations (Chapter 4), and preservatives used in food (Chapter 6).

2. *Biological* contaminants include various disease organisms that may be present in food and water (Chapters 6 and 7), those that can be transmitted by animals and insects, and those that can be transmitted by person-to-person contact (Chapter 10).

3. *Physical* factors that influence health and well-being range from injuries and deaths occurring as a result of accidents, to excessive noise, heat, and cold, to the harmful effects of ionizing and non-ionizing radiation (Chapter 12).

4. *Socioeconomic* factors, though perhaps more difficult to measure and evaluate, significantly affect people's lives and health. Statistics demonstrate compelling relationships between morbidity and mortality and socioeconomic status. People who live in economically depressed neighborhoods are less healthy than those who live in more affluent areas.

Clearly, illness and well-being are the products of community, as well as of chemical, biological, and physical forces. Other factors contributing to the differences range from the unavailability of jobs, inadequate nutrition, and lack of medical care to stressful social conditions, such as substandard housing and accompanying high crime rates. The contributing factors, however, extend far beyond socioeconomics. Studies have shown that people without political power, especially disadvantaged groups who live in lower-income neighborhoods, often bear a disproportionate share of the risks of environmental pollution. One common example is increased air and water pollution due to nearby industrial and toxic waste facilities. Disadvantaged groups also suffer more frequent exposure to lead paint in their homes and to pesticides and industrial chemicals in their work.

None of the preceding definitions of the environment is without its deficiencies and, as noted in the section that follows, the list is by no means complete. Classification in terms of inner and outer environments or the gaseous, liquid, and solid environments, for example, fails to take into account the significant socioeconomic factors cited earlier or physical factors such as noise and ionizing and nonionizing radiation. As a result, consideration of the full range of existing environments is essential for understanding the complexities involved and controlling the associated problems.

THE URBAN ENVIRONMENT

Another environment that is assuming increasing importance is that of our large cities, the so-called urban environment. One of the primary reasons is that whereas an estimated 50 percent of the global population lived in urban centers in 2007, this figure is projected to increase to 60 percent by 2030 (Figure 1.3). Although many view the urban environment as offering a better life, particularly for people living in the developing countries, its quality,

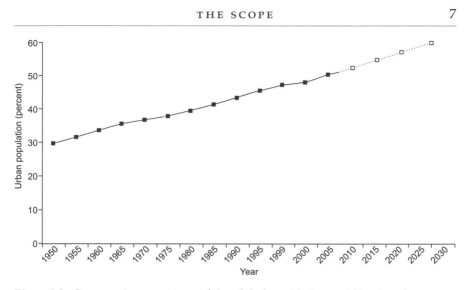

Figure 1.3 Increase in percentage of the global population residing in urban areas, 1950–2030

in truth, has been declining (United Nations, 2009). In fact, many urban environments today are increasingly noisy, congested, frustrating, and unhealthy. Exacerbating the situation, wildlife habitat is scarce, streams flow in artificial channels, wetlands are being filled, and aquifers are being depleted. Adding to these problems is the fact that the heat islands created by urban centers increase both the costs for cooling and the concentrations of air pollutants (DeKay and O'Brien, 2001). Another factor that is often overlooked is that as people move to urban areas, deforestation may be reduced but consumption in other facets (i.e., increased use of the Internet, mobile telephones, motor vehicles, and airplanes) will increase. Only with the application of a systems approach, and the consideration of all facets of environmental sustainability (i.e., pollution, overpopulation, resource depletion, and mass consumption), will the outcome of such an assessment be an accurate reflection of what the outcome (i.e., sustainability) of a proposed action will be (Chad, 2009).

If these problems are to be ameliorated, methods must be found to make urban areas environmentally and socially sustainable. Recognition of this need is not new. Leonardo da Vinci proposed some 500 years ago that pedestrian and vehicular travel within cities be separated by placing them on two different levels (Bugliarello, 2001). This has been accomplished in the city of Boston, where, by moving a major vehicular transportation artery underground, officials have created large open spaces in the heart of the downtown metropolitan area. Another approach is to develop what might be called a hybrid city by making gardens an integral part of the urban area. This has been common practice in the People's Republic of China for centuries, as has the reserving of adjoining areas for agriculture and using city-generated waste as fertilizer. In a similar, but more modest effort, officials in New York City now promote the development of ad hoc urban gardens. Through this step, they have, in a sense, made urban agriculture an explicit element in city planning.

Other measures being used to revitalize cities are typified by officials in Bogotá, Colombia, who have constructed riding pathways to encourage people to use bicycles. In Copenhagen, Denmark, they have actually replaced curbside parking spaces for automobiles with bicycle lanes and walkways. As a result, that city has changed from being car oriented to being people oriented. The success of this latter effort is demonstrated by the fact that the total distance traveled by motor vehicles in Copenhagen is 10 percent less today than it was in 1970. Similar changes are occurring in Washington, DC, through the revitalization of portions of the downtown area that were left abandoned during the 1970s and 1980s (Sheehan, 2002). In multitudes of other U.S. cities, abandoned commercial properties are being revitalized through the brownfields program (Chapter 9).

Another approach is to incorporate a range of environmental features into the design, construction, and landscaping of city buildings. This approach, which is called "green architecture," includes the installation of systems for treating wastewater from toilets so that it can be recycled, and the use of windows designed not only to open but also to admit sunlight while concurrently reducing the addition or loss of heat through radiation. Still other revitalization steps include shifting to smaller, decentralized sources of energy, such as wind and solar power, while taking advantage of cogeneration and conservation (Chapter 16). The planting of trees is also being encouraged to provide both shade and sinks for stormwater runoff. Adding to the benefits is that the leaves of trees absorb airborne gases, such as sulfur dioxide, carbon monoxide, nitrogen dioxide, and ozone, and also serve as sticky surfaces for the removal of airborne particles (Fields, 2002).

Defining Environmental Health and its Associated Problems

In its broadest sense, *environmental health* is the segment of public health that is concerned with assessing, understanding, and controlling the impacts of people on their environment and the impacts of the environment on them. Even so, this field is defined more by the problems it faces than by the approaches it uses. These problems include the treatment and disposal of liquid and airborne wastes, the elimination or reduction of stresses in the workplace, the purification of drinking-water supplies, the provision of food supplies that are adequate and safe, and the development and application of measures to protect hospital and medical workers from being infected with diseases such as acquired immune deficiency syndrome (AIDS) and severe acute respiratory syndrome (SARS). Concurrently, there are challenges of protecting the global population from other diseases, such as the so-called swine flu (H1N1) virus.

Environmental health professionals also face long-range problems that include the effects of toxic chemicals and radioactive wastes, acidic deposition, depletion of the ozone layer, climate change, resource depletion, and the loss of forests and topsoil. The complexity of these issues requires multidisciplinary approaches. Thus a team coping with a major environmental health problem may include scientists, physicians, epidemiologists, engineers, economists, lawyers, mathematicians, and managers. Input from experts in these and related areas is essential to the development, application, and success of the control strategies necessary to encompass the full range of people's lifestyles and their environment.

Just as the field of public health involves more than disease (for example, health care management, maternal and child health, epidemiology), the field of environmental health encompasses the effects of the environment on animals other than humans, as well as on trees and vegetation and on natural and historic landmarks. While many aspects of public health deal with the "here and now," many of the topics addressed within the subspecialty of environmental health are concerned with the previously cited impacts of a long-range nature.

ADDRESSING THE FULL RANGE OF ENVIRONMENTAL PROBLEMS

Adopting a global view, one of the fundamental sources of the environmental problem is the ever-increasing population. For this reason, the problems of the environment are inextricably intertwined with those of the size of the population. For example, the projected worldwide population in 2011 will be 7 billion people (Figure 1.4). By 2045, it is projected to reach 9 billion (Kunzig, 2010). To date, however, only two nations have addressed this problem in a direct and effective manner. The first was Taiwan, which, in 1979, adopted a policy that one child per family was insufficient to maintain the

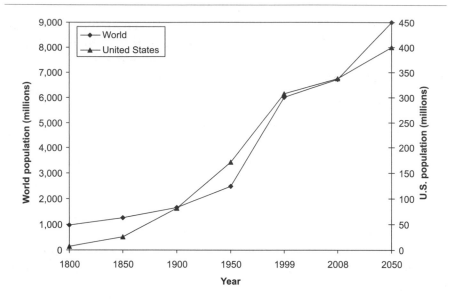

Figure 1.4   Trends in United States and global population with projections to 2050

population and that three would lead to an unacceptable increase. The successful implementation of this policy is documented by the fact that their 2008 population increase was 0.63 percent, compared to a 2007 U.S. population increase of 0.90 percent. Although the U.S. increase was partially due to immigration, it is nonetheless a factor of more than 1.4 times as high. The second to take action was the People's Republic of China, where, following the death of Chairman Mao Tse-tung in 1976, a "one child policy" was instituted by incoming Chairman Deng Xiaoping. Although originally designated a "temporary measure," it was recently announced that this policy would be continued through the 2006–2020 planning period. To enforce compliance, fines are imposed on couples who exceed this limit, pressures are applied for parents to abort a second pregnancy, and sterilization is imposed on couples following additional pregnancies. Interestingly, the increasing population has also been connected with the increasing number of global conflicts (Hayden and Potts, 2010).

In addition to the increasing population, there are environmental challenges due to multiple other sources, including changes in the demographics and the movement of people into the cities. As of 2009, 20 percent of the global population was between 18 and 24; in the sub-Saharan African nations, two-thirds of the members of the population were in that age range.

That this will not change soon is exemplified by the fact that in Mali 50 percent of the women become mothers during their teens, and their average families include almost six children (Holden, 2009). In contrast, as of 2009, within the United States, people under 20 years of age represent only about 28 percent of the population, and those 65 or older less than 13 percent. In 2008, the corresponding average number of children was 2.1 per woman—which is roughly the replacement level. Nonetheless, the U.S. population growth rate was among the highest of any of the industrialized countries due to the fact that the vast majority of such countries have below-replacement fertility rates and the United States has higher rates of immigration.

Cancer and the Personal Environment

One of the stated objectives of the National Cancer Institute (NCI) is to translate the results of its research into ways of saving lives (Kaiser, 2002). A similar approach has been adopted by other groups, including members of the Harvard Center for Cancer Prevention (Colditz et al., 2002). In fact, it is the judgment of the latter group that sufficient information is now available on the causes and prevention of cancer to enable the burden of this disease in the United States to be reduced by more than half during coming decades. Examples of the major contributors to cancer in this country are summarized in Table 1.2. At the same time, the NCI staff is quick to acknowledge that this does not imply that everything about cancer is known. Rather, it is to emphasize that the information that is available today can be effectively used to reduce the incidence of these types of diseases. What is needed is the initiation of programs that relay this information to members of the public and encourage them to change their lifestyles and personal habits in a beneficial manner. If long-term progress is to be achieved, however, the proposed programs must be supported by the enactment of a series of public policies that are designed to make risk-reduction behaviors easier for individuals to choose and maintain. Therefore, the recommendations of the Harvard Center focus on the following five major behavioral risk factors.

TOBACCO USE
More than 160,000 people (44 percent of whom were women) died of lung cancer in the United States during 2008. It is the number one cause of cancer deaths in this country (American Cancer Society, 2008). Of these deaths,

an estimated 85 to 90 percent (i.e., 140,000) were caused by smoking cigarettes and 3,000 were attributed to the effects of secondhand smoke (EPA, 2009). This, however, does not account for the total number of deaths caused by cigarettes. If heart disease and other effects are included, smoking is estimated to have caused an average of 443,000 deaths and 5.1 million years of potential life lost in the United States each year from 2000 through 2004 (CDC, 2008; Pell et al., 2008). For more than half a century, however, cigarette manufacturers have skillfully continued to present their products in a favorable manner. One early example is the fact that tobacco companies gave a free carton of cigarettes each week to every member of the U.S. armed forces throughout World War II, thus ensuring that a large percentage of the public would be smokers for life.

Concurrently, efforts to encourage a reduction in smoking have slowly gained momentum. One of the first major victories was the prohibition of smoking on passenger airliners. This was followed by similar prohibitions in private and public buildings and outdoor spaces. More recently, legislators, even in "tobacco" states such as North Carolina and Virginia, passed laws banning smoking in restaurants. Equally encouraging is the passage by Congress, and signing into law by President Barack Obama in 2009, of a mandate that cigarette manufacturers cease glamorizing smoking, especially to women and adolescents. In addition, manufacturers will be required to list all the ingredients of cigarettes on each package (U.S. Congress, 2009). As a result of such efforts, the percentage of the U.S. population who smoke has been reduced from almost 50 percent in the early 1950s to 21 percent today (CDC, 2009; Moeller and Sun, 2010).

In 2009, the U.S. Department of Defense proposed halting the sale of tobacco on military property and eliminating its use by members of the military services within the next 20 years. This action is based on the fact that 32 percent of active-duty military personnel are currently tobacco users and the accompanying continuing increase in health-care costs related to tobacco use (Zoroya, 2009).

PHYSICAL ACTIVITY

Physical activity has numerous health benefits, including reductions in the risk of colon and breast cancer, and possibly a reduced risk of lung and prostate cancer (Chapter 6). It also reduces premature mortality, cardiovascular disease, hypertension, diabetes, and osteoporosis. Yet, surveys show that less than one-third of the people in the United States meet the federal recommendations on physical activity, and 40 percent of adults engage in no leisure-time physical activity at all (U.S. Public Health Service, 2001). This is not surprising, since most communities in the United States are not structured to accommodate or encourage physically active lifestyles.

What is needed is the development of policies and environmental approaches that enhance and facilitate opportunities for young people to participate in physical activity (Colditz et al., 2002). A first step is the inclusion of physical education and fitness as part of the curriculum in public schools. As of 2002, only one U.S. state (Illinois) required daily physical education for kindergarten through twelfth grade, and, even there, waivers were available to permit students to replace physical education with other activities, such as band or choir. Nationwide, only 8 percent of elementary schools, 6.4 percent of middle schools, and 5.8 percent of high schools provide daily physical education for the entire school year. Twenty-five percent of school children are not provided an opportunity for any form of physical education at all (Brink, 2002).

While exercise is important during childhood and adolescence, parents and school officials must also recognize that exposure of these age groups to ultraviolet radiation while participating in outdoor activities plays a role in the future development of skin cancer. The extent of this problem is illustrated by the fact that the number of new cases of melanoma, the most serious form of skin cancer, has increased in this country by 150 percent since 1975; at the same time, the number of related deaths has increased by more than 40 percent. Since more than half of a person's lifetime exposure to the sun occurs during the younger years, it is particularly important that these age groups be protected. Care must be taken not to exacerbate one problem while protecting people from other types of cancers and diseases (CDC, 2002).

OBESITY AND HEALTHY DIETS

Obesity is a dramatically increasing problem in the United States. In fact, data collected in early 2010 showed that more than 23 million U.S. children and teenagers in the age range of 2–19 years old were obese or overweight. This represented a fourfold increase in the last few years. According to health and medical experts, this meant that nearly one-third of U.S. children were at early risk of type 2 diabetes, cardiovascular disease (i.e., heart disease and high blood pressure), some forms of arthritis, several types of cancer, and even stroke—conditions usually associated with adulthood. In addition, the associated health risks impose increased costs on families, the national health-care system, and the economy. In response, President

Obama in 2010 established a task force to address the problem. Their report was that five steps should be taken: (1) Create a healthy start on life for children from pregnancy through early childhood; (2) empower parents and caregivers to make healthy choices for their families; (3) serve healthy food in schools and ensure access to healthy, affordable food; and (4) increase opportunities for physical activity (Gerber, 2010). Another consideration that has been suggested is to limit the proximity of fast-food establishments near public schools (Davis and Carpenter, 2009). There is clear and convincing evidence, for example, that a diet rich in plant foods and moderate in animal products reduces the risk of cardiovascular disease and diabetes (Chapter 6). The estimated associated annual costs of these and other impacts exceed $150 billion and are escalating. Foreseeing these problems, the latest "food pyramid" published by the U.S. Department of Agriculture (2008) stresses not only a proper diet but also that it be supplemented by a planned program of daily exercise.

ALCOHOL

On the basis of epidemiological evidence, the International Agency for Research on Cancer concluded in 1988 that alcohol is a carcinogen and an independent risk factor for cancers of the liver and upper aerodigestive tract. Subsequent evidence has confirmed that alcohol consumption also increases the risk of breast cancer and possibly colon cancer. Although this evidence, as well as the role that alcohol plays in increased vehicular deaths and injuries, clearly calls for a reduction in its use, other data show that such a reduction would lead to a higher rate of cardiovascular disease. On the basis of this evidence, it would appear that the only ethical course of action is to conduct public educational campaigns that focus on reducing the abuse of alcohol and encouraging those who drink to do so moderately (Colditz et al., 2002). Such campaigns should include efforts to make people aware that the same benefits against cardiovascular disease can be obtained through the consumption of certain types of grape juice. Concurrently, there are increasing problems related to depression, which is often the result of alcohol abuse. Recent information indicates that more than 8 million members of the U.S. population consider suicide each year and 32,000 actually follow through (Newswatch, 2009).

The overriding message is that actions initiated by people on an individual basis represent an extremely effective method for controlling cancer. As noted, it is important that these actions be facilitated by government and private institutions through the development and implementation of policies that encourage and support the required behavioral changes. A

Table 1.3   Leading health indicators—Healthy People 2010 [a]

1	Physical activity
2	Overweight and obesity
3	Tobacco use
4	Substance abuse
5	Responsible sexual behavior
6	Mental health
7	Injury and violence
8	Environmental quality
9	Immunization
10	Access to health care

a. The listings are not intended to represent the relative importance of the various indicators.

good example of such support is the "Healthy People 2010" initiative of the U.S. Department of Health and Human Services (USDHHS, 2009). This initiative emphasizes the need for individuals to choose healthy lifestyles for themselves and their families. It also challenges communities and businesses to support health-promoting policies in schools, work sites, and other settings. Ten leading health indicators have been designated for measuring success in achieving the goals of the initiative (Table 1.3). As may be noted, the first five involve individual choices. While the second five relate primarily to what might be called system-wide issues, at least two, namely, injury and violence, and immunization, clearly depend on personal choices to some degree.

Specific Problems of Different Age Groups

While the environments in which people live and work are important, there are different factors that must be considered in assessing the problems of each specific age group. Examples of several of these are discussed here.

THE ELDERLY

For the elderly, one of the major sources of potential hazards is the home and the safety of the environment it provides. Specific problems include areas that are poorly lighted, combined with light switches that are either not clearly marked or cannot be seen in the dark; pathways that have obstructions, such as cords, loose throw rugs, or carpet edges that are curled; chairs and tables that are not sturdy and/or move easily; chairs and toilet seats that are low and difficult to get out of or off of; areas that are slippery, particularly in bathtubs and showers; and tubs and showers that are not equipped with grab bars. Even though these hazards are well known, in many cases they are not addressed even in homes that are supposedly designed for the elderly. Surveys show, for example, that two or more such deficiencies exist in almost 60 percent of the bathrooms and in 23 to 42 percent of the other rooms in such homes. Nearly all homes in one survey had at least two potential hazards (Gill et al., 1999).

While these problems can readily be solved, others represent a more formidable challenge. One is to ensure that routine actions of the elderly do not cause them injury and/or death. Many approaches being applied in this case are based on sophisticated electronic systems, examples being those that determine if ill people have taken their medication, recognize if they have become immobile, or have fallen and injured themselves. Other systems can regulate the temperature of the water in the bathtub and even jog the memory of an occupant if a kettle on the stove has been left unwatched too long.

A primary reason for addressing these types of problems is the escalating life expectancy. Between 1950 and 2009, the number of centenarians in the world increased from an estimated few thousand to more than 340,000. The highest numbers reside in the United States and Japan, followed by Italy, Greece, Monaco, and Singapore. By 2050, the rate of increase in this age group is projected to be more than 20 times that for the total population, making it the fastest-growing segment of the population. Demographers attribute this to decades of medical advances and improved diets that have reduced heart disease and stroke. Also contributing are genetics and lifestyle, and the fact that doctors are more willing to treat in an aggressive manner the health problems of people once considered too old for health care. In the United States, it is estimated that by 2017 there will be more people 65 years of age and older than there will be children younger than 5. In fact, by 2040 the population of people 80 years of age and older is expected to increase by more than 230 percent, compared with less than a 35 percent increase for the total population, and the median age of the population will increase from 33 to 39 years old. Similar increases are taking place throughout the developed world (Yen, 2009).

YOUNG PEOPLE

As has been emphasized in the preceding sections, data consistently demonstrate that lifestyles and personal habits have major influences on the health of individuals. This includes their behavior in transportation vehicles, their choice of diets, and their decision on whether to smoke. Since patterns of adult behavior are largely established during youth, it is imperative that this age group be a primary audience for the receipt of information on these matters. As will be noted in the data presented below, the situation with respect to young people in the United States is particularly disturbing.

Recent surveys reveal, for example, that almost 20 percent of young people 10–14 years of age in this country have rarely or never worn a seat belt while in a car. Furthermore, during the preceding 30 days more than a third of them had ridden in a car with a driver who had been drinking alcohol, more than half had consumed alcohol, and more than a third had smoked cigarettes. Also indicative of such lifestyle choices was that less than a third of them had eaten the recommended number of servings of fruits and vegetables during the day preceding the survey (CDC, 1998). Although these revelations are due to a range of factors, one of the most significant is that they reflect an apparent lack of understanding, on the part of parents and caregivers, of the importance of helping children develop healthy living habits. Unfortunately, they may also reflect a lack of communication between parents and their children.

The types of problems being faced by young people, however, do not end here. As often is the case, unexpected and more subtle problems are discovered. One example is the harm caused by the backpacks used by young students to carry their books and other personal items to school. Noting that she was suffering back pain, a 14-year-old female student in Texas interviewed her classmates and found not only that a number of them were experiencing similar discomfort but also that some of them were suffering shoulder discomfort. On the basis of her study, she recommended that students carry no more than 10 percent of their body weight in such devices, that they carry them using both straps (not by slinging one strap over one shoulder), and that they place the heaviest items in the bottom of the pack so that they are close to the body (Guyer, 2001). Follow-up studies showed that backpack injuries are sufficiently painful to cause an estimated 3,000 to 4,000 U.S. school students to report to emergency rooms each year. Possible solutions include replacing hardbacks with paperbacks; printing slimmer, two-volume sets (one for each semester); or issuing smaller textbooks supplemented by CD-ROMs.

Other concerns include the quality of the air inside schools (Chapter 5). In many cases, poor air quality is caused by inadequate ventilation rates. In fact, studies show that the air in 30 to 40 percent of the nation's schools contains molds and other pollutants, such as volatile organic chemicals emitted from cleaning products, photocopiers, and classroom furnishings. Trailer units, used to provide additional space in overcrowded schools, have been found to have high airborne concentrations of formaldehyde and benzene. Noise is also a problem in teaching facilities located near airports. Such problems represent not only a risk to health but also a detriment to the learning process (Wakefield, 2002).

These types of problems are primarily those of children in the developed nations. Worldwide, there are many additional factors and activities to consider. According to the International Labour Office, for example, an estimated 100 million or more children aged 5 to 14 years work full time. Many of them, the vast majority of whom live in the developing countries, particularly those in Asia and Africa, are employed in tasks in which the accident rates are high. In addition, they are frequently subjected to harmful substances, physical agents, and psychosocial hazards (Forastieri, 1997).

INFANTS AND SMALL CHILDREN

A close examination of the behavior and biological characteristics of infants and young children shows that there are multiple reasons that they are more susceptible to certain types of environmental stresses. For example, the metabolic pathways of the young, especially during the first few months of life, are immature; children are in the growing stage of life, a time during which their development processes are easily disrupted, and after exposure, they have more years in which to develop the range of chronic diseases that may be initiated. Another contributing factor is that infants and small children spend a considerable amount of time crawling either on the floor indoors or on the ground outdoors. This not only exposes them to higher levels of environmental toxicants, but it also exposes them to the possibility of absorbing toxic chemicals through the skin and ingesting them through hand- and object-to-mouth activities (Suk, 2002). Even if these factors are taken into account, additional, frequently surprising problems are often discovered. One was that the activation of a safety air bag in a motor vehicle could be fatal to infants and small children (Chapter 11).

Even so, the need to address the environmental health problems of children was late in being recognized, especially in the United States. Fortunately, this problem is now being corrected. The primary stimuli for these changes were (1) the issuance in 1993 of the National Research Council (NRC) report "Pesticides in the Diets of Infants and Children" and (2) the convening in 1994 by the Children's Environmental Health Network of the first

Table 1.4   The ten most significant public health achievements in the United States during the 20th century [a]

1	Vaccination
2	Motor-vehicle safety
3	Safer workplaces
4	Control of infectious diseases
5	Decline in deaths from coronary heart disease and stroke
6	Safer and healthier foods
7	Healthier mothers and babies
8	Family planning
9	Fluoridation of drinking water
10	Recognition of tobacco as a health hazard

a. The listings are not intended to represent the relative importance of the individual advances.

scientific conference on this subject. In rapid sequence thereafter, the U.S. Congress passed the Food Quality Protection Act of 1996, which incorporated the major recommendations of the NRC report, including a requirement that pesticide standards be set at levels that are protective of the health of children. The same year, the EPA established an Office of Children's Health Protection, and the following year (1997) the president issued an executive order requiring that all federal agencies reduce environmental threats to children; Congress followed with passage of the Children's Health Act in 2000.

In a similar manner, there has been increasing emphasis on these problems worldwide. For example, in 2002 the World Health Organization convened the International Conference on Environmental Health Threats to the Health of Children: Hazards and Vulnerability. One of the highlights of this conference was the issuance of "The Bangkok Statement," a pledge to protect children against environmental stresses (ATSDR, 2002). Also significant was the designation of "Healthy Environments for Children" as the theme of World Health Day for 2003 (Eskenazi and Landrigan, 2002). The ten most significant public health achievements within the United States during the twentieth century are summarized in Table 1.4.

Assessing Problems in the Ambient Environment

Among the many tasks that confront environmental health professionals is to understand the various ways in which humans interact with the ambient (indoor or outdoor) environment. In fulfilling this task, a primary step is to study the process or operation that leads to the generation of a problem and to determine how best to achieve control. Components of such an analysis include (1) determining the source and nature of each environmental contaminant or stress, (2) assessing how and in what form that contaminant comes into contact, or the stress impacts, with people, (3) measuring the resulting physical and economic impacts, and (4) applying controls when and where appropriate. In the case of air and water pollution, experience shows that instead of focusing on one or more individual sources within a given facility, every effort should be made to gather data on all the discharges from the facility, all the sources of each specific pollutant, and all the pollutants being deposited in the adjoining region, regardless of their nature, origin, or pathway (Chapters 5, 8, and 13).

Even though tracing the source and pathways of each contaminant is important, an essential part of the process is to determine the effects on human health and the environment. When a pollutant is being evaluated for the first time, and exposure limits have not been established, such efforts may entail establishing relationships between the exposure, the resulting dose, and the associated effects (Chapter 3). Armed with this information, appropriate governmental bodies, often in concert with various professional societies and organizations, can then move forward to establish standards for limiting exposures to the contaminant or stress (Chapter 13).

To assess the effects of exposures correctly, care must be taken to account not only for the fact that they can derive from multiple sources and enter the body by several routes, but also because elements in the environment are constantly interacting. In the course of transport or degradation, agents that were not originally toxic to people may become so, and vice versa. If the concentration of a contaminant in the environment (for example, a substance in the air) is relatively uniform, local or regional sampling may yield data adequate to estimate human exposure (Chapter 14). If concentrations vary considerably over space and time (as is true of certain indoor pollutants) and the people being exposed move about extensively, it may be necessary to measure exposure of individual workers or members of the public by providing them with small, lightweight, battery-operated portable monitoring units. Development of such monitors and the specifications for their use requires the expertise of air pollution engineers, industrial hygienists, chemists and chemical engineers, electronics experts, and quality-control personnel.

At the same time, advances in technology have produced highly sophisticated and sensitive analytical instruments that can measure many environmental contaminants at concentrations below those that have been demonstrated to cause harm to health or the environment. For example, techniques capable of measuring contaminants in parts per billion are common. The mere act of measuring and reporting the presence of certain contaminants in the environment often leads to concern on the part of the public, even though the reported levels may be well within the acceptable range. The accompanying fears, justified or not, can lead to expenditures on the control of environmental contaminants instead of on other, more urgent problems. Those responsible for protecting people's health must be wary of demands for "zero" pollution: it is neither realistic nor achievable as a goal in today's world. Rather, given the host of factors that are an integral part of our daily lives, the goal should be an optimal level of human and environmental well-being.

The General Outlook

In the course of their work, medical and public health personnel have achieved remarkable success in reducing human morbidity and mortality. One major benefit has been a significant increase in the average human life span. One important consequence has been a dramatic growth in the world's population and an accompanying heavier burden on the environment. In fact, a large share of the social, economic, and environmental decline in many parts of the world today results from the increased production of materials and wastes, and higher consumption of resources in order to meet the expanding expectations of an ever increasing number of people. Many of these practices have global ecological effects, and the combination of local and global effects will inevitably affect human health. While advanced technologies can help control some of the environmental impacts, the problem of population growth needs to be vigorously addressed. Fortunately, this need is being recognized, as exemplified by the third United Nations International Conference on Population and Development, held in Cairo in September 1994. Other indications of this increasing awareness are the Rio Declaration of 1992, which created the UN Commission on Sustainable Development, and the World Conference on Sustainable Development, held in Johannesburg, South Africa, in 2002. Limitations on population growth will of necessity be a strong component of any plans for long-range sustainable development. More details on the nature of population growth and its impacts are provided in Chapter 18.

While advances in modern science and technology have given humans the capability to control much of the natural world, choices will nonetheless have to be made to ensure that the controls, as applied, result in an optimal level of health for both the environment and the public. The overall goal should be to achieve the maximum good for the maximum number of people. As part of this exercise, those people living in the developed countries must decide what changes in their lifestyles they are willing to make to ensure the "greatest good" for the majority of the world's population, a vast number of who live in the developing countries.

2

TOXICOLOGY

A s Defined by the Society of Toxicology (2009), *Toxicology is the study of the adverse effects of chemical, physical, or biological agents on living organisms and the ecosystem, including the prevention and amelioration of such adverse effects.* Another definition is that it is *the science that deals with the effects, antidotes, and detection of poisons* (Random House, 2009). As the discussion that follows confirms, the challenges in this field are enormous. Globally, it is estimated that more than 70,000 chemicals are in common use. Adding to this total are from 200 to 1,000 new synthetic compounds that the chemical industry markets each year. Although these materials are manufactured and distributed so that society can take advantage of their benefits, the accompanying processes result in releases of many such materials into the environment. These include a variety of prescription and over-the-counter drugs that are discharged into the environment as components of human and animal wastes. The complexity of the situation is exemplified by the fact that this last group contains antimicrobials, anticonvulsants, antidepressants, and anticancer compounds. Obviously, if these gain access in sufficient quantities to streams and rivers, they can represent a danger to fish and other aquatic life (Service, 2002). As a result of these and other activities, humans and other species are exposed to a wide range of chemicals in the general environment, as well as in the home and in the workplace. In fact, trace quantities of toxic chemicals are present in our food, our air, and our drinking water.

In response, the Environmental Protection Agency (EPA, 1986) and especially the State of California (2009), are taking action to increase their regulation of such products. The Board on Toxicology and Environmental

Health Hazards of the National Research Council also periodically recommends the addition of other chemicals to the list (NRC, 1995b). These choices are based on evaluations of those known to cause cancer, birth defects, or other reproductive harm. Based on these and other sources, the EPA total now includes approximately 775 chemicals. Furthermore, the NRC, the Food and Drug Administration, and the National Institute for Occupational Safety and Health conduct related studies and impose other restrictions (OSHA, 2007).

These guidelines apply to carcinogens, such as styrene, marine diesel fuel, a category of herbicides, vinyl acetate (used in the production of plastics, paints, and adhesives), and wood dust. Those of reproductive concern include chloroform, toluene, several types of ethers, ethylene oxide (an industrial gas used in the production of chemicals for medical sterilization), carbaryl (a common pesticide), and the refrigerant methyl chloride. Also being considered is bisphenol-A (BPA), a compound used in hard plastics and the lining of food cans that has been linked to developmental disorders. The California State Senate had previously passed a bill that bans the use of BPA in children's food and drink containers. Other states are following their lead (State of California, 2009). A readily available source of updated information on this topic can be found in the following reference (Lewis, 2004).

The Field Encompassed by Toxicology

As will be demonstrated repeatedly in the discussions that follow, the efforts of toxicologists involve both science and art. The science lies in the observational or data-gathering aspects, and the art is in the projection of these data to situations where there is little or no information (Doull and Bruce, 1986). When the evaluations address the presence of chemicals in the environment, the situation is far more complicated. In these cases, specially trained environmental toxicologists must expand the work of their coworkers, who traditionally deal with the effects of a single chemical in a single animal species, to include assessments of the effects, both direct and indirect, of combinations of chemicals on total ecosystems. This is what is known as *environmental toxicology*. The outcomes of such efforts are increasingly used by regulatory agencies to assess chemical risks, assign priorities to the cleanup of hazardous waste sites, establish government policies, and set levels of allowable exposure (Gochfeld, 1998).

For years, laboratory toxicological studies followed a rather standard format. Today, such studies have entered a completely new era, namely, the development of the field of *molecular toxicology*. Applying modern experimental technologies, it is possible to explore the effects of toxins on the organism at all stages of development, on organ systems, tissues, and cells, as well as on enzymes, receptors, hormones, and genes. This includes studies of the biochemical and molecular aspects of uptake, transport, storage, excretion, lactivation, and detoxication of drugs, agricultural, industrial, and environmental chemicals, natural products, and food additives. Of particular interest are aspects of molecular biology related to biochemical toxicology. These include studies of the expression of genes related to detoxication and activation enzymes, toxicants with modes of action involving effects on nucleic acids, gene expression and protein synthesis, and the toxicity of products derived from biotechnology. As a result, this has become the most rapidly expanding specialty in the field of toxicology. This is confirmed by the increasing number of articles appearing in the international *Journal of Biochemical and Molecular Toxicology.*

Pathways of Exposure and Excretion

Although protection of other species is important, the discussion that follows will emphasize the impacts of toxic chemicals on humans. As previously discussed (Chapter 1), the major routes of intake in this case are the lungs (inhalation), the gastrointestinal tract (ingestion), and the skin (absorption). In the case of the respiratory tract, the primary site of uptake is through the alveoli in the lungs—especially for gases such as carbon monoxide, nitrogen oxides, and sulfur dioxide and for vapors of volatile liquids such as benzene and carbon tetrachloride. The capacity of the lungs for absorbing such substances is facilitated by the large surface area of the alveoli, and the high flow and proximity of the blood to the alveolar air. Liquid aerosols and airborne particles may also be absorbed through the lungs. In contrast, the deposition of airborne particles is heavily influenced by their size, the particles of primary interest today being those in the size range 2.5 micrometers or less (Chapter 5). Chemicals that are foreign to the human body are known as xenobiotics. Such substances can be either naturally occurring or of human origin.

Once a chemical is absorbed, the nature and intensity of its effects depend on its concentration in the target organs, its chemical and physical form, what happens to it after it is absorbed, and how long it remains in the tissue or organ in question (following the central tenet that "the dose makes the poison"). After being taken up in the blood, a toxic chemical will be rapidly distributed throughout the body. As part of this process, it may be transferred from one organ or tissue to another, and it may be converted into a new compound or metabolite. This process is known as *biological transformation*. Metabolic processes in the cytoplasm (i.e., the portion of a cell exclusive of the nucleus), for example, can alter toxic substances through various chemical reactions, including oxidation and reduction. In general, these reactions tend to result in new products that are less absorbable and more polar (charged) chemically and thus are more readily excreted in the urine. The removal of toxic chemicals from the body is thereby enhanced. In certain cases, the new product or metabolite may be more toxic than the parent compounds. In most cases, however, the newly formed compounds tend to be less toxic (Smith, 1992).

Such reactions are known as *bioactivation*, the details of which can be elucidated through mass spectrometry and computer modeling. In fact, bioactivation is being applied in the design and development of new drugs, gaining insights into the mechanisms of cell death, toxicity and carcinogenicity, identifying metabolite-specific biomarkers of exposure and effect, and extrapolating the results from animal models to humans (Elfarra, 2009).

The principal pathway for excretion of chemicals from the human body is the urine, but the liver (via absorption from the bile into the blood and excretion through the bowels) and the lungs (via various clearance mechanisms—Chapter 1) can also be important excretory organs. In general, the GI tract is not a major route of excretion of toxicants. Among the less significant routes are the sweat glands (Lu, 1991).

Toxic chemicals may cause injuries at the site of first contact, or they may be absorbed and distributed to other parts of the body where they exhibit their effects. Those effects may be considered reversible or irreversible. In general, reversible effects are observed for short-term exposures at low concentrations; irreversible effects are more commonly observed following long-term exposures at higher concentrations. Toxic agents may also produce either immediate or delayed effects. A notable example of the latter is carcinogenesis; many types of cancer do not appear in humans until a decade or more after exposure. The effects of a toxic agent may be influenced by previous sensitization of the exposed person to the same or a similar chemical, a notable example being the case with beryllium. Such effects are often classified as *allergic* reactions (Lu, 1991).

Expressing the Effects of Chemicals

There are a variety of ways for expressing effects of toxic chemicals. Summarized in Table 2.1 is a comparison of various chemicals on the basis of the quantities required to produce death in 50 percent of an exposed population of animals. Some, such as botulism toxin, produce death at concentrations of only nanograms (10^{-9} gram) per kilogram of body weight. Others, such as ethyl alcohol, may have relatively little effect even after doses of several grams per kilogram. These types of data are often then used to classify various toxic agents in terms of the probable amount required to produce death in humans (Table 2.2). Under this categorization, botulism toxin would be classified as supertoxic, whereas ethyl alcohol would be classified as slightly toxic. Although primarily qualitative, such a classification scheme serves a useful purpose in providing laypeople with answers to the question: How toxic is this chemical? (Klaassen, 1986). Toxic chemicals can also be classified in terms of their target organ (liver, kidney), their use (pesticide, food additive), their source (animal or plant toxin), and their effects (cancer, mutations).

The presence of toxic chemicals in various media within the environment and their uptake by different species can lead to a variety of interesting

Table 2.1 Approximate concentrations of various chemicals required to produce death in 50 percent (i.e., the LD_{50}) of an exposed population of animals

Chemical	LD_{50} (mg/kg of body weight)
Ethyl alcohol	10,000
Sodium chloride	4,000
Ferrous sulfate	1,500
Morphine sulfate	900
Phenobarbital sodium	150
Picrotoxin	5
Strychnine sulfate	2
Nicotine	1
d-Tubocurarine	0.5
Hemicholinium-3	0.2
Tetrodotoxin	0.10
Dioxin (TCDD)	0.001
Botulinum toxin	0.00001

Table 2.2 General classification of chemicals based on the probable dose necessary to cause death in humans

Toxicity rating	Probable lethal dose for humans	
	Dose	For average adult
Practically nontoxic	>15 g/kg	More than 1 quart
Slightly toxic	5–15 g/kg	Between 1 pint and 1 quart
Moderately toxic	0.5–5 g/kg	Between 1 ounce and 1 pint
Very toxic	50–500 mg/kg	Between 1 teaspoon and 1 ounce
Extremely toxic	5–50 mg/kg	Between 7 drops and 1 teaspoon
Supertoxic	<5 mg/kg	A taste (less than 7 drops)

situations. The concentrations of certain heavy metals, such as mercury, in plankton, for example, will be higher than in the water in which they live, and the concentrations in fish will be higher still. The concentrations in birds that feed on the fish will be even higher, perhaps by as much as several hundredfold. These phenomena are known as *biological magnification* or *bioaccumulation*. Bioaccumulation refers to how pollutants enter a food chain, and biomagnification refers to how much pollutants tend to concentrate more as they move from one trophic level to the next. Biomagnification, for example, led to the harmful effects of DDT [1,1,1-trichloro-2,2-bis (p-chlorophenyl)ethane] on pelicans via a thinning of the shells of their eggs. For these and other reasons, it is unlikely that procedures for the establishment of an acceptable level of intake of a chemical by humans can be directly applied in setting a corresponding limit for the environment. In a similar manner, DDT will concentrate in a human mother's milk to the extent that her baby's intake of this pesticide per unit of body weight may be more than 20 times that in the mother's diet (Croteau, Luoma, and Stewart, 2005).

Conventional Tests for Toxicity

Depending on the dose, the effects of toxic chemicals on animals may range from rapid death to sublethal effects and to situations in which there are apparently no effects at all. Often the first step in the prediction of effects is to conduct a series of laboratory studies involving a single chemical and a single animal species. Because of legal and ethical limitations, most such studies are conducted using rats or mice rather than humans. To examine the effects associated with exposure over various time periods, toxicological studies have generally been divided into two categories, acute and long-term. Long-term studies are further divided into short-term tests and long-term tests (Lu, 1991). Obviously, in these latter two cases the amounts of the chemicals administrated must be below the lethal level, the goal in this case being to simulate environmental exposures of humans.

ACUTE TOXICITY STUDIES

Acute toxicity studies may require only hours to conduct and may involve only a single administration of the chemical being tested. If death is the endpoint being observed, the data are generally analyzed by beginning with a plot showing the relationship between the dose and the percentage of the animals that die. Such a curve often exhibits the pattern shown in Figure 2.1. The portion of the curve between "Minimum" and point "B" represents the range of doses in which the most susceptible animals respond; the portion between "B¹" and "Maximum" represents the range in which the

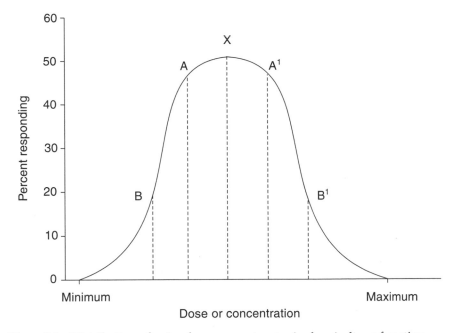

Figure 2.1 Distribution of animal responses to a toxic chemical as a function of dose

most resistant animals respond. The peak of the curve (directly below point "X") indicates the dose that causes 50 percent of the exposed animals to die. This is designated as the LD_{50} and, in the case of humans, is often expressed as the LD_{50} at different times, for example, at 30 and 60 days following a single acute exposure. In the case of small animals, the LD_{50} is generally expressed in terms of much shorter time periods after exposure. Since the curve follows a normal (or Gaussian) distribution, statistical procedures can be used to evaluate the resulting data (Loomis, 1968).

Although the Gaussian distribution is interesting, data resulting from toxicological studies are generally plotted in the form of a curve relating the dose or concentration to the *cumulative* percentage of animals exhibiting the given response. The curves in Figure 2.2 show this type of plot for two different chemicals, A and B. The curve to the left represents the more toxic of the two compounds, since the dose (or concentration) required to cause death in 50 percent of the exposed population (LD_{50}) is lower. Such graphs are commonly referred to as dose-response curves and are plotted using an arithmetic scale on the vertical axis and a logarithmic scale on the horizontal axis. One advantage of this format is that a major portion of the

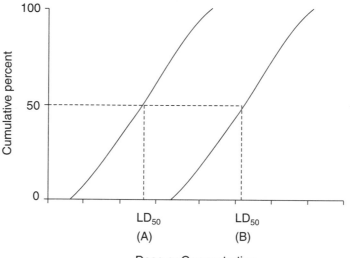

Figure 2.2   Cumulative percentages of animals showing responses to toxic chemicals. The LD_{50} designates the dose that is lethal to 50 percent of the exposed animals. The curve to the left represents the more toxic of the two chemicals.

curve is linear; for this portion, the response is directly related to the dose or concentration of the chemical agent (Smith, 1992).

Figure 2.2 also illustrates another approach for determining the LD_{50} for exposed animals. As in the previous case (Figure 2.1), this applies to those deaths that occur within a specified period of time after exposure. One of the advantages of this approach is that the endpoint is easily measurable; it either occurs or it does not. In fact, in previous years determination of the LD_{50} was one of the primary goals of many acute toxicity studies. This is far less true today, particularly in light of the diminished need for this type of information by regulatory agencies. Another contributing factor is the increased interest in both cancerous and noncancerous diseases, as well as possible behavioral effects, that may be caused by chemical exposures. Nonetheless, studies using the LD_{50} as an endpoint provide an excellent illustration of the differences in the dose required to reach this endpoint for one animal species versus another, and as a function of age in the same animal (Tables 2.3 and 2.4). As will be discussed later, such differences increase the complexity of extrapolating toxicological information from various animals to humans.

Other benefits of acute toxicity studies are that they can provide information on the probable target organs for the chemical and its specific toxic effect, as well as guidance on the doses to be used in more prolonged (long-term) studies. Acute toxicity studies can also provide information on the synergistic and antagonistic effects of certain combinations of chemicals. An interaction is described as *synergistic* when exposure to one chemical causes a dramatic increase in the effect of another. One example is the enhanced toxicity of ketone in combination with haloalkane. Another is the

Table 2.3   Effects of species differences on LD_{50} for TCDD (tetrachlorodibenzo-p-dioxin)

Species	LD_{50} ($\mu g/kg$)
Guinea pig	2
Mink	4
Rabbit	50
Monkey	70
Mouse	200
Rat	350
Hamster	2,000

Table 2.4   Influence of age on LD_{50} for DDT [1,1,1-trichloro-2,2-bis(*p*-chlorophenyl) ethane] in rats

Age	LD_{50} (mg/kg)
Newborn	>4,000
10 days	730
2 weeks	440
1 month	360
2 months	250
4 months	190
Adult	220

enhanced effects of the chemical and radiological carcinogens in tobacco (Chapter 15). In contrast, an interaction is described as *antagonistic* when exposure to one chemical results in a reduction in the effects of another. The protection that selenium provides against mercury is an example of this type of interaction. Such information is very important in the evaluation of environmental exposures.

CHRONIC TOXICITY STUDIES

Chronic toxicity studies are conducted on both a short- and long-term basis. Short-terms studies generally involve repeated administrations of a chemical, usually on a daily basis, over a period of about 10 percent of the life span of the animal being tested (for example, about 3 months in rats and 1 to 2 years in dogs); however, shorter durations such as 14-day and 28-day treatments have also been used by some investigators. Long-term studies involve repeated administrations over the entire life span of the test animals (or at least a major fraction thereof). For mice, the time period would be about 18 months; for rats, 24 months; for dogs and monkeys, 7 to 10 years.

For short-term studies, generally two or more species of animals are used, the objective being to have them biotransform the chemical in a manner essentially identical to the process in humans. It cannot be assumed, however, that this will be the case. In fact, differences in the abilities of various species to biotransform chemicals are the basis for the effectiveness of many of the pesticides that have been developed to be selectively toxic to only one insect, plant, or animal (Smith, 1992). Under normal circumstances, the animals selected are the rat and the dog because of their appropriate size, ready availability, and the preponderance of toxicologic information on their reactions to a wide range of chemicals (Lu, 1991). Differences in response by gender require that equal numbers of male and fe-male animals be used, and a control group must be maintained for com-parison purposes. In addition, the chemical should be administered by the same route of exposure that is anticipated for humans. Other factors that must be considered include the possibility that exposed population groups may include some people who are unusually susceptible, and that effects may have occurred but were not observed (Moriarty, 1988).

Endpoints for Toxicological Evaluations

As indicated in the previous discussion, acute and short-term tests served as the principal approaches used in earlier toxicological studies. In these cases, only death or tissue damage served as recognized endpoints. As toxicolo-gists sought to obtain information for evaluating a fuller range of effects in humans, the laboratory studies were expanded and new and different end-points were adopted. Today the evaluation of human exposures tends to be directed to studies involving a full range of endpoints or effects, including those on behavior and other noncancer endpoints. The more prominent of these are discussed below.

CARCINOGENESIS

Chemical carcinogenesis is recognized today as a multistage process, in-volving at least three steps: initiation, promotion, and progression. Although formerly it appeared that various chemical compounds and physical agents were either purely initiators or purely promoters, more recent interpreta-tions suggest that some chemicals and agents are both initiators and pro-moters. Current theory posits that the development of cancer involves the activation or mutation of oncogenes, or the inactivation of suppressor genes, and that this causes a normal cell to develop into a cancerous cell.

Due to the time and expense required for related tests using animals, toxicologists have for years experimented with the development of short-term, in vitro tests (i.e., conducted outside the body). One of the most widely applied is the Ames test (Ames, 1971), which is a measure of the mutagenic-ity of chemicals in bacteria. It is based on evidence that deoxyribonucleic acid (DNA) is the critical target for most carcinogens and on the fact that mutagenic chemicals are often also carcinogenic. Although the Ames test provides an indication of the ability of a chemical compound to induce mutations or stimulate other types of biological activities, it does not reflect the complex patterns of uptake, metabolism, detoxification, and excretion that occur in the whole animal, or the gene or target organ specificity—information that can be critical in evaluating cancer responses (Butterworth et al., 1999).

REPRODUCTIVE TOXICOLOGY

Toxic effects on reproduction may occur anywhere within a continuum of events ranging from germ-cell formation and sexual functioning in the parents through sexual maturation in the offspring. For this reason, and because exposure of the mother, father, or both may influence reproductive outcome, the determination of the relationship between exposure and these types of effects is highly complex. In addition, critical exposures may include those to the mother long before or immediately prior to conception, as well as to the mother and fetus during gestation.

DEVELOPMENTAL TOXICOLOGY

Developmental effects (commonly known as teratogenesis) can lead to the formation of congenital defects in the unborn. They have been known for decades and are an important cause of morbidity and mortality among newborns. Such effects encompass embryo and fetal death, growth retardation, and malformations, all of which can be highly sensitive to chemical exposures. For some years, no connection was suspected between such effects and chemicals. Toxicologists therefore had a tendency to assume that the natural protective mechanisms of the body, such as detoxication, elimination, and the placental barrier, were sufficient to shield the embryo from maternal exposure to harmful chemicals. These concepts changed dramatically after the clinical use of thalidomide, a sedative first employed in Germany in the late 1950s to relieve morning sickness in pregnant women, led to a host of developmental effects in their fetuses.

NEUROTOXICITY

Although fewer than 10 percent of the more than 70,000 previously cited chemicals in use have been tested, almost 1,000 have been identified as known neurotoxins (i.e., poisons that affect the brain or spinal cord, examples being rattlesnake venom and the poison injected by the black widow spider). The impacts on humans range from cognitive, sensory, and motor impairments to immune system deficits. Often there are major differences between the degree of neurotoxic response observed in animals and humans (Stone, 1993).

IMMUNOTOXICOLOGY

Various toxic substances are known to suppress the immune system, leading to reduced host resistance to bacterial and viral infections and to parasitic infestation, as well as to reduced control of neoplasms. The importance of these effects is well illustrated by the concern about AIDS, in which the infected person often dies due to his/her inability to resist an organism that would not be a problem in a healthy individual. Certain toxic agents can also provoke exaggerated immune reactions leading to local or systemic reactions.

Extrapolations of Small Animal Data to Humans

The application of animal bioassay data for estimating human responses to environmental exposures involves two types of extrapolations. One is to determine or estimate the relative responsiveness of humans and the animal species used in the bioassays—the so-called extrapolation from small animals to humans. The second is to extrapolate from the biological effects observed at relatively high exposures to the range anticipated in the ambient environment (Lippman, 1992). In the past, the general approach has been to assume that the dose-response relationship in the low-dose range is linear for carcinogenic agents and nonlinear (i.e., has a threshold) for non-cancer-producing agents. Another important influencing factor is that a chemical that has been tested and found to be carcinogenic may be so simply because the detoxification pathways in the animal being studied were overwhelmed (Schmidt, 2002).

Also to be considered is that the pathways of environmental exposures are often different from those applied in laboratory tests, and the nature of the chemicals and the rates at which they are metabolized may have been modified. This could lead to different lengths of time they are retained in the target tissues and thereby affect the sensitivities of these tissues to the exposures.

Dose-Response Relationships

During recent years, there have been multiple changes in the concepts related to the relation of a given dose from a toxic agent and its health impacts. Initially, the standard approach was to assume that the relationship was linear, beginning with the lowest dose and extending upward from there. This was done to ensure that protection standards were conservative, and it was described as the linear-non-threshold (LNT) concept. Although the NCRP (1993) emphasized that the relationships depicted by the LNT relationship should not be used for estimating the associated risk, this precaution was frequently ignored. For example, to quantify the health impacts to a population group that received an annual radiation dose equal to the limit for members of the public would require following more than 40,000 people throughout their lifetimes and comparing them to a similar group that had not been exposed, one of the reasons being that the anticipated increase in the rate of fatal cancers would be about 200, at most, compared to a normal expected number of cancer deaths, due to normal causes, of more than 7,000 (NRC, 1995a).

In other applications, an example being the protection of workers from toxic chemicals, the concept of the threshold limit value (TLV) was developed (Figure 4.3, page 81). As the term implies, this was based on the concept that the long-term health impacts from low doses of such agents would be sufficiently low that they could be tolerated without observable

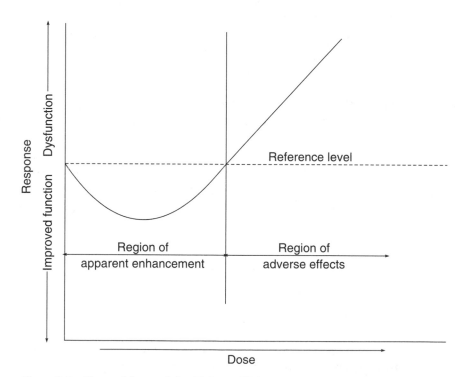

Figure 2.3   General form of the "J-shaped" dose-response curve

health effects. More recently, it has been documented that even radionuclides, such as ^{226}Ra and ^{239}Pu, which are documented carcinogenic agents at high intakes, have similar thresholds for observable health effects (Los Alamos Scientific Laboratory, 1995).

Another depiction of the relationship between dose and risk is what is called the "J-shaped" curve (Figure 2.3). This is based on the observation that some studies imply that a low dose of a toxic agent may be beneficial to health. This is commonly known as *hormesis.* An analogy is that vaccinations, which include the injection of a small dose of an infectious disease (e.g., mumps and measles), permit the body to develop immunity so that a person is not affected when he/she is later exposed to the disease (Calabrese, Baldwin, and Holland, 1999; Kaiser, 2003).

Providing Scientific Data and Regulatory Guidance

Even so, there are many other types of problems to be addressed. One is the need to develop systematic systems for summarizing and critically analyzing the existing toxicological information and assembling a database on which to assess trends in the amounts of toxic agents present in the environment. Another is to implement on a worldwide basis a uniform system for classifying the risks of toxic chemicals. By congressional mandate, the Agency for Toxic Substances and Disease Registry (ATSDR) produces toxicological profiles for hazardous substances present at National Priorities List (NPL) sites. These substances are ranked based on frequency of occurrence at such sites and their toxicity and potential for human exposure. For each such material, its toxicological and adverse health effects are succinctly characterized, an assessment is made of the adequacy of the information available for making such judgments, and the key references on which the assessments are based are identified. If the available information is determined not to be adequate, then additional testing needed to fill the voids is identified. Toxicological profiles are developed from a priority list of 275 substances. As of 2009, more than 300 profiles that cover more than 250 substances have been published (ASTDR, 2009). ATSDR also prepares toxicological profiles for the U.S. Department of Defense and the U.S. Department of Energy on substances related to federal sites. Acting in its role as a regulatory agency, the EPA establishes limits for these substances.

To assemble a database on which to assess environmental trends on toxic agents, the EPA maintains the Toxics Release Inventory. Through this program, more than 23,000 factories, refineries, mines, power plants, and chemical manufacturers self-report to the EPA on an annual basis the identities and quantities of toxic chemicals that they release into the environment. In addition to providing useful data on trends in the quantities of such releases, these reports often make industrial leaders more aware of how their activities are affecting the environment and stimulates them to be more careful. Data show, for example, that such releases were reduced by 10 percent between 1999 and 2000; the reduction from 1988 to 2000 was almost 50 percent. At the same time, however, the EPA has been careful to point out that while extremely useful, such data are not sufficient to determine exposure or to estimate the potential adverse effects of such releases on human health and the environment.

In an effort to meet this need, at least on a partial basis, the ATSDR periodically measures the quantities of certain chemicals in blood and urine of people throughout the nation. Toxic agents being assessed include lead, mercury, and uranium; the breakdown products (or metabolites) of several organophosphate pesticides (representing about half of all insecticides used in the United States); phthalate metabolites (additives found in plastics, particularly PVC); and cotinine (a metabolite of nicotine). One of the impetuses for this program was a study by the National Academy of Sciences suggesting that as many as one in four developmental and behavioral problems in children in this country may be linked to genetic and environmental factors. Neurotoxic compounds, such as lead, mercury, and organophosphate pesticides, were cited by the Centers for Disease Control and Prevention (CDC, 2001, 2003) as possibly playing a significant role. Although the latest data confirm continuing reductions in the concentrations of lead in children, almost a half million children between 1 and 5 years of age exceed the limit. Nonetheless, progress is being made. The levels of synthetic contaminants, such as DDT and dieldrin, for example, have declined by more than 90 percent during the last quarter of the twentieth century. At the same time, it is important to note that while the declines are continuing, the rate of decrease has slowed. In addition, there are some subpopulations that are still exposed to unusually high amounts of these and other contaminants (Kamrin, 2003).

Establishing Exposure Limits

Two basic principles should be applied in setting health-based exposure limits for human populations. The first is to use human data whenever possible; the second is to use surrogate chemicals or surrogate species only when the scientific evidence indicates that they provide an appropriate basis for such application. When these principles have been properly considered, the next step is to determine whether the information in the database is relevant and appropriate for estimating effects using the existing or anticipated exposure conditions. If these criteria are fulfilled, and there is a threshold or no-effect level for the specific adverse impact on which the estimates are to be based, the data can then be used to establish an appropriate exposure limit for humans. In so doing, scientists and regulators generally incorporate a safety factor into the threshold or no-effect level observed in animals. Selection of this factor should not only reflect the confidence of the evaluator in the quality and relevance of the data but also account for differences in the susceptibility and kinetics between test and target species and between individual members of the exposed population (Doull, 1992). The magnitude of these safety factors is illustrated by the values used by the Safe Drinking Water Committee of the National Research Council (NRC) in recommending no-response levels for various toxic agents in drinking water (NRC, 1983; 1986).

- A factor of 10 was used when valid chronic exposure data existed on humans and supportive chronic data were available on other species; the factor was added to ensure protection of the more sensitive individuals.

- A factor of 100 was used when there were no data on humans but satisfactory chronic toxicity data existed for one or more other animal species; the 100 includes a factor of 10 to protect sensitive individuals, plus a factor of 10 to account for interspecies extrapolations.

- A factor of 1,000 was used when the chronic toxicity data were limited or incomplete.

Regardless of how sound these safety factors are thought to be, a basic principle of health protection is to keep all exposures as low as reasonably achievable (ALARA) (NCRP, 1993).

Applying Toxicological Data to the Environment

Whereas the laboratory toxicologist is primarily concerned with the effects of toxic chemicals on individual organisms, evaluation of the effects of these same chemicals in the environment is far more complicated. The complications arise from several sources (Moriarty, 1988):

- Different species, and different groups and individuals within a single species, may react differently to identical exposures to the same chemical.

- Some pollutants may occur in more than one form, and the determination of either the details of exposure or the resulting biological effects may be difficult. Further complicating the situation is the fact that, in many cases, the structure of individual chemicals is being changed by interactions within the environment.

- Depending on the circumstances, the interactive effects of two or more toxic chemicals may be mutually additive, synergistic, or antagonistic.

- The indirect effects of the toxic chemical may be equally or more important than the direct effects. In fact, a chemical that kills no organisms but retards development may have more ecological impact than one that is lethal.

When multiple species are involved, additional complications arise. Even though predictions of the biological effects may be correct for the species under study, other species may be significantly more vulnerable and/or susceptible. In fact, the effects of many pollutants on wildlife may pass completely unnoticed. Ideally, the goal would be to identify the first-affected species. Adding to the complications is that even when an obvious effect is observed, identification of the chemicals that are responsible is often extremely difficult.

The problems of assessing the effects of chemicals within the environment do not end here. Alterations in the physical and chemical characteristics of the environment may have an impact on the ability of a species to survive: witness the releases of sulfur dioxide into the atmosphere that result in acid rain, and airborne discharges of carbon dioxide and other chemicals that affect global temperatures. In a similar manner, lakes and streams may be enriched through the release of sewage and agricultural chemicals, which in turn leads to eutrophication and detrimental impact on the survival of certain types of aquatic life. The analysis of indirect effects of these types must take into account not only the realization that the impacts of certain airborne emissions may be global in nature but also that their concentrations and resulting impacts can vary significantly from one region of the world to another.

As a result of these complications, it is quite probable that precise predictions of the effects of chemicals within the environment are unlikely to be achieved in the foreseeable future. Nonetheless, continuing guidance is needed to make sound judgments relative to the introduction and use of chemicals, and environmental toxicologists will undoubtedly continue to direct their attention to these problems. As a general guide, the chemicals that will be most important in terms of the environment are those that have known toxic effects, that are persistent, and that are biologically concentrated by various animals and/or plants.

The General Outlook

Since the early 1990s, a variety of sophisticated analytical technologies have become available for use by toxicologists. As the previous discussion indicates, these advances are enabling them to gather detailed information on the effects of toxic chemicals both on the functions of the body, as a totality, and at the molecular level. In the latter case, the information includes data on the content of the genes in our DNA, the proteins and regulatory molecules made from these genes, and the molecules that along with these proteins form the basis of normal biological function. One of the primary challenges facing toxicologists is to dissect and interpret what these data mean in terms of the normal functioning of the human body. Once this is accomplished, they should be able to provide environmental health specialists and regulators a vastly improved scientific basis for establishing permissible limits for toxic chemicals and implementing control measures in the most cost-effective manner (Greenlee, 2002). Concurrently, events are taking place that demonstrate that the leaders of increasing numbers of industrial organizations are recognizing the benefits of environmental stewardship. Even more encouraging is that they are responding by providing funds to academic and scientific institutions to support toxicological research on the effects of chemicals. This is exemplified by the U.S. chemical industry, which, in 1999, committed more than $100 million in support of such research. Specific areas being investigated include chemical carcinogenesis; endocrine, reproductive, and developmental toxicology; neurotoxicology; and respiratory toxicology (Henry and Bus, 2000).

Still to be developed is an international system for classifying the risks of toxic chemicals on a uniform basis. The need for such a system is illustrated in several ways. The International Agency for Research on Cancer, as well as the European Union, Germany, and Sweden, classify such materials on the basis of their carcinogenic potential in humans. The Netherlands and Norway, in contrast, do not explicitly differentiate between the effects in humans versus other animals. The Netherlands classifies carcinogens according to genotoxicity, that is, those that cause DNA damage. Norway, on the other hand, classifies carcinogens according to their potency. Other differences are exhibited by the classification approaches used in Germany, where rankings are based on data on malignant tumors only, and Norway, where data on both malignant and benign tumors are considered. Norway uses both published and unpublished data, while the IARC restricts the basis for its classifications to published data. Having noted these differences, it is not surprising that a review of the status of a group of eight chemicals, as classified by these countries, revealed a consensus for only two: benzene and vinyl chloride (Seeley, 2001). Nonetheless, the fact that efforts to achieve harmonization are under way is encouraging. The identification of differences such as these represents a sound first step.

3

EPIDEMIOLOGY

ACCORDING to the National Research Council (NRC, 1991), epidemiology is *the study of the effect on human health of physical, biological, and chemical factors in the external environment, broadly conceived. By examining specific populations or communities exposed to different ambient environments, it seeks to clarify the relationship between physical, biological, or chemical factors and human health.* In contrast to the field of toxicology, which is experimental in nature and involves laboratory studies ranging from those conducted at the molecular level to those involving animals, the field of environmental epidemiology is nonexperimental and involves studies of existing human population groups that have been inadvertently exposed to one or more chemical and/or physical agents. In the sections that follow, the general principles of environmental epidemiology will be outlined. The discussion will include a review of some of the precautions that must be taken both in the design of such studies and in the analysis and interpretation of the collected data.

A Classic Example

John Snow is often considered to be the founder of epidemiology. This was based on his classic studies of the transmission of cholera in London in the mid-1800s (Monson, 1990). These studies illustrate many of the principles of a valid environmental epidemiologic study. Snow, a practicing physician, observed that people working with cholera patients did not always contract the disease, and yet people who did not have contact with infected patients often did. In this case, the cause of the disease was known; what was not known was the pathway through which people were being infected. Snow postulated the existence of some vehicle that transmits the disease and, with support from other physicians and local laypeople, hypothesized that one possibility was the presence of sewage (fecal) contamination in their drinking water. To determine the source, Snow conducted a study of population groups in different parts of the city who obtained their drinking water from different suppliers. Recognizing that other factors could influence the spread of the disease, he analyzed the mortality rates in a single subdistrict, where the only observable difference was that one portion of the population obtained its drinking water from one supplier and the other obtained its water from a second supplier. Using a chemical test that took advantage of a difference in the chloride content of the two water supplies—chlorine (Chapter 7) is a disinfectant that will kill disease organisms in water—he was able to identify the supplier of each individual household. Using these data, he confirmed that the disease was transmitted by sewage in the drinking water supplied by one of the companies (Goldsmith, 1986; Monson, 1990).

As pointed out by Monson, several factors make Snow's study a model of environmental epidemiology:

- Snow recognized an association between exposure and disease—that is, between the source of the drinking-water supply and the incidence of cholera.

- He formulated a hypothesis—that fecal contamination of drinking water was the specific agent of transmission of the disease.

- He collected information to substantiate his hypothesis—in subdistricts where the drinking water was supplied by only one company, the association was stronger.

- He recognized that there could be an alternative explanation for the association—that social class or place of residence might influence transmission of the disease.

- He applied a method to minimize the effects of the alternative explanation—he compared cholera rates within a single district or neighborhood, rather than between neighborhoods, on the basis of their water supply.

- He effectively minimized the collection of biased or false information—since most residents were not aware of the name of the company that supplied their water, he applied a chemical test to make this determination in a positive manner (Goldsmith, 1986).

These criteria have withstood the test of time and are regarded today as fundamental to the design of all types of epidemiologic studies.

Modern Environmental Epidemiology

Based on the work of Snow, early epidemiologic studies were "disease centered," and the diseases primarily involved were infectious in nature. As a result, investigators at that time relied primarily on laboratory investigations with little attention to study design. Their basic principles were that a microorganism should be considered as causally related to a disease when it was present in all subjects affected. The implication that a given agent was the source of the disease was then confirmed by isolating it in the laboratory, inoculating it into animals, and demonstrating that the animals developed the disease (Terracini, 1992).

Today the trend is to employ epidemiologic studies that are "exposure centered." This approach is an outgrowth of the realization of a multitude of factors. One is that in the developed countries of the world, degenerative diseases such as cancer, whose etiology is multifactorial, have become the prevailing pathology. The result has been an increasing awareness of the need for a rational, systematic, explicit, and reproducible approach to evaluating the associations between various diseases and environmental agents. Meeting this need requires the consideration of certain basic criteria. These have been enumerated by Hill (1965):

- The strength and specificity of the association

- The consistency of the findings in different studies

- The existence of a dose-response gradient between the exposure and the occurrence of the disease

- The biological plausibility of the proposed association

- The coherence of the evidence with the natural history of the disease

- The supporting experimental, or quasi-experimental, evidence

Although subsequent investigators have expanded on these criteria, they continue to serve as one of the foundations of modern epidemiology, much as Snow's principles did during the early years. The primary changes have been to emphasize the control of confounding variables and to improve study design (Terracini, 1992).

Design of an Epidemiologic Study

One of the first considerations in the design of an environmental epidemiologic study is to define its objective and scope. As an extreme, one might consider monitoring the health records of the whole population and linking that information with as many data on environmental factors as possible. Basic to such a study would be national death statistics and records on morbidity. To extend this type of study to include inquiries into the "health and habits" of individual members of the population on a national scale, however, might be considered an intrusion on privacy—and the financial costs would be prohibitive. Nonetheless, if success is to be achieved, the World Health Organization (WHO, 1983) has emphasized that some form of additional data gathering may be required.

An alternative approach is to focus on small groups of people considered to be at risk. The objective in this case would be to consider a specific disease or effect and to compare the available information on exposures in this group to those in a "control group." Depending on the type of study, the control group is generally one that either has not been exposed to the agent in question or does not have the disease being investigated. Because it is unethical to expose people to potentially hazardous environmental agents solely for purposes of epidemiologic study, essentially all such studies are nonexperimental. As a result, it may be difficult to define or quantify the exposures received by the population group being evaluated.

Three of the multitude of ways in which environmental epidemiologic studies can be classified are described below.

COHORT STUDY

This is one in which a population that has received unusual exposures is followed over time to determine what diseases they develop and whether there is an increase in the incidence of those diseases that might be presumed to have been caused by the exposures. A cohort study may be either prospective, in which case the disease has not yet occurred at the time the exposed or nonexposed groups are defined, or retrospective, in which case the disease has already occurred. The epidemiologic studies of the survivors of the World War II atomic bombings in Japan exemplify a prospective cohort study (NRC, 2006). The purpose in this case was to follow those who had received radiation exposures (and survived) to determine whether, and to what degree, they later developed cancer. The study by John Snow was an example of a retrospective cohort study. Nonetheless, in both cases the health effects are identified over the follow-up period of interest and the data analyzed in a similar manner. The observed health effects are then compared to those expected based on an appropriate control group and/or related to variations in the estimated doses.

CASE-CONTROL STUDY

In this case, people who are known to have a specific disease are examined to determine what if any exposures that they are receiving now or have received with unusual frequency in the past might have been the source of the disease. Early epidemiologic studies of the relationship between cigarette smoking and lung cancer (Doll and Hill, 1950) are examples of this type of study. This approach has also been used to evaluate various diseases in occupational settings, one example being the associations between certain illnesses and pesticide exposures (Cantor et al., 1992). As might be anticipated, various combinations of approaches are often included within a single study.

ECOLOGIC STUDY

This is a third type of study that might be considered under some circumstances. In this case, disease rates are compared for groups of subjects for whom exposures are judged to differ, and the groups are usually defined based on geographic locations in which mortality and morbidity rates are known. This being the case, such studies can often be conducted quickly and inexpensively. The data sources in such studies, however, do not always provide the best delineation of exposure, particularly in cases where the current locations of segments of the population do not reflect the nature or extent of their past exposures (i.e., they have moved frequently, such as in the case for families of military parents). For these and other reasons, ecologic studies are generally regarded as hypothesis generating, at best, and the outcomes must be regarded as questionable until confirmed by either cohort or case-control studies (NRC, 1995).

The basic differences in the various types of epidemiologic studies can be summarized as follows (Monson, 1990):

- In a cohort study, individuals are included on the basis of whether they have been exposed; in a case-control study, individuals are included on the basis of whether they have the disease being evaluated.

- In a prospective cohort study, the disease has not occurred at the time the exposed and nonexposed groups are defined; in a retrospective cohort study, the disease has already occurred.

- In a prospective cohort study, the investigator usually compares the disease rates of two or more groups (for example, smokers and nonsmokers); in a retrospective cohort study, mortality rates among the exposed group are compared to mortality rates of some general population—no specific comparison group is identified.

- In a case-control study, the past history of exposure is the primary information that is collected. As a result, such studies can be completed relatively quickly. In contrast, because time must pass in order for the disease to develop, completion of a prospective cohort study often requires a relatively long period of time.

- Because of the length of time that prospective studies require, the general approach is to evaluate a number of exposures in relation to one disease. This is in contrast to a cohort study, in which one exposure is evaluated in relation to a number of diseases.

- In an ecologic study, the sources of the data are general, at best, and the results can always be challenged in terms of their scientific validity.

In the conduct of current environmental epidemiologic studies, the general approach is not to compare an exposed and a presumably nonexposed group. Instead, it is to compare the incidence of a given disease as a function of the degree, extent, or amount of exposure. This approach has been adopted because it is often difficult to identify persons who have not been exposed at all to a given physical or chemical agent.

Other Challenges

Environmental epidemiologists continue to face a variety of other challenges. Several of the more important of these are discussed below.

EXPOSURE ASSESSMENT

As mentioned earlier, assessment of the exposures to which a population study group has been subjected is a crucial but often inadequately addressed component of epidemiology. A difficulty is that assessments in the workplace (Chapter 4) require different approaches than those in the home or ambient (outdoor) environment (Chapter 14). The same is true in assessing exposures to different types of agents. Regardless of these challenges, valid environmental monitoring and accurate estimates of exposures, particularly those in the ambient environment, are essential if confidence is to be placed in the associations that are developed between exposures and observed adverse consequences to human health (NRC, 1991). The discussion here will apply primarily to assessments of environmental exposures.

An additional complication is that exposures to physical agents, such as noise or vibration, are generally transitory and the resulting assessments must therefore be made on a real-time basis. Unfortunately, in some cases (as with electric and magnetic fields), assessment personnel do not yet fully understand which parameters are indicative of exposure (Brain et al., 2003). Nor do they know whether it is average or peak exposures that are important. In the case of chemical and radioactive contaminants, the field of environmental monitoring and exposure assessment requires consideration of the source of the contaminant, its associated pathways of exposure, its avenues of transport through each medium, its routes of entry into the body, the intensity and frequency of contact with the contaminant of the persons exposed, and its spatial and temporal concentration patterns. The importance of such movement and interactions is exemplified by the fact that the composition as well as the physical form of chemical contaminants can be readily altered by any or all of these factors. The progression from the release of a contaminant, to its movement through the ambient environment and uptake by humans, to the production of associated health effects is depicted in Figure 3.1.

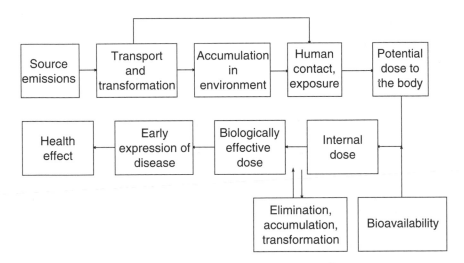

Figure 3.1 Progression of factors that influence the behavior of a contaminant within the environment, its uptake by humans, and the resulting health effects

For these reasons, an accurate assessment of exposures from airborne particulates requires, for example, not only identifying what the contaminant is, its physical form (amorphous, crystalline, discrete particulate, or fibrous), and particle-size distribution but also, in many cases, its physico-chemical surface properties. Especially complex is the assessment of exposures necessary for cross-sectional studies of the effects of air pollution. In earlier times, such studies typically involved a comparison of the health of populations in communities having different ranges of specific contaminants in the outdoor air. It is now recognized that people spend 90 percent or more of their time indoors. This represents another challenge. Thus, the assessment of their exposures must include determination of the concentrations of airborne contaminants both inside their homes and places of work.

As will be noted later (Chapter 5), factors that can contribute to airborne exposures within a home include personal or family eating habits, the type of cooking facilities (natural gas or electricity), personal hobbies and recreational activities, pesticide applications within the home and garden, and the nature of the domestic water supply. Contaminants released into the air during showering, bathing, and cooking may also represent sources of exposure through inhalation (NRC, 1991). The hierarchy of exposure data or surrogates, ranging from quantified measurements of individual exposures to simply knowing the person's residence or place of employment, is depicted in Table 3.1.

Table 3.1 Hierarchy of exposure data or surrogates

Types of data	Approximation to actual exposure
Quantified personal measurements	Best
Quantified area or ambient measurements in vicinity of residence or other sites of activity	
Quantified surrogates of exposure (e.g., estimates of drinking water and food consumption)	
Residence or employment in proximity to site of source of exposure	
Residence or employment in general geographic area (e.g., county) of site of source of exposure	Poorest

HEALTH ENDPOINTS

A second major challenge in the design and implementation of an environmental epidemiologic study is the selection of the health endpoints to be evaluated. Formerly it might have been adequate simply to determine whether the chemical or physical agent in question was causing an increase in the number of deaths (mortality) or hospital admissions (morbidity) among the exposed population. Subsequently, the potential increase in the incidence of cancer became the health endpoint (or indicator) of primary importance. Today environmental and public health officials, as well as the public at large, have a wider range of concerns, including the possible impacts of environmental agents on the quality of life, associated health-care costs, and the accompanying years of life lost.

STATISTICAL VALIDITY

Even if all the other problems were resolved, the successful application of the principles of epidemiology generally requires small population groups exposed to a relatively large dose of a very toxic material. In other cases, this is often next to impossible, the major limitation being to ensure that the resulting data are statistically valid. This is exemplified by the data in Table 3.2, which are based on the doses commonly encountered by radiation workers and members of the public. The unit of dose in this case is the millisievert (mSv). For purposes of perspective, the average annual dose to members of the U.S. public from natural background (i.e., cosmic and solar radiation, naturally occurring uranium in building materials, and radon in the home, etc.) is about 3 mSv. In case of a major nuclear accident, the doses can be much higher. For example, the doses to the workers who were involved in the recovery of the 1986 accident at the Chernobyl nuclear power plant in Ukraine (then part of the Soviet Union) ranged up to hundreds of mSv.

For those and other reasons, epidemiologists have never been able to confirm any differences in the health (i.e., cancer rates) among population groups such as those living in the mountains (i.e., areas of relatively high natural background) and those living at sea level (i.e., areas of low background). For purposes of perspective, the average annual dose to workers at U.S. commercial nuclear power plants is less than 1 mSv. This being the case, epidemiologists have not been able to document any ill effects of their radiation exposures. This subject will be discussed in more detail in Chapter 12.

Table 3.2 Size of exposed population group and radiation dose required to detect an increase in total cancer mortality, assuming lifetime follow-up

Mean whole body dose (mSv)	Excess cancers per 10,000 population	Required sample size
2.5	1.9	32,000,000
5.0	3.8	7,900,000
10.0	7.5	2,000,000
20.0	15.0	500,000
30.0	22.5	220,000
40.0	30.0	130,000
50.0	37.5	80,000
60.0	45.0	56,000
70.0	52.5	41,000
80.0	60.0	31,000
90.0	67.5	25,000
100.0	75.0	20,000
120.0	90.0	14,000
150.0	113.0	9,100
200.0	150.0	5,200

Other Complications

Other factors add to the complexity of epidemiological studies. If, for example, the only effect of an agent at a given intensity is a small change in bodily function, well within an individual's normal physiological range of variation, then its importance in comparison with other factors affecting health must be carefully weighed. Competing factors that must be considered include the duration of the effects and the number of persons likely to be affected. The relative importance of a minor immediate effect versus a potentially more serious but delayed effect must also be evaluated (WHO, 1983). A key criterion is whether the chemical or physical agent being evaluated has been demonstrated to be capable of causing the suspected effect. Unless it has, successful conduct of the proposed study may be seriously impaired.

Assessment of any of these endpoints requires some standardized measure of effects. The indicators that have been developed for measuring behavioral effects of noxious environmental agents, for example, fall into two broad groups: (1) measures of psychological and psychophysiological functioning and (2) measures of mental state and behavior. Psychological tests have proved effective in the detection and measurement of organic brain damage. In a similar manner, relatively simple techniques, such as Raven's progressive matrices and vocabulary and memory tests, have proved

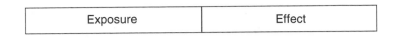

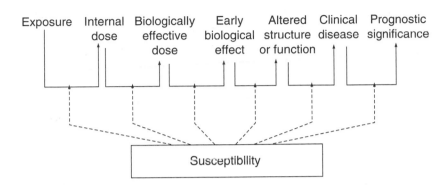

Figure 3.2 Relation of susceptibility, exposure, and effect

both reliable and practical in field studies involving the screening of large numbers of individuals (WHO, 1983).

An emerging development in resolving these problems is the use of biological measurements made at the molecular level as indicators (or biomarkers) of the effects of a particular environmental agent or stress. A distinguishing feature of these indicators is the increased analytical sensitivity they provide. In fact, they can be used to describe events along the continuum between exposure and clinical disease and help assess the relationship between susceptibility, exposure, and effect (Figure 3.2). This is an outgrowth of the field of molecular toxicology, discussed earlier. It has, in turn, led to what is called *molecular epidemiology*, an approach that is now being widely applied to the field of cancer epidemiology. Application of these techniques can, for example, be used to measure DNA damage, heritable genetic polymorphisms that influence susceptibility, and "cancer family" genes. Especially exciting is the discovery that it may be possible to identify environmental carcinogens, including radiation, based on the genetic fingerprint they leave on cells that become cancerous (Hande et al., 2003).

Conduct of an Environmental Epidemiologic Study

The practical problems in the organization of an environmental epidemiologic study include both the level of study to be conducted (simple to complex) and the resources required. One of the first objectives is to identify the population group to be examined. It is often helpful to consider the conduct of an initial study among workers who may be exposed to the same agent. One advantage of this approach is that exposures in occupational settings are often higher than in the general environment. At the same time, it should be borne in mind that a working population is preselected. It excludes children, the elderly, and those whose health is already impaired, as well as individuals who may be hypersensitive to certain agents. For certain occupations, the working population also frequently includes a disproportionately low number of women. Furthermore, exposures of workers are limited in most cases to 8 hours a day. As a result, caution must be exercised in the extrapolation of the resulting observations to the general population (WHO, 1983).

Once the study group has been identified, contacts need to be established with individuals within the group to guarantee their interest and cooperation. Where individuals decline to participate, care must be taken that their response does not bias the results of the study. If a number of people are engaged in collecting information, joint training sessions are required to ensure uniformity of approach, and it may be necessary and beneficial to interchange the teams periodically during the data-collection period. Experience has shown that the most effective approach, where the effects of common environmental agents on individuals within the general population are being studied, is to have the data collectors visit the subjects in their homes. Although this technique is labor intensive, it is often justified by the improved quality of the results. Any instruments used to collect data need to be calibrated on a regular basis, and all related methods should be standardized. If biologic indices of effects are used, it may be necessary to have all measurements performed in a single laboratory.

Ethical problems may also arise. If some tests are intrusive (for example, the collection of blood samples), prior permission will be required. Confidentiality is another issue. Thus it is common practice to include the names and addresses of those interviewed only on the original survey form.

As indicated by the World Health Organization (WHO, 1983) and Monson (1990), the computer has had a revolutionary impact on the conduct of environmental epidemiologic studies. In fact, the dramatic increase in the number of such studies since the 1950s is directly related to the development and wide availability of these devices. As Monson has stated, the ability to collect large amounts of data, to store them, and to conduct extensive analyses is "the hallmark of epidemiology today." This is especially true of data showing weak associations between exposure and effects. Unfortunately, however, the computer has separated many epidemiologists from the data they are analyzing. As a result, they may not be familiar with weaknesses inherent in the collection of the data or with limitations in the computer programs used.

Classic Case Studies

Many environmental epidemiologic studies have served as examples of the beneficial uses and applications of this methodology. One of the earliest documented the fact that the intake of fluoride in drinking water led to a reduction in dental caries (Terracini, 1992). Subsequent studies (Chapter 12) led a quantification of the relationship between the dose from ionizing radiation and the induction of a fatal cancer (NRC, 2006). Two others, the relationship between cigarettes and lung cancer and between airborne concentrations of extremely small airborne particles and population death rates, are discussed here.

CIGARETTES AND LUNG CANCER

The history of the documentation of a definitive association between cigarette smoking and lung cancer is a classic example of the useful application of environmental epidemiology. It is also an example of how the personal choices of individuals can have an extremely detrimental effect on their health and of how difficult it is, even when a relationship has been thoroughly demonstrated, to implement effective control measures.

In the middle to late 1940s, physicians in several of the industrialized countries of the world, including the United States and the United Kingdom, observed an increasing number of diagnoses of men with lung cancer. A decade earlier, such cancer had been a medical curiosity. Although cigarette smoking was immediately suspected as a cause, obviously the presumption had to be confirmed. Two types of studies were undertaken—case-control studies in which persons with and without lung cancer were asked about past habits, including smoking, and cohort studies in which smokers and nonsmokers were followed and the rates of development of a variety of diseases, including lung cancer, were measured (Monson, 1990).

One of the leading epidemiologists who conducted such studies was Richard Doll, working first with A. Bradford Hill and later with Richard Peto. On the basis of an initial case-control study, Doll and Hill (1950) concluded that "smoking is a factor, and an important factor, in the production of carcinoma of the lung." They admitted, however, that they had no evidence about the nature of the carcinogen. Based on a subsequent series of longer-term cohort studies, Doll and Peto (1976) concluded that the death rate from lung cancer in smokers was ten times that in nonsmokers. These studies and related research culminated with the issuance in 1964 of the surgeon general's report "Smoking and Health" (USPHS, 1964). Interestingly, one of the principal actions that finally brought about a noticeable reduction in cigarette smoking in the United States was the publication by Trichopoulos (1994) of the results of his long-term epidemiologic studies that documented that nonsmokers were harmed by "secondhand" (sidestream) smoke. Related studies conducted by the Environmental Protection Agency (EPA) led to the conclusion that secondhand smoke was responsible for approximately 3,000 lung cancer deaths annually among nonsmoking members of the U.S. public. Furthermore, infants and young children are especially sensitive to other types of health effects. These include an annual estimated 150,000 to 300,000 cases of respiratory infections, such as bronchitis and pneumonia, among infants and young children up to 18 months of age, and similar negative effects on up to 1 million children who have asthma (EPA, 1992). More recent studies at the University of Glasgow in Scotland showed that following a ban on smoking in enclosed public areas in 2006, 17 percent fewer people were admitted to the hospital for heart attacks and acute coronary problems and that the decline was especially dramatic in nonsmokers, suggesting that secondhand smoke is a risk for heart attack (Pell et al., 2008).

EFFECTS OF AIRBORNE PARTICLES

In 1974, researchers at the Harvard School of Public Health initiated a study of the relation between human respiratory health and the concentrations of particulate matter (PM) and sulfates (a component of smaller particles) in ambient air within the United States. The study involved a random sample of more than 8,000 people living in six eastern cities. One of the major findings, based on the results of 15 years of subsequent observations, was that death rates among the study populations correlated with the concentrations of fine particulate air pollution in the communities in which they lived (Dockery et al., 1993; Pope et al., 1995; 2002). The particles that proved to be most significant were those 2.5 micrometers ($PM_{2.5}$) or less in size. Airborne releases from motor vehicles and power plants were identified as the primary sources. Exacerbating the situation, such particles are also formed by photochemical transformations in the air (Chapter 5).

A similar study was undertaken by the American Cancer Society in 1980. In this case, the study population included more than 500,000 people residing in 154 cities. In this case, increased deaths were found as a result of exposures to particles in the same size range even though the concentrations of particles smaller than 10 micrometers (PM_{10}) in the air in the cities in which the people lived complied with the 1987 air-quality standards mandated by the U.S. Congress (1987). Even though the increase in mortality was small, nationwide it resulted in an estimated up to 60,000 deaths per year. Reacting to these findings, the American Lung Association sued the EPA seeking a review of the air-quality standards for particulate matter that are required to be designed to protect the public health. In 1997, the EPA announced new regulations to limit the concentrations of $PM_{2.5}$. Although these regulations were subsequently challenged, in 2001 the U.S. Supreme Court unanimously ruled in favor of the EPA's action.

Subsequent studies at the Johns Hopkins University not only confirmed these observations but also provided compelling evidence of the relationship between air pollution and lung cancer and heart disease. Based on an analyses of the data, it was estimated that there is a 6 percent increase in deaths from heart- and lung-related causes and an 8 percent increase in deaths from lung cancer for each 10 micrograms per cubic meter ($10\,\mu g/m^3$) increase in the concentrations of fine particulates ($PM_{2.5}$) in the air (Pope et al., 2002).

Revolutionary Change in Applications of Epidemiology

Prior to the observations described above, it had been assumed that the techniques of environmental epidemiology were generally not designed— nor should they be expected—to prove that a given environmental agent causes a specific disease or health effect. In most cases, the best anticipated outcome was to demonstrate a *relationship* or *association* between a given agent and one or more specific health effects. This was an outgrowth of the extended studies of the health effects of ionizing radiation. To "prove" that a given cancer was due to a specific environmental or occupational stress was considered extremely difficult, primarily due to the fact that (1) radiation is a very weak carcinogen and, for that reason, the health effects in the low-dose region are minimal, and (2) as previously discussed, the required magnitude of the exposed population, to detect an increase in total cancer mortality, is so large. The first challenge is illustrated by the limited usefulness of the information in the low-dose region (Clarke, 2008; Figure 3.3). To document this fact, he has stated that "in this zone the relationship is irrel-

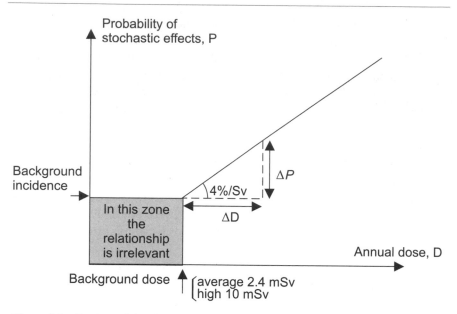

Figure 3.3   Region of the dose-response curve above where risk factors apply

evant." The second was documented by the information that was provided in Table 3.2 (page 52).

For these reasons, when decisions need to be made to control suspected agents within industry or the community at large, many aspects of the situation must be taken into account—the strength and consistency of the association, toxicological and clinical findings, and the economic and social implications of control measures. An ancillary consideration is whether there is a plausible mechanism through which the given chemical or physical agent can cause the suspected effect. With the exception of the health effects of cigarette smoking, this situation continues to exist today. Interestingly, the overwhelming effects of tobacco were due, in large measure, to the fact that there is a synergistic effect between the chemical and naturally occurring radioactive carcinogens in tobacco (Darby et al., 2005; EPA, 2003).

The General Outlook

The role of toxicological and epidemiologic studies in environmental and public health continues to expand. Initially, the primary outcome of such studies was to provide information on the relationships between a given environmental stress and one or more diseases. This being the case, one possible response would have been to view this field as a "negative" science, that is to say, the outcomes or information it provided was predominately "bad." As noted earlier, the use and application of such studies have broadened considerably. They now provide an important tool for the amelioration and perhaps the prevention of the impacts of environmental contaminants on public health and the environment. Equally important are the studies of the correlations between factors, such as diet and exercise and health, through which epidemiologists are illuminating steps that can be applied for enhancing personal health and well-being. Another outcome is the identification of voids in our knowledge, the net result being that the information can be used to identify areas in which additional research would be beneficial (Muirhead, 2001).

If the benefits of epidemiologic studies are to be realized, however, it is important that the outcomes of such studies be adequately communicated to the media and the public. This is especially true in terms of the benefits of the lifestyle choices, mentioned above, relative to diet and exercise (Chapter 1). Otherwise, these benefits may not be realized by major segments of the population.

4

THE WORKPLACE

As EARLY as the fourth century B.C.E., Hippocrates apparently observed adverse effects on miners and metallurgists caused by exposure to lead. In 1473, Ulrich Ellenbog recognized that the fumes of some metals were dangerous and suggested preventive measures. In the early 1500s, Georg Bauer (known as Georgius Agricola), a physician and mineralogist, attributed lung disease among miners in the Carpathian Mountains to the inhalation of certain kinds of mineral dusts, observing that so many miners succumbed to the disease that some women were widowed as many as seven times. In 1700, Bernadino Ramazzini published the first complete treatise on occupational diseases, *De Morbis Artificum Diatriba* (Diseases of Workers). As a result of this pioneering effort, Ramazzini is known as the "father of occupational medicine" (Franco, 2001). In the mid-1880s, Karl Bernhard Lehmann, whose work continues to serve as a guide on the effects of exposure to airborne contaminants, conducted experiments on the toxic effects of gases and vapors on animals. During the same period, the first occupational cancer—scrotal cancer in chimney sweeps—was observed in England (Clayton and Clayton, 2001).

In the United States, occupational health problems received little attention until the twentieth century. Although the U.S. Bureau of Labor was created in 1885, and even when it became the Department of Labor in 1913, its stated goals included no mention of the health of workers beyond "promoting their material, social, intellectual, and moral prosperity" (U.S. Congress, 1913). Alice Hamilton's classic work, *Exploring the Dangerous Trades*, now perhaps the most widely quoted book on this field in the world, was not published until 1943 (Hamilton, 1943).

The Field of Occupational Hygiene

Today, the profession that has primary responsibility for recognizing, evaluating, and controlling hazards in the workplace is known in most countries of the world as *occupational hygiene*. The exception is the United States, where it is identified as *industrial hygiene*. In either case, the primary responsibility of those working in this field is to address the full range of chemical, biological, and physical hazards, including the musculoskeletal problems that are becoming increasingly common in the modern technological world. As the scope of the challenges imply, if industrial hygienists are to address these problems effectively, they must have the abilities to combine the skills and knowledge of people working in the physical sciences and engineering, as well as in the health sciences and medicine. They must also be able to apply relevant information being generated in the fields of toxicology and epidemiology (Herrick, 1998). If successful in these efforts, industrial hygienists have an opportunity of demonstrating the effectiveness of the public health intervention model (Chapter 1), which emphasizes the prevention of the development of disease as contrasted to delaying until people become ill and then seeking to cure them.

Protective Legislation

As exemplified by the establishment of the U.S. Department of Labor, protective legislation came piecemeal and slowly. In 1908, the federal government provided limited compensation to civil service employees injured on the job. In 1911, New Jersey became the first state to enact a workers' compensation law. Although many other states rapidly followed suit, it was not until 1948 that all the states in the United States required such compensation (Clayton and Clayton, 2001).

The first significant federal legislation for workers not involved in government projects, however, was not enacted until 1969, when the Federal Coal Mine Health and Safety Act was passed. This legislation was followed by the landmark Occupational Safety and Health Act of 1970, whose principal purpose was to prescribe the standard "which most adequately assures, to the extent feasible, on the basis of the best available evidence, that no employee will suffer material impairment of health or physical capacity" (U.S. Congress, 1970). Among its provisions were the establishment of the Occupational Safety and Health Administration (OSHA) and the creation of the National Institute for Occupational Safety and Health (NIOSH). The

Table 4.1 Significant federal legislation pertaining to occupational health and safety

Year	Act	Content
1908	Federal Workers' Compensation Act	Granted limited compensation benefits to certain U.S. civil service workers for injuries sustained during employment
1936	Walsh-Healey Public Contracts Act	Established occupational health and safety standards for employees of federal contractors
1969	Federal Coal Mine Health and Safety Act	Created forerunner of Mine Safety and Health Administration; required development and enforcement of regulations for protection of mine workers
1970	Occupational Safety and Health Act	Authorized federal government to develop and set mandatory occupational safety and health standards; established National Institute for Occupational Safety and Health to conduct research for setting standards
1976, 2000	Toxic Substances Control Act	Required data from industry on production, use, and health and environmental effects of chemicals; led to development of "right- to-know" laws, which provide employees with information on nature of potential occupational exposures
1990	Pollution Prevention Act	Established policy to ensure that pollution is prevented or reduced at source, recycled or treated, and disposed of or released only as last resort; led to substitution of less toxic substances in wide range of industrial processes, with significant reductions in worker exposure
2000	Energy Employees Occupational Illness Compensation Program Act	Provides monetary compensation and medical benefits to energy employees who have developed certain types of cancer that have been determined to have resulted from occupational radiation exposure covered under the act

stipulated purposes of OSHA were to encourage the reduction of workplace hazards, to provide for occupational health research, to establish separate but dependent responsibilities and rights for employers and employees, to maintain a reporting and record-keeping system to monitor job injuries, to establish training programs, to develop mandatory safety and health standards, and to provide for development and approval of state occupational safety and health programs. The responsibilities of NIOSH were to conduct the research necessary to establish a scientific foundation on which to base such standards and to implement education and training programs to ensure the availability of adequate numbers of qualified people to implement and enforce the Occupational Safety and Health Act (U.S. Congress, 1970).

In a later development, Congress incorporated "right-to-know" provisions in amendments to the 1976 Toxic Substances Control Act (U.S. Congress, 1976) that required employers to provide workers with information about the health hazards of their occupational environments. This stipulation has made it much easier for workers to be aware of the hazards they face and to raise questions about the protection being provided. Another law that has significantly reduced occupational exposures is the Pollution Prevention Act of 1990 (U.S. Congress, 1990). This law, which established a national policy to encourage the prevention of pollution at the source, with disposal to the environment acceptable only as a last resort, has led to the substitution of less toxic substances for use in a wide range of industrial processes. These actions, in turn, have significantly reduced workplace exposures. The latest development was the passage of the Energy Employees Occupational Illness Compensation Program Act of 2000, which provides monetary compensation and medical benefits to energy employees who have developed certain types of cancer related to the nature of their employment. A summary of the more important occupational health laws being applied in the United States is shown in Table 4.1.

Identification of Occupational Health Problems

Records maintained by the National Safety Council (NSC, 2009) show that almost 150 million men and women are gainfully employed in the United States. To some degree, all of them are exposed to occupational hazards and are at risk of job-related adverse health effects. The largest number of nonfatal occupational injuries and illnesses are distributed among the following age groups: 35 through 44 years of age (28 percent), 25 through 34

(24 percent), and 45 through 54 (21 percent). These impacts are compounded by the fact that about 19 percent of U.S. workers are employed in businesses that have fewer than 20 employees, and about 36 percent in companies with fewer than 100 employees. Unfortunately, smaller companies often lack the capabilities to identify occupational health hazards and the funds to finance associated control programs; moreover, many (for example, agricultural firms with fewer than 11 workers) are usually exempt from state and federal occupational health and safety inspections (Chapter 11). The ten leading work-related diseases and injuries are summarized in Table 4.2.

The effects of occupational exposures range from lung diseases, cancer, hearing loss, and dermatitis to more subtle psychological effects, many of which are only belatedly being recognized. Workplace exposures include those to airborne contaminants, ionizing radiation, ultraviolet and visible

Table 4.2   The ten leading work-related diseases and injuries, United States, 2010

Type of disorder or injury	Examples
Occupational lung diseases	Asbestosis, byssinosis, silicosis, coal workers' pneumoconiosis, lung cancer, occupational asthma
Musculoskeletal injuries	Disorders of the back, trunk, upper extremity, neck, lower extremity; traumatically induced Raynaud's phenomenon
Occupational cancers (other than lung)	Leukemia, mesothelioma, cancers of the bladder, nose, and liver
Severe occupational injuries	Amputations, fractures, eye loss, lacerations, traumatic deaths
Cardiovascular diseases	Hypertension, coronary artery disease, acute myocardial infarction
Reproductive disorders	Infertility, spontaneous abortion, teratogenesis
Neurotoxic disorders	Peripheral neuropathy, toxic encephalitis, psychoses, extreme personality changes
Noise-induced loss of hearing	Widespread and insidious; has impaired 10 or more million U.S. workers
Dermatologic conditions	Dermatoses, burns, chemical burns, contusions
Psychological disorders	Neuroses, personality disorders, alcoholism, drug dependency

light, electric and magnetic fields, infrared radiation, microwaves, heat, cold, noise, extremes of barometric pressure, and stress. Each of these may also interact with and exacerbate the effects of other chemical, physical, or biological agents. For example, cardiovascular diseases may be related to a combination of physical, chemical, and psychological job stresses. The workplace can also be the source of a wide range of infectious diseases. Hospital workers in particular must be concerned with protection against hepatitis B, tuberculosis, influenza, and other viral infections, including acquired immune deficiency syndrome (AIDS) and severe acute respiratory syndrome (SARS). In addition, they face exposures to waste anesthetic gases, both during operations and exhaled into the air by patients recovering from anesthesia (NIOSH, 2007b).

During 2007, 3.5 million disabling injuries occurred among workers in the United States. An additional 0.5 million suffered work-related illnesses, bringing the total to about 4 million. These injuries led to a total of about 75 million person-days of lost time, which combined with continuing losses due to injuries during prior years led to a total loss of about 114 million person-days. Significant progress, however, is being made. From 1990 through 2005, the total annual number of occupational deaths in the United States was reduced from 7,500 to about 5,200. About 1,100 of these were in construction; 750 in transportation and warehousing; 550 in agriculture, forestry, fishing, and hunting; 400 in professional and business services; and another 400 in government. The highest death rates (deaths per 100,000 workers) were in agriculture (26.3), mining and oil and gas extraction (24.4), transportation and warehousing (14.5), and construction (10.0). Even more significant, the overall death rate per 100,000 workers during this same period was reduced from 7.5 to about 3.2. This represents a reduction of almost 60 percent within a time span of 25 years (Table 4.3). Worldwide, an estimated 270 million work-related injuries occur annually, with an accompanying death toll of 350,000 (NSC, 2009).

Even so, the true magnitude of the health and economic impacts of occupational disease and injury in the United States is not known. This is due to a variety of factors: *First*, the recording of data on workers' illnesses and deaths is often incomplete or erroneous, and physicians frequently fail to relate observed diseases to occupational exposures. This is particularly true for neurologically based illnesses and chronic degenerative diseases (i.e., atherosclerosis and chronic obstructive respiratory ailments). In other cases, the diagnosed cause of death may not be coded onto the death certificate. Even when the required information is available, it may not be used to

Table 4.3 Deaths and disabling injuries in various industries, United States, 2007

Industry	Workers (10³)[a]	Deaths	Change (%) since 2006	Deaths per 10⁵ workers	Change (%) since 2006	Disabling injuries
Agriculture[b]	2,045	552	−13	26.3	−9	70,000
Mining[c]	730	178	−6	24.4	−12	20,000
Transportation & warehousing	5,265	766	−5	14.5	−7	280,000
Construction	11,416	1,140	−5	10.0	−6	420,000
Wholesale trade	4,357	183	−13	4.2	−9	110,000
Utilities	851	32	−38	3.8	−39	20,000
Professional & business services	15,219	407	0	2.7	−5	180,000
Manufacturing	16,204	358	−15	2.2	−15	410,000
Government	22,207	412	1	1.8	−2	510,000
Other services[d]	6,934	113	−20	1.6	−19	130,000
Leisure & hospitality	11,972	130	−4	1.1	−6	250,000
Retail trade	16,478	156	−18	0.9	−17	450,000
Financial activities	10,249	69	−16	0.7	−16	100,000
Educational & health services	19,904	121	−15	0.6	−16	500,000
All industries[e]	147,203	4,689	−8	3.2	−9	3,500,000

a. Includes persons of all ages.
b. Includes forestry, fishing, and hunting.
c. Includes oil and gas extraction.
d. Excludes public administration.
e. Includes 2 deaths that could not be attributed to a specific industry.

promote worker protection. *Second,* because the appearance of the health effects caused by chronic exposures in the workplace is delayed, and because many workers change jobs frequently, by the time a disease manifests itself it may be difficult to relate it to a specific exposure or combination of exposures. *Third,* even if an association between a specific disease and a given toxic agent is known to exist, it is often difficult to quantify the concentration of the toxic agent to which the worker was exposed and to estimate the intake and the accompanying dose.

For these reasons, it is not surprising that workplace exposures to toxic and hazardous substances annually cause the premature deaths of an estimated 40,000 people in the United States. These result primarily from cancer, neurological disease, cardiopulmonary disease, and other maladies (CommonDreams.org, 2009). This is more than eight times the number killed in workplace accidents (NSC, 2009). Nonetheless, the number of inspections of industrial facilities is relatively small. During 2008, for example, the Occupational Health and Safety Administration (OSHA) collected about 50,000 samples in assessing such exposures. This was an estimated one-third less than those collected in 1988. In a similar manner, records show that OSHA inspectors annually visit only about 30,000 of the more than a million workplaces in the United States. In fact, three-quarters of the facilities in which workers were injured by serious accidents in 1994 had not been inspected during the 1990s. One meliorating factor is that many state and local agencies are also involved in the inspection of industrial facilities.

Economic considerations also tend to delay or reduce attempts to address occupational health problems. Sensing the urgency to maintain production and fearing the loss of their jobs, workers may disregard controls designed to enhance health and safety, especially in cases when such measures slow production or interfere with comfort. Similar economic problems lead to the reduction of the funding for inspection agencies, such as OSHA. The situation, however, is not as bad as it may seem. Companies that insure workplace establishments conduct hundreds of thousands of inspections annually. If they discover that good practices are not being observed, they immediately cancel the insurance policies or increase the premiums. This provides a strong economic incentive for industrial organizations to pay close attention to worker safety. Additional inspections are conducted by state and local agencies involved with fire, elevators, food establishments, and motor vehicles.

Types and Sources of Occupational Exposures

Years ago, most workers were employed in manufacturing. Today, only about 20 million of the workers in the United States are employed in this category; the remainder are in service industries. Even so, both types of employment have associated occupational health problems. One of the most common in manufacturing is the presence of contaminants in the air that result from industrial processes. Other problems include noise, vibration, and ionizing radiation. Common problems in the service industries include inadequate indoor air quality, low-back pain, and cumulative trauma disorders. In certain situations, problems not heretofore recognized are assuming importance. These include the need to protect workers from potential exposures to biological agents and to provide them with safe (nonslip) floors and stairs and comfortable, employee-friendly workstation environments. Three of the primary agents or factors to which workers are exposed today are discussed in the sections that follow.

TOXIC CHEMICALS

As would be anticipated, toxic chemicals play a major role in occupationally related diseases. Their two primary portals of entry are the skin and the respiratory tract (Chapter 1). Once inside the body, such agents can affect other organs, such as the liver and kidneys. The ideal way to ensure that chemical exposures are properly controlled is to ensure that the techniques necessary for assessing their toxicological risk are available and applied before they are introduced into the workplace (Burgess, 1995). Typical of the chemicals that can gain access to the body through the respiratory tract are those that are released into the air as a result of activities associated with metal fabrication, machining, welding, and brazing, as well as follow-up operations involving the cleaning, electroplating, or painting of the finished product. Included in such releases are mineral dusts, metal fumes, and resin systems used in sand bonding agents, plus carbon monoxide. Specific examples of airborne contaminants produced in the construction industry, their effects, and methods for their control are summarized in Table 4.4. The last aspect, control, will be discussed in more detail in a later section. The operations that produce some of these contaminants also generate a host of physical hazards, such as noise, vibration, and heat stress, as well as dermal exposures to cutting fluids and coolants. The last two items produce upwards of a half-million cases of dermatitis in this country each year.

Table 4.4 Types, sources, effects, and control of typical airborne contaminants in the construction industry

Contaminant	Examples	Sources of exposure	Effects	Methods for Control
Volatile organic compounds	Aromatic hydrocarbons, chlorinated solvents, formaldehyde, toluene diisocyanate	Use as solvents and additives in paints to enhance color and spreadability	Headaches, respiratory problems, allergic reactions	Ventilation, reduce or eliminate use in paints
	Urea formaldehyde	Use as binder in particleboard and hardwood plywood paneling	Brain impairment, lung cancer, and naso-pharyngeal cancers	Add scavengers to prevent formaldehyde from volatilizing
Toxic metals	Lead	Renovation and demolition of old buildings and metal structures, use of lead-base paints	Nausea, fatigue, aches and pains, damage to central nervous system	Ban use of lead paint indoors, respiratory protection, periodic tests for lead levels in blood
	Cadmium, chromium, copper, nickel, and zinc	High temperature welding of metals such as stainless steel	Metal fume fever, and chemical pneumonia	Ventilation, use of air-supplied respirators
Silica	Sand, flint, agate, and quartz	Cleaning buildings and bridges using sand-blasting equipment	Silicosis	Use of non-silica containing abrasives, respiratory protection
Asbestos	Floor tile, pipe insulation, fire-proofing materials	Refinishing tile floors, maintenance of heating systems	Asbestosis, mesothe-lioma, and lung cancer	Isolation, personal protective equipment

NANOMATERIALS

An emerging problem is the development of and widespread application of so-called nanomaterials. Their most significant characteristic is their extremely small size, which, in turn, dramatically increases the ratio of their surface area to their volume. This alters their physical and electronic properties, both of which can significantly increase their absorption through the skin and lungs. Although their associated health impacts are only beginning to be explored, they are components of more than 800 consumer products. These include food, cosmetics, clothing, toys, and scores of other everyday products. They range from sunscreen that contains tiny particles of titanium dioxide, to odor-eating socks made with atoms of germ-killing silver, to drinks containing nanosolutions that are promoted as providing amazing health effects, to stain-repellant clothing. The risk to workers is due to the fact that nanomaterials are small enough to evade the body's defenses. If inhaled, they can readily deposit in the lungs (Helman, 2009) and migrate to the surrounding tissue, where they might cause diseases similar to mesothelioma. They can also move into the bloodstream and deposit in various organs (NIOSH, 2009). Although about $1.5 billion in federal funds is annually being spent on issues related to nanotechnology, only between 1 and 2.5 percent is being applied to the study of the associated health, safety, environmental, and safety risks. Unfortunately, NIOSH currently has no regulatory power in this regard (Bass, 2009).

INDUSTRIAL ILLNESSES

More than 200,000 occupational illnesses were recognized or diagnosed by private U.S. employers in 2007, a reduction of 10 percent since 2006. The overall incidence was 21.8 per 10,000 full-time workers, with manufacturing having the highest rate (50.5 per 10,000 workers); utilities, with 38.9, had the second-highest rate, followed by those in agriculture, forestry, fishing, and hunting, which had a rate of 33.7 (NSC, 2009). The release of bioaerosols into the air of the workplace is increasingly recognized as a common problem. These include microorganisms (i.e., culturable, nonculturable, and dead microorganisms) and fragments, toxins, and particulate waste products from all varieties of living things. Exposures are especially common in the health-care industry, where respirable aerosols, which contain blood, are routinely produced in the operating room during surgical procedures. These include pathogens, such as hepatitis B virus and the human immunodeficiency virus (HIV), which causes AIDS. Workers in funeral services, medical equipment repair, correctional facilities, and law enforcement, plus those at hazardous waste sites are similarly at risk. The accompanying exposures are estimated to lead to over 9,000 infections and more than 200 deaths in the United States each year.

PHYSICAL FACTORS

Overall, slips and falls (often involving inadequate lighting) and lifting and moving heavy objects, all of which are related to these physical factors, represent almost half of all workplace injuries. Three common sources are discussed below.

> *Ergonomics.* Nearly two-thirds of the illness cases reported among U.S. workers are associated with factors involving problems of the human-machine interface, a specialty area that is designated by this term. The accompanying injuries are due to inadequate attention to the complex relationships among people, machines, job demands, and work methods, particularly during their design (Seeley, 2008). Such relationships include repetitive motions, forceful motions, static or awkward postures, mechanical stresses, and local vibration (Figure 4.1). If these are properly addressed, the performance and health of the involved workers will not be jeopardized. Nonetheless, about 1 million people in the United States miss work each year due to injuries due to such ergonomic injuries, the most common being carpal tunnel syndrome. Another is low-back pain, which

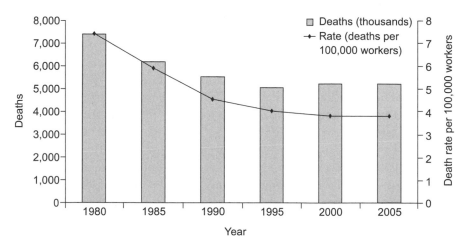

Figure 4.1   Trends in occupational injuries and deaths in the United States

can occur simply due to sitting in an office chair. Experience shows that if a worker who has suffered a low-back injury has not returned to work within 6 months, he or she will probably never return. While many such problems are chronic in nature, they can also result on an acute basis. Reaching for toilet tissue in a poorly designed hotel room can cause pain in the shoulder and arm; driving a car for several hours in heavy traffic can cause cramps in the legs, ankles, and feet. Ergonomic injuries are also a common problem among construction workers (NIOSH, 2007a). Although discussions over the past decade would imply that the recognition of such problems is new, in reality it is not. In the previously cited *De Morbis Artificum Diatriba*, Ramazzini noted that a variety of common occupational diseases were caused by prolonged, violent, and irregular motions and prolonged postures (Franco, 2001).

Noise. This is one of the most common of all occupational problems. Because hearing loss occurs gradually, invisibly, and often painlessly, many employers and employees do not recognize the problem early enough to provide protection. It can not only interfere with communication, disturb concentration, and cause stress, but it can also lead to vehicular accidents. The effects, however, do not end here. They also include elevated blood pressure and an increased pulse rate and respiratory rate that, in turn, increase fatigue. There is also evidence that excessive noise exposure to the abdomen of pregnant women during the fifth month of pregnancy can cause hearing loss in the fetus (NSC, 2009).

Heat stress and heat strain. Heat stress is the net load to which a worker may be exposed from the combined contributions of metabolic heat, environmental factors (i.e., air temperature, humidity, air movement, and radiant heat), and clothing requirements. It is a pervasive problem, especially among workers who wear protective clothing. As body temperature increases, the circulatory system seeks to cool the body by increasing the heart's pumping rate, dilating the blood vessels, and increasing blood flow to the skin. If these mechanisms do not provide sufficient cooling, the body perspires; the evaporation of sweat will cool the skin and the blood and reduce body temperature. Because sweating causes a loss of both water and electrolytes, some form of heat stress, including heat stroke, may develop if the body temperature is not reduced (ACGIH, 2008).

Occupational Exposure Standards

The American Conference of Governmental Industrial Hygienists (AC-GIH), established in 1938, has played a major role in reviewing and assessing the literature and recommending limits for the control of workplace exposures in the United States. One of its early contributions was the development of what are called *threshold limit values* (TLVs). These are based on the assumption that there is a threshold relationship between the dose from toxic chemicals and the appearance of health effects (Figure 4.2). TLVs for limiting occupational exposures to selected airborne contaminants are summarized in Table 4.5. Through the efforts of the ACGIH (2008), TLVs have been established for more than 600 chemical substances. After it was established in 1939, the American Industrial Hygiene Association joined these efforts. Today, OSHA and NIOSH also provide standards or limits, the primary difference being that they have regulatory authority, while the ACGIH and AIHA do not. A related source of information on the toxic properties of various airborne materials is provided by the "Toxicological Profiles" published by the Agency for Toxic Substances and Disease Registry (ATSDR, 2007).

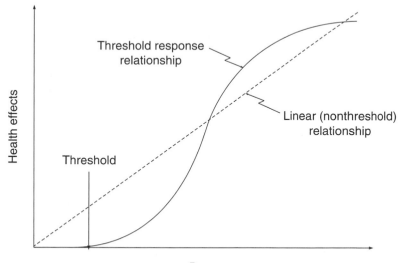

Figure 4.2 Assumptions on relationships between doses from toxic chemicals and their health effects for purposes of establishing intake limits

Table 4.5 Threshold limit values (TLVs) for selected airborne chemical substances [a]

Substance	Typical industrial uses or sources	Time-weighted average[b] ppm[d]	(mg/m^3)	Short-term exposure limit[c] ppm[d]	(mg/m^3)
Ammonia	Coke ovens	25		35	
Benzene	Gasoline refining, organic chemical synthesizing	0.5		2.5	
Carbon monoxide	Blast furnaces, coal mines	25		—	
Chlorine	Fabric bleaching, water purification	0.5		1	
Formaldehyde	Embalming, pathology	—		0.3[e]	
Lead (elemental and inorganic compounds)	Battery manufacturing		0.05[f]		—
Mercury (elemental and inorganic forms)	Fungicide applications		0.025[f]		—
Sulfur dioxide	Production of sulfuric acid, and as a preservative	2		5	
Trichloroethylene	Metal degreasing	10		25	
Vinyl acetate	Artificial leather manufacturing	10		15	

a. Depending on the chemical, the TLVs are listed in the units in which they were expressed by the American Conference of Governmental Industrial Hygienists. As may be noted, all except lead, which is listed in mg/m^3, are listed in ppm.
b. For normal 8-hour day, 40-hour workweek.
c. Not to exceed 15 minutes more than four times per day.
d. Parts per million.
e. Respirable fraction.
f. Ceiling limit.

To supplement the TLVs, the ACGIH also publishes biological exposure indices (BEIs) for more than three dozen chemicals. Recommended limits and indices for the control of selected chemicals are shown in Table 4.6. Through providing these two sets of recommendations, the ACGIH offers a two-step approach for assessing the importance of chemicals in the workplace: first, monitoring the air being breathed; second, monitoring the chemicals themselves or their metabolites in biological specimens (such as urine, blood, and exhaled air) collected from the exposed workers at specified intervals. The first step provides data on exposures of workers. The second provides data on the amount of a given contaminant that has been taken

Table 4.6 Biological exposure indices (BEIs) for selected chemicals

Chemical	Sampling time	Biological exposure index
Acetone Acetone in urine	End of shift	50 mg/L
Arsenic, elemental and soluble inorganic compounds Inorganic arsenic plus methylated metabolites in urine	End of workweek	35 µg As/L
Cadmium and inorganic compounds Cadmium in urine Cadmium in blood	Not critical Not critical	5 µg/g creatinine 5 µg/L
Carbon monoxide Carboxyhemoglobin in blood Carbon monoxide in end-exhaled air	End of shift End of shift	3.5% of hemoglobin 20 ppm
Lead Lead in blood	Not critical	30 µg/100 ml
Mercury Total inorganic mercury in urine Total inorganic mercury in blood	Prior to shift End of shift at end of workweek	35 µg/g creatinine 15 µg/L
Trichloroethylene Trichloroacetic acid in urine Trichloroacetic acid in blood	End of shift at end of workweek End of shift at end of workweek	15 mg/L 0.5 mg/L

Table 4.7 Threshold limit values for noise

	Duration/day	Sound level (dBA)[a]
Hours	34	80
	16	82
	8	85
	4	88
	2	91
	1	94
Minutes	30	97
	15	100
	7.5	103
	3.75	106
	1.88	109
	0.94	112
Seconds	28.12	115
	14.06	118
	7.03	121
	3.52	124
	1.76	127
	0.88	130
	0.44	133
	0.22	136
	0.11	139

a. No exposure should be permitted to be continuous, intermittent, or impact noise in excess of a peak C-weighted level of 140 decibels.

into the body, which, in turn, can be used to estimate the accompanying dose.

In addition, the ACGIH has developed TLVs for physical agents, including noise, heat and cold, vibration, lasers, radiofrequency/microwave radiation, magnetic fields, and ultraviolet and ionizing radiation (ACGIH, 2008). The TLVs for noise are of special interest because the latest recommendations are based on noise as a generic stress, not on the type of work or conditions under which it is generated (Table 4.7). The limits for ionizing radiation are based on the recommendations of the International Commission on Radiological Protection (Chapter 12).

Monitoring the Workplace

Workplace monitoring can be done to assess exposures of workers under routine conditions, to alert them to abnormal (accident) situations, or to design a control strategy. The type of monitoring program depends to a major extent on the nature of the stress being evaluated.

AIRBORNE CONTAMINANTS

If the source of exposure is airborne, air sampling may be the only approach necessary for assessing exposure. This is particularly true if a technique having the necessary sensitivity is readily available. If, however, both the measurement and its interpretation are difficult, a combination of monitoring techniques may be required. Although analyses of excreta, primarily urine and sometimes feces, generally provide more accurate information on such exposures, the information they provide is "after the fact." In contrast, air monitoring provides a warning of potentially unacceptable conditions. If the contaminants are particulate in nature, the information obtained should include their concentrations, size, chemical form, and solubility, since these factors affect where they will be deposited within the lungs, how effectively they will be retained, and their rate of uptake and metabolism by the body (Chapter 5).

Essentially all air samplers include a filter or sorbent collector, a fan to move the air and associated contaminants through the collector, and a means for controlling the air flow rate. The system selected depends on the purpose of the monitoring program and the type (particulate or gas) and concentration of the contaminant. The collection medium depends on the physical and chemical properties of the materials to be collected and analyzed. Particles are generally collected by means of various types of filters; gases and vapors are collected via solid sorbents and liquid reagents. The air mover may be small and serve only one sampler, or it may be a central vacuum system that serves a number of stations. Once a sample has been collected, its identification and quantification commonly require laboratory chemical or physical analysis.

A variety of sampling schemes are used. The most common are small, lightweight units that are battery powered and can be worn by individual workers. Although this permits them to be attached to the lapel or collar of a worker and collect samples representative of the air actually being breathed, it also limits the sensitivity since such samplers are often limited by the relatively low flow rate of air flow and the length of time they can be operated on battery power. As an alternative, passive samplers that collect the contaminant through diffusion or direct absorption without ancillary air flow have been developed. Another approach, but used less often, is to place samplers at fixed locations in the workplace, the positions having been selected so as to be as representative as possible of the breathing zones of the workers.

Because they need not be portable, these units can be provided sufficient power to sample at a much higher rate.

As implied by the discussion of BEIs, airborne monitoring programs may be supplemented by a variety of measurements of biological indicators of contaminants within the bodies of the exposed workers. Generally, this method of monitoring requires the collection of prescribed samples of urine, blood, sputum, hair, and body fluids and/or tissues that are analyzed for specific contaminants or their metabolites. To promote the use of the latest techniques, NIOSH publishes a manual of analytical methods, which is updated periodically (Schlecht and O'Connor, 2003). Since obtaining a representative sample is as essential as its analysis, especially in the case of urine, the required collection times are also specified (within 24 hours for urine). One of the advantages of such measurements is that the resulting data can be used to complement and/or confirm the adequacy of other types of workplace monitoring programs (ACGIH, 2008).

BIOLOGICAL AGENTS

For biological agents that are airborne, monitoring techniques closely parallel those for gases and particulates. Due to the many different types of bioaerosols that must be evaluated, however, no single sampling method or analytical procedure is optimal. Once a sample is collected, the contaminants must be identified. This generally requires microscopic examination for contaminants such as pollen grains, fungal spores, and house dust mites and culturing for microorganisms. Such techniques are continuously being expanded to incorporate newer technologies, such as gene probes and DNA amplification.

PHYSICAL AND PSYCHOLOGICAL FACTORS

Multiple measuring instruments are available for collecting real-time data for exposures to heat and noise. Assessments of the risks of ergonomic factors present an entirely different challenge. This is due to the multitude of settings in which workers are employed, the large number of interfaces between them and the equipment they use, and the increasing recognition that organizational and psychological factors may be as important as physical factors in terms of the impacts on their health. Further complicating the situation is the scarcity of data for quantifying the dose-response relationships for the associated physical, psychosocial, and organizational factors. The seriousness of these deficiencies is illustrated by observations that tasks that place high psychological demands on workers and allow them little control over the work process are causally related to atherosclerosis of the coronary arteries (Fine, 1996).

Control of Occupational Exposures

A complete and effective control program requires process- and workplace-monitoring systems, education, and the commitment of both workers and management to appropriate occupational health practices. Obviously, steps must be taken to ensure that protection is provided not only under normal operating conditions but also under conditions of process upset or failure, particularly in systems for controlling airborne contaminants. Although a majority of the problems associated with toxic chemicals can be controlled by ventilation, those associated with biological agents, particularly in the case of health-care workers, often require personal protective equipment. The situation is similar when protecting workers in the presence of physical stresses, such as noise and heat. To ensure that the best available technologies are applied, supervisory personnel must be knowledgeable of the full range of control measures available.

TOXIC CHEMICALS

Since the 1990s, the primary approach for the control of exposures from toxic chemicals has been to design each element in the manufacturing or production process to eliminate the generation of the contaminant. If this proves impossible, the second or supplementary approach is to prevent dispersal of the contaminant. If this cannot be achieved or if the degree of control is inadequate, the backup is to collect and remove the contaminant by exhausting the air into which it is released. As will be noted in the discussion that follows, there are six basic approaches that can be used to implement one or more of these goals.

Proper work practices and housekeeping. One of the best methods to prevent and control occupational injuries is to "design out," or minimize, hazards and risks early in the design process (Seeley, 2008). One of the most important components of this approach is to develop operating and maintenance procedures that minimize exposures and emissions. Examples include the use of handheld quick-response instruments to conduct periodic leak-detection surveys, the requirement that safe-work permits be obtained before a task is begun, and the use of "lockout" systems, which prevent operation of a facility except when conditions are safe. Appropriate housekeeping practices include chemical decontamination, wet sweeping, and vacuuming.

Elimination or substitution. This involves control at the source by eliminating the use of a toxic substance or substituting a less toxic one. Examples include discontinuing the use of mercury in Leclanché-type batteries and using toluene or xylene instead of the more toxic benzene in paint strippers.

Process or equipment modification. The goal in this case is to design processes so that, as far as is practical, the hazardous materials involved are contained within sealed or enclosed equipment and maintenance requirements and associated exposures are minimized. This is frequently applied to older processes that do not meet existing or proposed occupational health standards and can effectively be modified and upgraded.

Isolation or enclosure. Operations involving highly toxic materials can be isolated from other parts of the facility by constructing a barrier between the source of the hazard and the workers who might be affected. The barrier can be a physical structure or a pressure differential. Another approach is to isolate the process from a worker through the use of robots.

Local exhaust ventilation and air cleaning. Airborne gases or particulates produced by essentially all industrial operations can be captured at the point of generation by an exhaust ventilation system. Two possible types of equipment are a glove box (Figure 4.3) and a laboratory hood. Before the exhaust air is released to the environment, however, it should be passed through an air-cleaning device (such as a filter, adsorber, or electrostatic precipitator) to remove any contaminants present.

Personal protective equipment. Controls can also be applied to individual workers. The concept is to isolate the worker rather than the source of exposure. People working with heavy equipment, for example, should be provided with protective helmets, goggles, and safety shoes. Those working with corrosive and toxic chemicals should be provided with respirators, face shields, and protective clothing (with appropriate care taken to avoid the potential for heat stress). Even so, unless the use of such equipment is limited to

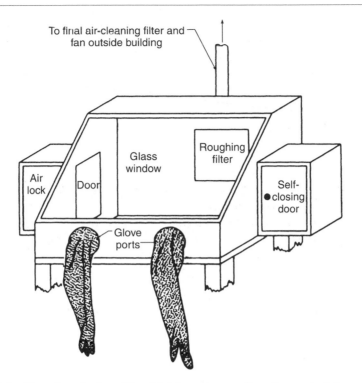

To final air-cleaning filter and
fan outside building

Glass
window

Roughing
filter

Air
lock

Door

Self-
● closing
door

Glove
ports

Figure 4.3   Glove box for handling highly toxic or radioactive materials

situations in which control at the source is not practical or has
failed, procedures for maintaining the workplace free of contaminants may either not receive adequate attention or be ignored
(OSHA, 2006).

BIOLOGICAL AGENTS

One of the best approaches to controlling airborne biological agents in the
workplace is to limit the types of environments—namely, wet spots and
pools of water—that promote the growth of organisms. Another key step is
proper maintenance of the air-handling system, especially the humidifier.
When exposures to biological agents arise primarily through puncture
wounds from contaminated needles, as in the health-care setting, the principal controls are to ensure that used needles are placed in puncture-resistant
containers, hands are washed to reduce contamination, and appropriate personal protection, such as gowns, gloves, and goggles, are worn. Control

in this case is also dependent on careful housekeeping, with specific requirements for discarding contaminated needles and other sharp instruments and proper handling of the accompanying wastes.

PHYSICAL FACTORS

As noted earlier, this category involves a range of factors. Four of these are
discussed below, the first two of which involve ergonomics.

Desk-type operations. The resolution of these types of problems
continues to be a major challenge. The laptop computer represents a typical example, the primary concern being the development of carpal tunnel syndrome by the users of such devices.
One of the major obstacles in essentially all such cases is the lack
of obvious visual indications of a problem (Keyserling et al., 2005).
Nonetheless, steps are readily available to implement controls.
Guidance on steps that can be taken in the case of the laptop
computer is illustrated in Figure 4.4. One often neglected factor is
that there is a correlation between the accuracy of vision prescrip-
tions and neck, back, and shoulder muscular skeleton problems
among computer operators. For this reason, OSHA recommends
that all computer operators have regular vision examinations
(Daum, 2004).

Musculoskeletal disorders. A typical example of this type of problem is
low-back pain. Typical methods for resolving these problems include
the use of mechanical aids to lift heavy weights, rearranging the
workplace layout to help workers avoid unnecessary twisting and
reaching, modifying seat design to permit adjustments in the height
and lumbar support, and establishing new guidelines for the packaging of products so their weights are compatible with human capabilities. Although back supports or belts are popular, their use to
avoid workplace injuries is of questionable value (Wassell et al.,
2000). The better approach is to focus on job design.

Heat stress. Another example is the previously discussed use of protective clothing and the possibility of heat stress. Control measures in
this case include reducing humidity to improve evaporative cooling,
increasing air movement via natural or mechanical ventilation,
providing radiant-reflecting shields between workers and the heat
source, reducing demands in terms of workload and duration, or
some combination of these elements.

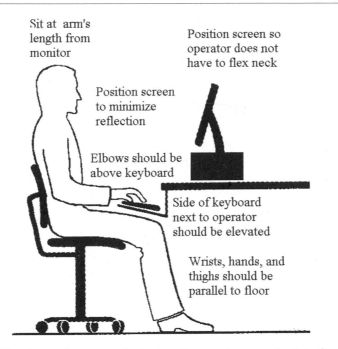

Sit at arm's length from monitor

Position screen so operator does not have to flex neck

Position screen to minimize reflection

Elbows should be above keyboard

Side of keyboard next to operator should be elevated

Wrists, hands, and thighs should be parallel to floor

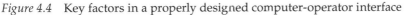

Figure 4.4 Key factors in a properly designed computer-operator interface

Noise. This problem can be controlled at the source by damping, reducing, or enclosing the vibrating surface that produces it. For instance, low-speed, high-pitch fan blades can be substituted for high-speed, low-pitch ones; sound absorbers can be placed between the source and the employees; and hearing protection can be provided to individual workers. One approach that is becoming popular is the use of headsets that contain a small computer capable of analyzing incoming noise. The information is fed to an electronic controller, which generates an opposing sound wave that in essence cancels out a portion of the incoming noise.

Special Groups

Although progress is being made in the United States in the overall reduction of workplace injuries and illnesses, as is often the case certain age and occupational groups appear not to have received the attention they deserve. Three such groups are discussed below.

TEENAGERS

Some 70 to 80 percent of all teenagers in the United States work at some time during high school (NIOSH, 2004). Typical places of employment include the services and retail trades. During an average summer, upwards of 4 million people in this age group will be involved in such activities. Because they frequently lack work experience, safety training, and an appreciation for the need to observe safe practices, this group is at a particularly high risk for injuries. For those employed in eating establishments, the most common sources of injuries are burns due to exposures to hot oil and grease, hot water and steam, and hot cooking surfaces. These and related events lead to more than 200,000 injuries each year. Even more troublesome is that 70 to 80 members of this age group die due to their injuries (CDC, 2001a).

OUTDOOR WORKERS

Many people work outdoors. These include those involved in abrasive blasting to remove surface coatings, scale, and rust in preparing large metal structures for finishing operations, such as in the repair of bridges, buildings, and ships; those employed at hazardous waste sites who may be exposed to toxic chemicals; airport workers who are exposed to air pollution and noise; building and highway construction workers subjected to the hazards of lifting cranes and earth-moving machinery; and farm workers who are exposed not only to higher rates of injury from accidents involving moving machinery but also to a wide range of toxic chemicals and pesticides (CDC, 2001b). This is particularly the case in terms of injuries to young people working on farms (NIOSH, 2007c).

HEALTH-CARE WORKERS

One of the most hazardous occupations in the United States is in the health-care industry. From the late 1980s to the late 1990s, the rates of injuries and illnesses in this industry doubled. In addition to the multiple sources of exposure described earlier, analyses have shown that, in one health-care facility, *Staphylococcus aureus* was present on more than half of the cell phones of staff members (Chapter 6). Likewise, workers in the health-care industry showed an unusually high rate of lost work days. In fact, nurses and nurses' aides had the highest claims rate for back injuries of any occupation. In fact, the overall injury and illness rate for health-care workers was higher than that for mining. Exacerbating the problem was the revelation that the number of occupational health and safety professionals employed in this profession was low compared to what would have been anticipated (Levine, 2001).

The General Outlook

As noted earlier, there has been an ongoing shift in employment in the United States from manufacturing to the service industries. This is due primarily to the outsourcing of so much of our industrial production to other countries. Other significant changes are also occurring. These include the increasing tendency of people to work at home, supported through technological innovations such as computers and e-mail. Another important factor is the increasing number of women workers, two-thirds of whom have children. In fact, at the end of 2009, the number of women in the U.S. workforce exceeded that of men. At the same time, the numbers of men running the household is increasing. Consider also that, on average, the salaries of women are only slightly more than three-quarters of that of men, and that women may be more vulnerable to certain toxic substances in the workplace. All of these changes could indicate the need to review and revise the standards for protection of workers. Additional developments in electronic communications and robotics have further decentralized our workforce. Also relevant is the fact that the U.S. workforce includes an increasing number of minorities, is increasing in age, and faces ever-stronger competition from overseas. Witness, for example, the impacts of the North American Free Trade Agreement (NAFTA).

Concurrently, increasingly complex arrays of materials, processes, equipment, and technologies are being introduced into industrial operations. Fortunately, most of these involve the use of new and less toxic chemicals and/ or the introduction of less hazardous processes and equipment. Recognizing the potential benefits, NIOSH is actively seeking to identify and promote these types of activities. Other proposed changes in industrial operations, however, have the potential of introducing new risks. To ensure that these are adequately evaluated and that effective measures are available and will be applied to reduce their workplace and environmental impacts to acceptable levels, the EPA has mandated that manufacturers document the types of controls that will be applied in what are called "pre-manufacturing notifications." This activity, in concert with the earlier cited Pollution Prevention Act, holds the promise of significantly reducing toxic chemical exposures to both workers and the public. Another encouraging sign is that corporate leaders are increasingly recognizing that an effective occupational health program can have long-term economic benefits. In fact, improvements in protecting the health of the workforce often depend more on the manner in which corporate management views this subject than on technical factors related to the control measures that must be applied. Still another change is the increasing awareness of many corporate leaders of the important role of physical activity, nutrition programs, stress management, and other positive lifestyle behaviors on the health and productivity of their workforce.

5

INDOOR AND OUTDOOR AIR

IN ITS BROADEST SENSE, air pollution may be defined as *the presence in the air of substances in concentrations sufficient to interfere with health, comfort, safety, or the full use and enjoyment of property*. From a more encompassing sense, it is *the presence in the outdoor atmosphere of one or more contaminants such as dust, fumes, gas, mist, odor, smoke, or vapor in quantities and of characteristics and duration such as to be injurious to human, plant, or animal life or to property, or to interfere unreasonably with the comfortable enjoyment of life and property* (*West's Encyclopedia of American Law*, 2010). Interestingly, records reveal that problems of air pollution have occurred for centuries. During the Roman Empire, for example, from 80,000 to 100,000 metric tons of lead, 15,000 tons of copper, 10,000 tons of zinc, and over 2 tons of mercury were used annually in industrial operations. Uncontrolled smelting of large quantities of related ores resulted in substantial emissions of these materials into the atmosphere. In the 1300s, authorities in England, recognizing that airborne contaminants were being generated by some of the industrial operations, banned silver and armor smithing. Nonetheless, it was not until 1881 that regulatory control of air pollution was initiated in the United States, when the cities of Chicago and Cincinnati, recognizing the need to control emissions of smoke and soot from furnaces and locomotives, passed the first air-pollution statutes in this country. This was followed by the passage of similar ordinances in Pittsburgh in 1895 due to emissions from local steel mills. An ordinance passed in Boston in 1911 was the first to acknowledge that air pollution has regional and national as well as local effects. Shortly thereafter, a number of county governments in the United States passed similar laws (Garth, 2001).

As is often the case, it required the occurrence of several major, acute episodes to demonstrate conclusively to policymakers and the public that air pollution could have significant health effects. The *first* occurred in 1930 in Belgium's Meuse River valley, during which high concentrations of air pollutants held close to the ground by a thermal atmospheric inversion during a period of cold, damp weather led to the deaths of 60 people. The principal sources of pollution were industrial operations, including a zinc smelter, a sulfuric acid plant, and glass factories. Most of the deaths occurred among older people with a history of heart and lung disease. The *second* occurred in 1948 in a river valley in Donora, Pennsylvania, where about 20 people died due to exposures to air pollution from iron and steel mills, zinc smelters, and an acid plant (Helfand, 2001). The *third* occurred under similar circumstances in London in 1952. This caused about 4,000 deaths and was due to airborne releases as a result of domestic coal burning (Figure 5.1). Most of those admitted to hospitals were elderly or already seriously ill and were affected by shortness of breath and coughing. Half a century later, controversy remains on whether the estimated number of people killed was accurate, the primary point being that over the next several months additional thousands may have succumbed due to delayed effects (Stone, 2002). Similar episodes occurred in London in 1959 and 1962, and analyses of death records have shown that additional episodes took place in 1873, 1880, 1882, 1891, and 1892.

More recently, concern is mounting over the effects of decades of environmentally blind industrial development in Eastern Europe and the former Soviet Union, which appear to have produced widespread threats to health and life from air pollution. Although specific data are lacking, reports indicate that high concentrations of airborne contaminants may have caused tens of thousands of people to develop respiratory and cardiovascular ailments. In some cases, air pollution was so severe that drivers had to use their headlights in the middle of the day. In many industrial areas, 75 percent of children now have respiratory disease. Outrage over environmental pollution is even said to have been a catalyst in the 1990 revolution against Communist rule in Poland (Boyes, 2009).

Today the effects of air pollution on human health and on the global environment are widely recognized. Most industrialized nations have taken steps to prevent the occurrence of acute episodes and to limit the long-term, or chronic, health effects of airborne releases. All the same, estimates suggest that up to 8 percent of Americans suffer from chronic bronchitis, emphysema, or asthma either caused or aggravated by air pollution. Newer epidemiologic data suggest that tens of thousands of members of the U.S.

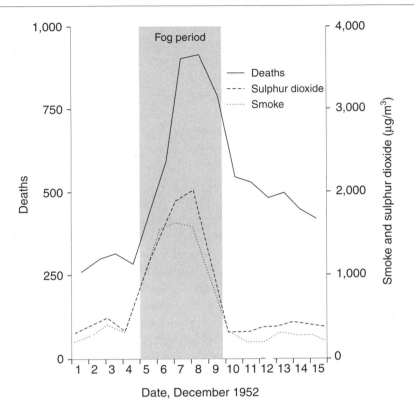

Figure 5.1 Daily mean pollution concentrations and number of deaths during the London fog episode of 1952

public may be dying annually, even though the concentrations of most airborne contaminants are within federal limits (Pope et al., 2002). The costs to society are enormous: a lower quality of life for the affected individuals, shorter life spans, and less productivity and time at work.

Human Responses to Air Pollution

The intake of pollutants into the lungs and retention at potential sites of injury depend on the physical and chemical properties of the pollutant as well as the extent of activity of the subject exposed. Gases that are highly water soluble, such as sulfur dioxide and formaldehyde, are almost completely removed in the upper airways. Less-soluble gases, such as nitrogen dioxide and ozone, penetrate to the small airways and alveoli.

The ease of entry and the sites for deposition of particulates are heavily influenced by their aerodynamic size and the anatomy of the space through which they are moving. Relatively large particles are susceptible to inertial impaction in the airways, where the flow rate is high and the passageways change direction frequently. Particles that penetrate to the small bronchiolar and alveolar region can rapidly deposit in the lungs through settling and diffusion. Fractional depositions in various regions of the respiratory tract of inhaled particles within a range of sizes are shown in Figure 5.2. As may be noted, the total collection efficiency is lowest for particle sizes of about 0.5 micrometer. The reason for this is that such particles do not settle rapidly and are too large to diffuse effectively. Another factor that influences particle delivery and deposition is the aerodynamics of respiration. Total deposition is higher and is more uniformly distributed with slow, deep breathing, as contrasted to rapid, shallow breathing.

As with all kinds of environmental stresses, the human respiratory system has a variety of mechanisms to reduce the effects of airborne pollutants. Particles ranging in size from 5 to 10 micrometers or more are effectively removed by the nose, which acts as a prefilter. Particles that are inhaled

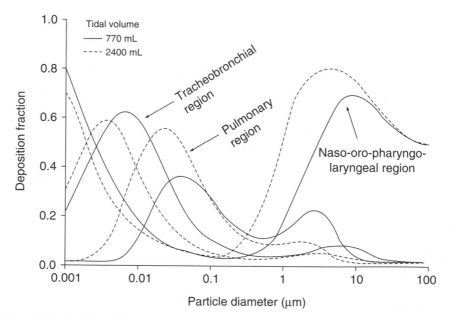

Figure 5.2   Relation of particle diameter to calculated regional deposition in the lungs for spherical particles of density 1 gram per cubic meter

and deposited in the upper respiratory tract can be removed by mucociliary action. Those that are deposited in the lower parts of the lungs can be engulfed and destroyed by cells called macrophages. Usually the cilia sweep the macrophages, along with dirt and bacteria-laden mucus, upward to the posterior pharynx, where they are expectorated or swallowed. Exposure to airborne gaseous irritants may cause sneezing or coughing and thus prevent their entry into the deeper parts of the lungs. Even if gases are taken into the lungs and absorbed, the body has mechanisms that detoxify most of them, a notable exception being carbon monoxide. Where detoxification takes place depends on how soluble the gas is in various tissues and organs and how and with what it reacts chemically. Despite these mechanisms, some pollutants will still be deposited in the body. If they remain in the lungs, they may cause constant or recurrent irritation and lead to long-term illnesses. If transported by the bloodstream to other parts of the body, they can cause chronic damage to organs such as the spleen, kidneys, or liver.

The likelihood of an adverse response to an inhaled pollutant depends on the degree of exposure, the site of deposition, the rate of removal or clearance, and the susceptibility of the exposed person. Recent epidemiologic data suggest that particulate matter (PM) 2.5 micrometers or less in size (PM$_{2.5}$) has far more impacts on health than heretofore recognized. Although the reasons for this are yet to be confirmed, one major factor could be, as noted above, the relatively high deposition in all regions of the lungs of particles in the smaller size ranges. Another factor may be the larger amounts of toxic and carcinogenic compounds that smaller particles can adsorb per unit mass. Although the mass of one 10-micrometer particle is equivalent to that of one thousand 1-micrometer particles, the total surface area of the thousand smaller particles will be at least one hundred times that of the larger particle. This is important because the chemical contaminants that attach to these smaller particles may be more important than the particles themselves (Weinhold, 2002).

Ambient (Outdoor) Air Pollution—Sources and Methods for Control

As noted in Chapter 1, there are many components of the environment, two of these being the outdoor and indoor environment. In a similar manner, air pollution can be divided into that which is outdoors and that which is indoors. The first of these will be discussed in the sections that immediately follow.

SOURCES

Sources of the major pollutants vary significantly in terms of their relative contributions. Although the combustion of fossil fuels (primarily coal) in electricity-generating plants is a primary source (85 percent) of the releases of sulfur dioxide, other activities, such as paper and pulp mills, smelters, and food-processing plants, can be significant contributors. The combustion of fuel also represents a major source of oxides of nitrogen (41 percent) and volatile organic compounds (37 percent). Petroleum refineries, solvent manufacturers, and distributors and users of their products, such as gasoline stations and dry cleaners, are also important sources. As will be noted in the discussion that follows, motor vehicles, especially diesel-powered vehicles, are a significant contributor of volatile organic compounds, plus carbon monoxide and oxides of nitrogen. As in addressing any such problem, there are many factors to be considered.

Diesel engines have long been recognized for their power and durability. Since the fuel they burn has a higher energy content, this type of engine is more efficient than those fueled by gasoline. On the negative side, airborne emissions associated with the operation of diesel engines have a host of adverse effects on human health. These include the fact that the particulate matter they release contains a number of contaminants, such as oxides of sulfur, volatile organic compounds, and aromatic hydrocarbons. The extent of the impacts of these contaminants on human health was confirmed by the EPA in 2002 when they concluded that, although uncertainties remain, diesel exhausts from large trucks and other sources probably cause lung cancer. This conclusion agrees with that of various world health agencies. Taking such factors into account, diesel engines are estimated to be the source of at least 70 percent of the total toxic risk posed by air pollutants in the United States. Exacerbating the situation, the rate of increase in the number of diesel-powered vehicles in use in this country during the last few years has been dramatic. While such vehicles consumed about 29 billion gallons of fuel in 1996, this had increased to almost 36 billion gallons by 2000. A continued increase of 2 percent per year is projected for the next several decades (Weinhold, 2002).

GENERIC METHODS FOR CONTROL

Experience has repeatedly demonstrated that it is better both economically and scientifically to prevent or reduce the production of a pollutant than to control the amount that is being released. Regardless of the care that is taken, some amounts of any toxic substance that is produced will escape through one route or another. Even if a contaminant is captured prior to its release, it still must be isolated and destroyed.

Some controls can be incorporated during the manufacture of a product. This is exemplified by the installation of emission-control devices in automobiles. In a similar manner, emissions controls can be specified nationwide for major industrial operations (such as power stations, solid-waste incinerators, and metallurgical plants) that have uniform characteristics. In other cases, however, controls must be tailored to the specific characteristics of a particular industrial operation. These include the size of the operation, the processes used, and the age and condition of the facility. On a generic basis, approaches for controlling releases of air pollutants from industrial and commercial operations can be categorized as follows. As will be noted, many of these are similar to those used to control airborne contaminants in the workplace (Chapter 4).

Atmospheric dilution. This minimal form of control is designed to take advantage of the diluting capacity of the local atmosphere to reduce the concentrations of the pollutants to acceptable levels. In so doing, a common approach is to discharge the releases through a very high stack. This, in reality, simply spreads the risk over a larger area. Due to the rapid growth in the population and the ever-increasing number of air pollution sources, this approach is being increasingly restricted. With the increasing global population and the accompanying impacts on the environment, the age-old adage *Dilution is not the solution to pollution* is definitely no longer applicable!

Substitution or limitation. This involves either eliminating a pollutant by substituting materials or methods that do not produce it, or restricting the amounts of key chemical elements available for pollutant production. Examples are using substitutes for lead to improve the octane rating of gasoline and limiting the permissible sulfur content in coal and oil burned in electric power plants. One application of the latter approach is the use of biodiesel fuel, which contains less sulfur, in diesel-powered vehicles. Adding to its benefits is that it also reduces emissions of heavy hydrocarbons, particulate matter, carbon monoxide, and carbon dioxide (Dooley, 2002).

Reduction in quantity produced and/or released. This includes improving the combustion efficiency of furnaces and adding exhaust and emission controls to motor vehicles. To ensure that such reductions are accomplished on a systematic basis, the EPA, under the New Source Review (NSR) requirements of the Clean Air Act, has applied more stringent requirements for the control of emissions from new power plants, and certain other sources, than for existing facilities. The concept was that as industries replaced their existing plants with more modern facilities, there would be reductions in the accompanying airborne emissions. As will be noted later, this goal has unfortunately not been achieved.

Process or equipment change. Typical approaches include the use of fully enclosed systems for processes that generate vapors. Examples are floating covers on tanks that store volatile fluids and the use of electric motors instead of gasoline engines as a source of power.

Air-cleaning technologies. Common examples are the applications of filters, electrostatic precipitators, scrubbers, adsorbers, or some combination of these to remove pollutants from airborne exhaust systems.

Administrative, economic, and regulatory approaches. These include programs that incorporate the provision of economic incentives to promote mass transportation and carpooling and that promote land use management to ensure that designated areas are restricted to residential, commercial, or industrial use. Another economic incentive is the concept of tradable emission credits through which the operators of an industrial facility who reduce emissions below the standard or prior to the timetable set by the law can earn credits that can be applied to future emissions or sold to an operator of another facility (Chapter 13).

Applications of essentially all such strategies or approaches involve trade-offs. Reductions in one type of pollutant, for example, frequently lead to increases in other types of environmental and public health problems. While the use of electric-powered automobiles reduces airborne emissions in metropolitan areas, certain aspects of their operation—for example, the use of lead-acid batteries—can increase an existing or create a new source of contamination. In addition, the provision of electricity to recharge such vehicles can increase the quantity of contaminants discharged from nearby electricity-generating stations. In a similar manner, the installation of scrubbers to remove sulfur from power-plant emissions produces large quantities of solid waste.

Case Studies—Control of Ambient Air Pollution

Three case studies are discussed here. These illustrate both the complicated nature of such activities and the multiple factors that must be considered and evaluated.

MOTOR VEHICLES

A review of some of the steps that have been undertaken to control emissions from motor vehicles illustrates the wide range of problems entailed in such efforts. For years, lead was added to gasoline to improve its octane rating. After multitudes of studies showed that lead was extremely toxic and that its use in gasoline was leading to an alarming increase in its concentrations in the ambient air and other components of the environment, legislation was passed requiring that its use be phased out. The results of this action, supported by the development of suitable substitute additives, were dramatic. Airborne emissions of lead into the atmosphere were reduced by 99 percent between 1980 and 2008. Additional details will be provided later.

The success in applying similar approaches to reduce other types of motor vehicle emissions, however, has been far less successful. Faced with the need to reduce the emission of compounds that serve as precursors to the production of smog, Congress incorporated into the Clean Air Act a requirement that distributors in nonattainment areas add oxygen-rich compounds to gasoline to help it burn more cleanly. Initially, one of the more widely used additives was methyl tertiary-butyl-ether (MTBE). Leaks and spills soon led to the contamination of groundwater supplies in an estimated 5 to 10 percent of the areas in which MTBE-treated gasoline was being used. Adding to the disappointment was that because MTBE is a relatively stable compound, any natural biodegradation that might take place proved to be very slow. The selection of a replacement compound, however, proved to be equally challenging. One that was considered was dimethyl carbonate (DMC). Although DMC burns even more cleanly than MTBE, its manufacture involves the use of phosgene, a poisonous gas, whose handling encompasses a host of environmental problems (Service, 2002).

Another approach for reducing the discharge of toxic compounds from motor vehicles was the installation of emission controls, a prime early example being the catalytic converter. If this and other similar devices are to be effective, however, it is important that they continue to perform as designed. To ensure that this is the case, inspection programs have been established in all regions of the United States in which certain of the federal clean air standards are being violated. Pollutants that are analyzed in such inspections typically include carbon monoxide, hydrocarbons, and, in some cases, nitrogen oxides. Owners of vehicles that fail to meet the standards are required to have their pollution-control systems repaired. Even this approach, however, is not without its challenges. Evaluations soon revealed that the inspection programs were targeting the wrong vehicles, the origin of this problem being that most state regulatory agencies devote too much of their resources to the inspection of newer cars, which have the latest control technologies and are therefore far less polluting. Instead, they should have been concentrating on the older models, which, although representing only about 10 percent of those on the road, emit about 50 percent of the most harmful pollutants. The *Cash for Clunkers* program, which encouraged drivers to replace older vehicles, proved very effective in this regard (U.S. Congress, 2009). The wisdom of this approach was also confirmed by the fact that the owners of older vehicles are generally people of limited economic means (NRC, 2001).

AIRBORNE MERCURY EMISSIONS

Airborne releases of mercury from chlorine-manufacturing plants and coal-fired electricity-generating stations had produced environmental and public health problems for many years (Lydersen, 2009b). Once in the atmosphere, this contaminant deposits into surface waters and gains access to the food chain through fish and shellfish. Interestingly, after it was mandated that chlorine-producing plants convert to a mercury-free technology, hundreds of tons of mercury they had on hand appears to have reached the global market, where it could be used in small-scale unregulated operations, such as gold mining (Lydersen, 2009b). In a similar manner, after the EPA ruled that coal-fired plants remove mercury from their airborne emissions, the control systems that were installed removed the mercury from the air, only to be subsequently discharged in their liquid waste streams. The impacts of such releases are widespread and long-lasting. Nearly half of the U.S. lakes and reservoirs contain fish with potentially harmful concentrations of mercury, as well as polychlorinated biphenyls. In fact, these two contaminants were present in all fish samples collected from 500 U.S. lakes and reservoirs during 2000 and 2003. Studies also showed that 49 percent of these waters contained mercury concentrations that exceeded the limits considered safe for people eating average amounts of fish (Associated Press, 2009). Concerned about the situation, the Environmental Protection Agency conducted public hearings in North Carolina in September 2010 on its plans to develop a proposal to regulate coal ash. The selection of North Carolina was based on the fact that it has more high-risk storage sites than any other state (State News, 2010).

SHORT-HAUL DIESEL LOCOMOTIVES

A significant source of air pollution, particularly for people living near railroad yards in metropolitan areas, are "short-haul" locomotives. Their primary task is to provide power for shifting railroad cars from track to track. Although they constitute only a small percentage of the total locomotives in service, the nature of their task means that they often remain in place while idling. As a result, they affect far more people than a train moving rapidly on a track outside a city. Compounding the problem, railroad companies typically retire their oldest, most heavily polluting locomotives to fill the "short-haul" need. Fortunately, pressure during recent years from the EPA and other regulatory agencies has led to the development of new, less polluting switch engines. As is sometimes the case, the leader in this field is a small, family-operated business, located in Illinois. Based on an innovative design and operational procedures, these new units are quieter and release no visible signs of emissions or odors. Tests show, for example, that they release up to 90 percent less particulate matter and oxides of nitrogen than the units now in use. One source of this reduction is that the new locomotives have three smaller engines, rather than one large one, thus enabling them to use only as much power as is required. The older units now in operation run at full power even when idling. As of 2009, 300 such new units have been built and are in use in the Washington, DC, metropolitan area. While this represents progress, much work remains. A total of about 5,000 short-haul locomotives are in operation in the United States (Lydersen, 2009a).

Criteria Pollutants

Airborne contaminants are considered potential pollutants in terms of their effects not only on human health but also on agricultural products and on buildings, statues, and other public landmarks. In fact, concentrations of some air pollutants considered acceptable for avoiding damage to agricultural products (so-called *secondary* standards) are lower than those considered acceptable for humans (so-called *primary* standards). That standards are needed to protect property is confirmed by many instances of damage, one being extensive discoloration of the marble of the Taj Mahal in India (Architecture Week, 2007).

Under the requirements of the Clean Air Act amendments of 1970, 1977, and 1990 (U.S. Congress, 1990), the Environmental Protection Agency (EPA) is required to identify a set of *criteria* pollutants that play key roles in terms of their effects on human health and the environment. To date, six such pollutants have been identified. Their identities and physical and biological characteristics are summarized below (CEQ, 1997; Findley and Farber, 2000).

Carbon monoxide (CO) is a colorless, odorless, poisonous gas that is slightly lighter than air. It is produced through the incomplete combustion of carbon and is primarily generated by the operation of internal combustion engines, primarily by the operation of motor vehicles. CO enters the bloodstream and reduces the delivery of oxygen to the body's organs and tissues. The health threat is most serious for people who suffer from cardiovascular disease, particularly those with angina or peripheral vascular disorders. Exposures to elevated carbon monoxide concentrations are associated with impairment of visual perception, work capacity, manual dexterity, learning ability, and performance of complex tasks.

Lead (Pb) is a heavy, comparatively soft metal that for years was used as an additive to gasoline and household paint, as well as in shotgun pellets and stained-glass windows. When inhaled, it accumulates in the blood, bone, and soft tissues. Because it is not readily excreted, it also affects the kidneys, liver, nervous system, and blood-forming organs. Excess exposure may cause neurological impairments such as seizures, mental retardation, and/or behavioral disorders.

Nitrogen dioxide (NO$_2$) is produced when fuels are burned at high temperatures. The main sources are transportation vehicles and power plants. When inhaled, NO$_2$ and other oxides of nitrogen can irritate the lungs and lower resistance to respiratory infections such as influenza. Although the effects of short-term exposure are not yet clear, continued or frequent exposure to high concentrations causes increased incidence of acute respiratory disease in children. Nitrogen oxides are also an important precursor of both ozone and acidic precipitation, and they may affect both terrestrial and aquatic ecosystems. As such, the limit for nitrogen dioxide is also designed to support the control of ozone.

Ozone (O$_3$) is formed in the atmosphere as a result of chemical reactions between oxides of nitrogen and volatile organic compounds, such as hydrocarbons (HCs). If it is inhaled, it damages lung tissue, reduces lung function, and sensitizes the lungs to other irritants. Scientific evidence indicates that ambient levels of ozone not only affect people with impaired respiratory systems, such as asthmatics, but healthy adults and children as well. Specific effects, particularly at elevated concentrations, include eye and lung irritation. Ozone is also responsible for several billion dollars of agricultural crop loss in the United States each year.

Ambient small airborne particles were, in the past, primarily produced by the combustion of fuel in stationary power plants, diesel-powered vehicles, and various industrial processes. They are also produced by the plowing and burning of agricultural fields. If inhaled, they can lead to respiratory symptoms, eye irritation, fatigue, and cardiovascular disease, as well as aggravate existing respiratory conditions, alter the defenses of the body against foreign materials, damage lung tissue, and produce latent cancers and premature mortality (Keady and Halvorsen, 2000). An emerging issue is the nanoparticle, which is vastly smaller. In fact, it has the same dimensions as biological molecules, such as proteins, which can absorb onto their surface some of the large molecules that enter the tissues and fluids of the body. Inhaled nanoparticles can deposit in the lungs and be transported through the blood into other organs such as the brain, liver, heart, and spleen, and possibly into the fetus. Oddly enough, nanoparticles are not rare. They are released by common kitchen appliances—gas and electric stoves and electric toasters (Biswas and Wu, 2005). Nonetheless, their estimated health impacts are extremely high. This is why so much increasing attention is being directed to the health effects of airborne particles in the size range of PM_{10} and $PM_{2.5}$ (Pope et al., 2002).

Sulfur dioxide (SO$_2$) is a corrosive, poisonous gas that is produced in power plants, particularly those that use high-sulfur coal as a fuel. SO_2, and oxides of nitrogen, after being released into the atmosphere can be chemically converted into sulfates and nitrates. These, in turn, may later be deposited on the Earth in the form of so-called "acid" rain or snow. At high concentrations, SO_2 affects breathing and produces respiratory illness, alterations in the defenses of the lungs, and aggravation of existing respiratory and cardiovascular disease. Sulfur dioxide can also produce foliar damage on trees and agricultural crops. It is also a greenhouse gas.

National Ambient Air Quality Standards

As guidance in the control of air pollutants, the EPA is required to establish and revise, when deemed necessary, national ambient air quality standards (NAAQS) for each of the six *criteria* pollutants (EPA, 2009). These are tabulated in Table 5.1. Although the benefits of the NAAQS are obvious, it is important to recognize that the regulated pollutants serve primarily as surrogates for many other more toxic materials in the air. These include carcinogens, mutagens, and reproductive toxins. Specific examples not covered by the NAAQS are acid aerosols, polynuclear aromatic hydrocarbons, many

Table 5.1   National Ambient Air Quality Standards for six criteria pollutants

Pollutant	Primary standard	
	Averaging time	Level
Carbon monoxide (CO)	8-hour[a]	9 ppm (10 mg/m^3)
	1-hour[a]	35 ppm (40 mg/m^3)
Lead (Pb)	Rolling 3-month	0.15 μg/m^3
Nitrogen dioxide (NO$_2$)	Annual	0.053 ppm (100 μg/m^3)
Particulate matter (PM$_{10}$)	24-hour[b]	150 μg/m^3
Particulate matter (PM$_{2.5}$)	Annual[c]	15.0 μg/m^3
	24-hour[d]	35 μg/m^3
Ozone(O$_3$)	8-hour[e]	0.075 ppm (2008)
	1-hour[f]	0.12 ppm
Sulfur dioxide (SO$_2$)	Annual	0.03 ppm[g]
	24-hour	0.14 ppm

a. Not to be exceeded more than once per year.

b. Not to be exceeded more than once per year on average over 3 years.

c. To attain this standard, the 3-year average of the weighted annual mean PM$_{2.5}$ concentrations from single or multiple community-oriented monitors must not exceed 15.0 μg/m^3.

d. To attain this standard, the 3-year average of the 98th percentile of 24-hour concentrations at each population-oriented monitor within an area must not exceed 35 μg/m^3.

e. To attain this standard, the 3-year average of the fourth-highest daily maximum 8-hour average ozone concentrations measured at each monitor within an area over each year must not exceed 0.075 ppm.

f. The standard is attained when the expected number of days per calendar year with maximum concentrations above 0.12 ppm is less than 1.

g. Has a 3-hour secondary standard of 0.5 ppm that cannot be exceeded more than once per year.

toxic metals, and volatile organic compounds. Some of these, such as mercury, asbestos, beryllium, vinyl chloride, benzene, arsenic, and radioactive materials, are controlled by emission standards, that is, through the establishment of limits on the amounts of these substances that can be released into the atmosphere through industrial operations. In a similar manner, the amount of ozone in the ambient air is regulated primarily through the establishment of limits on releases of oxides of nitrogen and volatile organic compounds (VOCs). Playing a major role in such controls are the New Source Performance Standards and limits on the amounts of VOCs permitted in, for example, varnishes and paints (Larsen, 2002).

It is anticipated that the list of criteria pollutants will be expanded as scientists identify additional pollutants that have significant impacts on human health and the environment or are deemed as being capable of serving as surrogates for other pollutants that do. In fact, the Clean Air Act Amendments of 1990 require that the EPA review and evaluate some 200 additional air pollutants. For each such pollutant, the agency must identify and quantify its major sources and specify the control technologies that are acceptable.

Also under way is a range of activities being conducted in response to the Regional Haze Rule, which was promulgated by the EPA in 1999. As the name implies, this rule was designed to remedy the effects on visibility of man-made air pollution in so-called Class I areas. These include all national parks, wilderness areas, and memorial parks larger than 5,000 acres and national parks larger than 6,000 acres that were in existence as of August 7, 1977. Specific goals of the rule, which is being implemented by the states, requires that (1) visibility be improved during the 20 percent most-impaired days, (2) there is no degradation in visibility during the 20 percent clearest days, and (3) the annual rate of visibility improvement that would lead to "natural visibility" conditions within 60 years (i.e., by 2064) be determined. The rule also requires that all major stationary sources subject to best available retrofit technology (BART) requirements be identified, using as a basis the collective benefits to visibility that would accrue through the control of such sources in Class I areas. In 2002, however, a three-judge panel of the U.S. Court of Appeals for the District of Columbia issued an opinion that the portion of the rule dealing with BART contravened the language of the Clean Air Act. It was also judged that the rule did not provide enough discretion to the states in applying the BART requirements. The latter judgment was based on the fact that the rule applied to the effects of a combination of sources. Although the court's decision left the precepts of the rule generally intact, the decision to vacate the BART provisions leaves several aspects of the rule yet to be confirmed (Jezouit and Frank, 2002).

Progress in Outdoor Air-Pollution Control

TRENDS

There are two primary indicators that are applied to evaluate progress in the control of ambient air pollution: (1) the improvement in air quality and (2) the reduction in emissions. Trends in the control of emissions of the six criteria pollutants during the last two to three decades are summarized in Table 5.2. Although they have separate standards, the data for $PM_{2.5}$ and PM_{10} particles are assessed as a combination, accounting for the differences

Table 5.2 Percent reductions in air quality improvements and pollutant
emissions, United States

	Air quality		Emissions	
Pollutant	1980–2008	1990–2008	1980–2008	1990–2008
Carbon monoxide (CO)	79	68	56	46
Ozone (O_3) (8-hour)	25	14	—[a]	—[a]
Lead (Pb)	92	78	99	79
Nitrogen oxides (NOx)	46	35	40	35
Volatile organic compounds (VOC)	—[a]	—[a]	47	31
Particulate matter	—[a]	—[a]	—[a]	57
PM_{10}	—[a]	31	68[b]	39[b]
$PM_{2.5}$ (annual)	—[a]	19[c]	—[a]	58
Sulfur dioxide (SO_2)	71	59	56	51

a. Trend data were not available.
b. Based on data since 1985; 24-hour value.
c. Based on data since 2000; 24-hour value is the same.

in their biological effects. As may be noted, there have been significant improvements both in air quality and reductions in airborne emissions. Excellent examples are those for carbon monoxide, lead, and sulfur dioxide, which have shown improvements in air quality of 68, 78, and 59 percent, respectively, between 1990 and 2008 (EPA, 2009). The corresponding reductions in airborne emissions have been 46, 79, and 51 percent. Also of note is that airborne emissions of $PM_{2.5}$ particles have been reduced by 58 percent (EPA, 2009).

Even so, many people are primarily interested in the potential effects of these materials on their health. In response, the EPA is developing an air quality index (AQI) to provide this type of information. The goal is to develop a classification system that will make it possible to express the quality of the air in terms ranging from "good" to "hazardous" on a nationwide basis. Since the system can be tied into data generated by continuously operating air monitors, the AQI is to be reported on a real-time basis (Bortnik, Coutant, and Hanley, 2002).

Indoor Air Pollution

Although few people realize it, concern for indoor air pollution has been ongoing for centuries. This is exemplified by the controls that were imposed on cigarette smoking. For example, a Mexican ecclesiastical council banned the use of tobacco in any church in Mexico in 1575; Pope Urban VII banned smoking in the Catholic Church in 1590, and Pope Urban VIII continued the ban in 1624 (Gilman and Xun, 2004). The earliest cities to do so were in Bavaria and certain parts of Austria in the late 1600s. Similar bans were promulgated in Berlin in 1723, Konigsberg in 1724, and in Stettin in 1744 (NPO Staff, 2009). The first modern smoking ban was imposed by the Nazi Party in every German university, military hospital, and post office in 1941. This last action was based on research on the hazards of smoking (Proctor, 1997). In 1975, the state of Minnesota banned smoking in most public places, making it the first state to do so. This was made possible under the Freedom to Breathe Act, passed by the State Legislature (Lohn, 2007). The first city in the world to ban indoor smoking in all public places was San Luis Obispo, California, in 1990. Interestingly, this type of action had its genesis in the restrictions placed on passengers in commercial aircraft (U.S. Congress, 1987). In 2005, San Francisco promulgated even more restrictive legislation, the goal being to prohibit smoking in some popular parks and outdoor spaces (*USA Today*, 2005).

Nonetheless, it was not until several decades ago that attention to indoor air pollution was addressed in an effective manner. That this was clearly justified is documented by two major observations. *First,* average members of the U.S. public spend from 87 to 90 percent of their time in the home or some other type of building. In fact, urban populations and some of the most vulnerable people (the young, the infirm, and the elderly) typically spend more than 95 percent of their time indoors. *Second,* indoor air pollutants can not only encompass a range of toxic materials but, in many cities, the concentrations of compounds, such as nitrogen oxides, carbon monoxide, airborne particulates, and other volatile organics, exceed those outdoors. Even if indoor concentrations proved to be lower, the longer duration of indoor exposures could render them significant when evaluated on an integrated time-exposure basis.

Adding to this interest was the increasing number of complaints by office workers and the response of the media to what was called the "sick building syndrome." These conditions were due, in part, to the responses to the so-called U.S. energy crisis of the 1970s, wherein new homes and buildings were tightened up as a means for conserving energy. In many cases, these responses included reducing the amount of fresh air brought in from the outside being circulated within such homes and buildings. Also contributing to the problem was an increase in the use of synthetic building materials and furnishings inside homes and offices. Concurrent with these developments, advances in measurement techniques increased the number of indoor contaminants being identified and evaluated (Long, 2002).

SOURCES

As noted above, airborne contaminants are generated by a variety of activities inside buildings. In a broad sense, there are six major types and/or sources of such pollutants (Spengler and Sexton, 1983).

Combustion byproducts—These are generated through the burning of wood, natural gas, kerosene, wax candles, or any similar materials.

Microorganisms and allergens—Sources include molds, spores, mycotoxins, glucans, animal dander, fungi, algae, and insect parts. The growth and presence of many of these were enhanced by changes in the design of homes and office buildings to enhance energy efficiency, which, in some cases, also increased the moisture content of the indoor air.

Formaldehyde and other organic compounds—Sources include building materials (such as plywood and particleboard), furnishings (draperies and carpets), and some types of foam insulation. Other sources include unvented gas combustion units and tobacco smoke. Personal-care products, cleaning materials, paints, lacquers, and varnishes may also generate chlorinated compounds, acetone, ammonia, toluene, and benzene.

Asbestos fibers—Until 1980, asbestos was used in many building materials, including ceiling and floor tiles, pipe insulation, spackling compounds, concrete, and acoustical and thermal insulation. Fortunately, asbestos is a source of fibers only if it is friable (shedding). As such, exposures today are minimal.

Tobacco smoke—Cigarette smoking serves as a source of fine airborne particles and ^{210}Po and ^{210}Pb, two naturally occurring radionuclides commonly present in tobacco, plus more than 2,000 compounds that are known to be carcinogens and/or irritants. Compounding the problem, the combination of chemical and radioactive carcinogens has been shown to be synergistic (Chapter 3).

Radon and its decay products—Although drinking water can be a source of radon in homes where the supply is obtained from the ground (such as from wells), in most circumstances the primary origin is the diffusion of radon (a naturally occurring radioactive gas that is produced by the decay of radium) into the building from the ground. Buildings with basements are especially vulnerable. Once radon is present in the indoor air, it decays into solid airborne radioactive atoms (i.e., ^{210}Po and ^{210}Pb) that are deposited in the lungs of those who are exposed (Moeller, 1990).

Consideration of indoor environments, however, should not be restricted to homes and office buildings. Relatively high concentrations of nitrogen dioxide have been observed in the air at hockey rinks because of the use of gasoline- or propane-powered vehicles to resurface the ice. The combustion of gasoline can also lead to concentrations of volatile components inside a passenger car during rush-hour traffic that can be six to ten times higher than at standard urban outdoor monitoring sites. Related studies show that subways frequently contain relatively high concentrations of airborne particles. Concentrations of carbon dioxide (CO_2) in commercial transportation vehicles, such as trains and subways, can also be relatively high when passenger loads are high. Similar studies in commercial airliners showed that while CO_2 concentrations remained stable during the cruise portion of the flights, they were significantly higher during pre- and post-flight periods. This reflects what appear to be lower ventilation rates at those times (Dumyahn et al., 2000). These and other conditions, combined with the low relative humidity of the air, often lead to complaints of eye irritation and respiratory problems.

CONTROL

Effective control of indoor air pollution depends on an understanding of several factors. The *first* relates to the characteristics of the contaminant (concentration, reactivity, physical state, and particle size). All such characteristics affect its removal. The *second* is the nature of the emissions. Are they continuous or intermittent, from single or multiple sources, primarily inside or outside? A *third* is the quantitative relationship between the exposure and its health effects. Are individuals to be protected primarily from long-term chronic exposures to low concentrations, or from periodic short-term exposures at peak concentrations? A *fourth* is the nature of the facility. Some controls are more readily applied in residential buildings, others in commercial or office buildings. Also influential are the age and condition of the building.

Control measures for some of the more important indoor air contaminants closely parallel those previously described for the control of ambient air pollution. As is frequently the case, however, different pollutants require different control measures. Adding to the complexity is the fact that indoor air pollution often arises through the interaction of a host of factors that are constantly changing: the temperature and humidity of the air, as well as any contaminants it may contain. Various environmental factors may also impact the building occupants: improper lighting, noise, vibration, and

Table 5.3 Indoor air pollutants, sources, and acute effects

Pollutant	Source	Acute effects/symptoms
Gases		
NO$_2$	Gas stoves, malfunctioning gas or oil furnaces/hot water heaters, fireplaces, wood stoves, unvented kerosene heaters, tobacco products, vehicle exhausts (garages)	Respiratory tract irritation and inflammation; increased air-flow resistance in respiratory tract; increased risk of respiratory infection
CO	Garages, transfer of outdoor air indoors, malfunctioning gas stoves and heaters, tobacco smoke	Impairment of psychomotor faculties; headache, weakness, nausea, dizziness, and dimness of vision; coronary effects with high concentrations
SO$_2$	Kerosene heaters	Bronchoconstriction, often associated with wheezing and respiratory distress; impairment of lung function; increased asthmatic attacks
Volatile organic compounds		
Formaldehyde	Tobacco smoke, glues, resins	Irritation of eyes and respiratory tract; headaches, nausea, dizziness; bronchial asthma at high doses; allergic contact dermatitis and skin irritation (occupational)
Reactive chemicals		
Isocyanates	Paints, foams, structural supports	Upper and lower respiratory tract irritation; bronchoconstriction; contact dermatitis; pulmonary sensitization
Trimellitic anhydride	Plastics, epoxy resins, paints	Bronchial asthma, asthmatic bronchitis, rhinitis; contact dermatitis

(continued)

Environmental particulates		
Biologic allergens: dust mites, cockroaches, animal dander, protozoa, insects (dusts, fragments), algae, pollen	Pets, insects, plants	Hypersensitivity pneumonitis, causing cough, dyspnea, and fatigue; allergic rhinitis; asthma
Toxins: fungi (including molds) and bacteria (endotoxins)	Fungi and bacteria (especially in high-humidity environments)	Hypersensitivity pneumonitis, causing cough, dyspnea, and fatigue, allergic rhinitis; humidifier fever, causing flu-like illness with fever, chills, myalgia, and malaise
Airborne infectious agents		
Legionella pneumophila	Bacteria (in contaminated water sources such as humidifiers and cooling systems)	Pneumonia, Pontiac fever (flu-like symptoms including fever, chills, myalgia, and headache)
Complex mixtures		
Tobacco smoke	Indoor smoking	Eye, nose, and throat irritation; nasal congestion, rhinorrhea; inflammation of lower respiratory tract

overcrowding. In addition, ergonomic factors and job-related psychosocial problems (such as job stress) may be important. Each of these alone or in combination can readily produce symptoms that are similar to those associated with poor air quality.

Since the exposures may involve mixtures of pollutants, it is often difficult to relate complaints of specific health effects to a given indoor contaminant. This is further complicated by the fact that even small problems can have disruptive and potentially costly consequences if the building occupants become frustrated and mistrustful. Effective communication among facility managers, staff, contractors, and building occupants is the key to cooperative problem solving. Another key is recognition that the expense and effort required to prevent most indoor air-quality problems is much less than that required to resolve them after they develop. This is especially the case when seeking to control such problems on a longer-range basis. One important step in achieving such a goal is to provide adequate guidance to the people responsible for the construction of new buildings, as well as those who manufacture the machinery and appliances that are used within them. Many existing indoor air problems can be controlled by following common-sense recommendations: maintaining proper sanitation, providing adequate ventilation, and isolating pollutant sources. The sources and acute effects of these and other indoor air contaminants are presented in Table 5.3.

Assessments of Control Programs in the Ambient Air

As noted in the case study on motor vehicles, while MTBE was found not to be acceptable in the reformulation of gasoline to reduce emissions from automobiles, this type of approach has several distinct advantages. One is that the reformulated gasoline is automatically used by all vehicles and, as such, generally proves to be cost effective. In contrast, compliance with the standards for emissions from the tailpipes of cars has increased their cost and appears to have served as a stimulus for lengthening the time that owners continue to use older, higher-emitting vehicles. In a similar manner, vehicle maintenance and inspection programs have often led to far fewer emissions reductions than projected (Krupnick, 2002).

One of the most successful control programs is the concept of allowance trading. A prime example is the sulfur dioxide (SO_2) trading program being used by the electric utilities. It has yielded benefits well in excess of the expected costs. Although some analysts feared that such trades would lead to hot spots or unfavorable rearrangements in emissions, this has not proved to be the case. Buoyed by this experience, the EPA now considers market-based instruments to be equal in effectiveness to command-and-control methods as regulatory procedures for reducing air pollution from these sources (Krupnick, 2002).

In contrast to the success of allowance trading, the NSR requirements of the Clean Air Act have had almost the opposite effect. Under the regulations, older plants were, in effect, "grandfathered" and did not have to comply with the newer emission restrictions so long as they were not modified in a significant manner. That is to say, routine upkeep was permitted but improvements were not. Unfortunately, guidance on how to distinguish between normal upkeep and improvements was not clear. As a result, the NSR requirements proved to be both excessively costly and environmentally counterproductive, and investments in new, cleaner, power-generating technologies did not occur as anticipated. Another negative impact was that the NSR requirements impeded the adoption of cleaner and more efficient energy technologies, such as cogeneration, wherein the waste heat from one industrial operation is used in other processes at the same site (versus being vented to the atmosphere). The reasons for the delays were essentially the same as those in the case of applying for approval for improvements in existing plants.

In response to these problems, the EPA subsequently modified the NSR requirements to remove these impediments. Even so, some analysts have concluded that, in reality, the development of national and regional allowance trading programs has made NSR redundant. They suggest that the ultimate solution is to place a limit or cap on total pollution emissions and use an allowance trading system to ensure that emission increases at one plant are balanced by offsetting reductions at another. The SO_2 program, which, as noted above, has successfully achieved targeted emissions reductions with a minimum of litigation, can serve as an excellent model for implementing such a program (Gruenspecht and Stavins, 2002).

As implied by the earlier discussion, programs established to control indoor air pollution have, in general, suffered the same range of fates as those for outdoor pollution. Some have been highly successful; others have not. As noted in the case study on radon, progress has been slow, and what remediation is being achieved appears to be due primarily to actions stimulated by factors other than the concerns of homeowners. Although it is difficult to judge, progress in the control of mold may be better, primarily due to the fact that some of its effects are rather immediate. Another influencing factor is that the presence of mold is indicative of basic problems with construction and/or maintenance of a home. As a result, the home-owner has more than one incentive to initiate remedial actions. Due to the magnitude of assessing nationwide changes in the levels of contaminants in indoor air, similar data are not available on comparable trends.

The General Outlook

The preceding discussions of ambient air pollution show that significant progress is being made in reducing the concentrations of many ambient airborne contaminants within the United States. Nonetheless, much work remains. The latest EPA data show that millions of people continue to live in areas where air contaminants pose significant health concerns. For example, an estimated 120 million people reside in areas in which the NAAQS for ozone is exceeded; about 37 million live where the standard for size $PM_{2.5}$ particles is exceeded, about 15 million live where the standard for size PM_{10} is exceeded, and about 5 million live where the standard for lead is exceeded (EPA, 2000). On a worldwide basis, the problems are much larger in magnitude. This is exemplified by the "Asian brown cloud" produced by the combustion of wood and fossil fuels that covered southern Asia in 2002. This cloud was estimated to be more than 3 kilometers (2 miles) thick, and the accompanying pollutants may be producing hundreds of thousands of excess deaths annually due to respiratory illnesses. Although one would expect that the cloud would slowly dissipate, the production of new pollut-ants appears to be replacing the losses as rapidly as they take place. Also to be remembered is that the harmful effects of air pollution are not restricted to humans. It poses an equal, or perhaps higher, risk to forests, natural veg-etation, and agricultural crops, in addition to its harmful effects on build-ings, statues, and other types of physical structures. Adding to the problem is that air pollution readily crosses national boundaries to affect areas far distant from the emission sources (Wilkening, Barrie, and Engle, 2000). This is a critically important consideration in addressing long-term problems, such as acidic deposition and climate change (Chapter 18).

One other item worthy of comment is the leadership demonstrated by California in the control of airborne emissions from automobiles. Because of the unique nature of air pollution problems in the Los Angeles basin, regu-lators in that state continue to impose requirements more stringent than those proposed by the EPA. Although automobile manufacturers complain vociferously, the net result inevitably is that the proposed standards are met, and ultimately the whole nation benefits. The state of California has similarly been a leader in promoting the development of energy-efficient transportation vehicles (Chapter 16).

While increasing attention is being directed to the problems of indoor air pollution, these activities continue to be hampered by several factors. A major one is the lack of resolution of certain public policy and public health questions relative to the proper role of the government in safeguarding air quality inside public and private spaces. Progress, however, is being made. Even so, one could reasonably ask why members of the U.S. Congress have not addressed these problems in the same forceful and constructive manner that has been applied to the ambient environment. One step that might be considered would be to promulgate legislation that the air inside office buildings must be equal to, or of higher quality, than that outdoors.

Another area of primary interest is the quality of the air in the workplace. In this case, the Occupational Safety and Health Administration (OSHA) has the responsibility at the federal level for ensuring that workers are not unnecessarily exposed. Strongly supporting OSHA are multitudes of similar groups at the state and local level. This is not to imply, however, that no one is addressing similar problems with respect to buildings that are used for other purposes. The American Society for Heating, Refrigeration, and Air-Conditioning Engineers (ASHRAE), for example, has developed uniform practices for designing and installing the equipment necessary to ensure acceptable indoor air quality. In a related manner, the American Institute of Architects has issued guidelines for the design and construction of hospitals and health-care facilities. These include recommendations for acceptable air-exchange rates. Related reports and recommendations have been issued by the Underwriters Laboratories, the American Industrial Hygiene Association, the American Conference of Governmental Industrial Hygienists, the International Society on Indoor Air Quality and Climate, and the Association of Energy Engineers (Latko, 2000).

With advances in computers and systems of electronic communications, increasing numbers of people in the more industrialized countries are using their homes as a secondary, or even a primary, place of work. For these and other reasons, the importance of indoor air pollution in homes is likely to increase. One approach would be to implement requirements similar to those that have proved successful in controlling radon in homes. That would be to establish regulations that all new homes and commercial buildings must meet requirements such as those recommended by ASHRAE and related groups to ensure acceptable indoor air quality. The same requirements might also be applied each time an existing building is being sold. Although care would need to be exercised, and a wide range of factors would need to be considered in developing such regulations (especially in terms of private residences), this could represent a place to begin. Such actions could be initiated at the local level and gradually be expanded to the state and federal level.

From the standpoint of private residences, a major consideration is that many people, as noted earlier, view their home as their "castle" and would object to any group, particularly a governmental agency, mandating that they must spend money to correct a problem that, in many cases, appears to have little or no impact on their health. Another, almost overwhelming challenge is the sheer magnitude of the problem. There are literally tens of millions of houses in the United States that have indoor pollution problems. For these reasons, any type of legislative approach to control these problems is likely to be difficult.

6

FOOD

Two Of The Most important factors that influence our health are the food and the water we consume. In fact, because essentially all our food is prepared using water, the two are closely intertwined. Although one would assume that members of the public have control over the quality of their food, this is far from the case. Its production, preparation, and handling continue to present new and novel challenges. These include the introduction of new agricultural and food technologies, such as genetically modified agricultural products; an increasing globalization of the food supply; changes in human demographics and food preferences; and intense public and media scrutiny on issues such as mad cow disease, H1N1 (swine flu), and disease and their relationship to food.

Food and Health—Role of Obesity

Food is related to health in multiple ways, one of the most important being its contribution to the current alarming increase in obesity in the United States. This encompasses all age groups, including children, adolescents, and young adults (Davis and Carpenter, 2009). In fact, not only are more than one in four members of the adult population in 31 of the U.S. states obese, but this percentage increased in 23 states during the past year. Interestingly, increases in people 65 years of age or older were lower than those in the 55 to 64 age range. Contributing to these increases is the consumption of so-called "fast foods" and meals consumed in restaurants where the portions are generally unusually large. In response, some state and local health officials now require that calorie intakes be posted on individual items listed in menus in restaurants and that the amounts of salt in foods, such as soups, pasta sauce, salad dressings, and bread, as well as canned products (i.e., tomato and cream soups), be reduced. Overall, about 75 percent of the salt intake by members of the U.S. public is through the consumption of prepared and processed food. One of the primary stimuli for these efforts is that excess salt intake leads to increased risk of heart disease. For every grain of salt intake reduction, it is estimated that as many as 250,000 cases of heart disease and 200,000 deaths could be prevented during the next decade. Interestingly, these actions have raised legal questions relative to such actions (Associated Press, 2009a).

In terms of its direct health effects, obesity has been shown to (1) reduce the life span of members of the U.S. public by an average of 6 to 7 years, (2) increase the risk of heart disease, stroke, and cancer, and (3) increase the incidence of obesity-related cancers, such as those of the colon, prostate, and breast (Begley, 2009). For these and other reasons, the American Cancer Society considers obesity to be the second largest cause of preventive cancer, the first being lung cancer caused by smoking (Chapter 5). In fact, cancer experts predict that within a decade, obesity may exceed smoking as an avoidable cause of cancer. From the standpoint of indirect effects, obesity (1) increases the risks associated with undergoing surgery and anesthesia, (2) increases the likelihood of developing asthma and sleep apnea (interrupted breathing during sleep), (3) increases the risk of developing gastrointestinal-associated problems such as gallstones, acid reflux, fatty liver disease, and pancreatitis, and (4) is associated with menstrual irregularities, polycystic ovarian syndrome, and increased difficulty of conceiving, as well as skin conditions such as fungal and bacterial infections, pressure sores, and psoriasis. In summary, obesity, which is above and beyond the health effects discussed in the section that follows, affects all aspects of life and represents one of the most important health challenges of U.S. public health professionals (Kava, Ross, and Whelan, 2009).

Sources of Guidance

To help prevent these potential harmful effects, the U.S. Department of Agriculture (USDA) has for several decades published what is called the *Food Pyramid* (Figure 6.1), the objective being to encourage the consumption of certain food groups (for example, grains, vegetables, and fruits) and to discourage the consumption of others (for example, meat, dairy products, fats, oils, and sweets). This is an outgrowth of the documentation of epide-

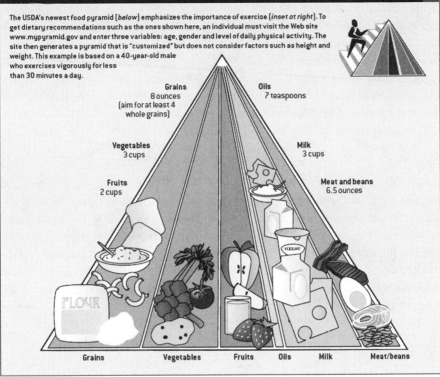

2005 FOOD PYRAMID

The USDA's newest food pyramid (*below*) emphasizes the importance of exercise (*inset at right*). To get dietary recommendations such as the ones shown here, an individual must visit the Web site www.mypyramid.gov and enter three variables: age, gender and level of daily physical activity. The site then generates a pyramid that is "customized" but does not consider factors such as height and weight. This example is based on a 40-year-old male who exercises vigorously for less than 30 minutes a day.

Grains
8 ounces
(aim for at least 4 whole grains)

Oils
7 teaspoons

Vegetables
3 cups

Milk
3 cups

Fruits
2 cups

Meat and beans
6.5 ounces

FLOUR

YOGURT

Grains   Vegetables   Fruits   Oils   Milk   Meat/beans

Figure 6.1   Food pyramid, U.S. Department of Agriculture

miological studies that revealed that the consumption of fruits, vegetables, and fiber protects the heart, while the consumption of whole grains reduces the risk of stroke and diabetes. More recently, studies have shown that the consumption of saturated fats in red meat, butter, and cheese contributes to coronary heart disease, and that foods containing trans fatty acids play a key role, particularly in relation to the types and quantities of cholesterol in the blood. In fact, some nutritional experts have postulated that these acids may be responsible for the epidemic of heart disease that began in the United States during the 1930s and the 1940s. Such acids are produced when partially hydrogenated oils are solidified in the production of margarine and shortening. Consequently, they were present for years in baked goods, chips, and fast foods. In contrast, grain foods are good sources of carbohydrates, plant oils are good sources of fats, and nuts and legumes, followed by fish, poultry, and eggs, are good sources of protein. Since the proper consumption of food groups is only one portion of maintaining a healthy body, the latest *Food Pyramid* emphasizes that one set of recommendations is no longer appropriate for everyone. As noted in Figure 6.1, the version shown is tailored for a 40-year-old male who exercises vigorously for less than 40 minutes a day. Related efforts are being explored by the U.S. Department of Health and Human Services (2009), which has initiated a website to provide information on food safety.

Contaminants in Food and Associated Illnesses

Although contaminants in food may include objectionable materials, such as rust, dirt, hair, machine parts, nails, and bolts, this is rare. For this reason, primary attention in the discussion that follows is directed to (1) infections caused by various biological agents, such as bacteria, viruses, molds, antibiotics, parasites, and their toxins that, if present, can cause a wide range of illnesses, and (2) poisonings, both acute and chronic, caused by chemicals, such as lead, cadmium, mercury, nitrites, nitrates, and organic compounds (Figure 6.2). Either group can gain access to the food chain at any of

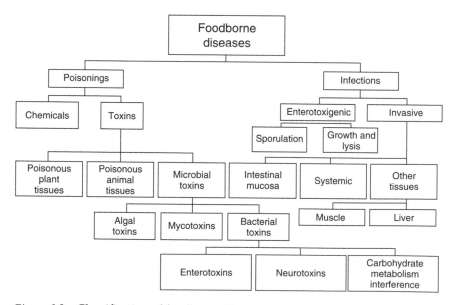

Figure 6.2   Classification of foodborne illnesses

a multitude of stages during growing, processing, preparation, or storage (Besser and Sischo, 2009). Microbial sources account for upwards of 95 percent of all reported outbreaks (97 percent of all cases). Accordingly, most of the attention in the discussion that follows is devoted to illnesses of this type. The major foodborne illnesses, their causative agents, the foods usually involved, and the incubation periods are summarized in Table 6.1. Additional details are available in the *Control of Communicable Diseases Manual*, published by the American Public Health Association (Heymann, 2008). Related guidance on methods for the microbiological examination of foods is available from the same organization (Downes and Ito, 2001).

Foodborne Illnesses and Their Causes

Foodborne illnesses are a major health burden in the United States. Overall, they cause about 13 million reportable illnesses, 325,000 hospitalizations, and about 1,800 deaths annually (Hoffman, 2009; Table 6.2, page 120). It will be noted, information in this table does not include the millions of cases in which the infections were asymptomatic (i.e., the infected person did not realize he/she was infected) or those illnesses that did not require treatment. Reports of recognized outbreaks show that such illnesses can be traced to food that has been improperly handled, contaminated with animal feces, and subsequently consumed either raw or insufficiently cooked. The primary cause of the contamination is that food handlers fail to wash their hands thoroughly after using the toilet. Ten rules for the safe preparation and consumption of food are listed in Table 6.3 (page 120) (Heymann, 2008). The food categories, most frequently involved are produce, seafood, and poultry, which account for an estimated 70 percent of all foodborne illnesses. These three, combined with a fourth category, luncheon and other meats, accounted for more than half the estimated U.S. foodborne-related deaths in 2009 (Hoffmann, 2009). Foodborne diseases are generally divided into four classes: (1) bacterial infections, (2) viral infections, (3) toxins produced by bacteria, and (4) those that are naturally occurring. Each of these is discussed below.

BACTERIAL INFECTIONS

One of the most important of these is *Campylobacter*, which annually infects an estimated 2.5 million people in the United States. The majority of victims suffer diarrhea, fever, and abdominal cramps 1 to 3 days after exposure. Although the illness generally lasts from 4 to 7 days, most of those affected will not require medical treatment. The causative agent, *Salmonella*,

Table 6.1 Examples of important foodborne illnesses, United States

Illness	Causative agent	Food usually involved	Incubation period
Foodborne parasites			
Amebiasis (amebic dysentery)	*Entamoeba histolytica*	Food contaminated with fecal matter	2–3 days to 1–4 weeks
Cryptosporidiosis	*Cryptosporidium parvum*	Vegetables, unpasteurized milk, fruits	2–28 days, 7 days average
Cyclosporiasis	*Cyclospora cayetanensis*	Imported berries, lettuce	1–11 days
Giardiasis	*Giardia lamblia*	Raw salads and vegetables	1–4 weeks
Trichinosis	*Trichinella spiralis*	Raw or under-cooked meat, usually pork or wild game	1–2 days to 2–8 weeks
Foodborne bacteria			
Dysentery	*Shigella dysenteriae*	Food contaminated with fecal material, raw vegetables, egg salads	Up to 1 week
Gastroenteritis, abdominal cramps, and diarrhea	*Shigella sonnei and Escherichia coli*	Food contaminated with fecal material, person-to-person contact, raw produce, parsley. Contaminated with human or animal fecal material	2–4 days
Salmonellosis	*Salmonella* spp.	Eggs, poultry, unpasteurized milk or juice, cheese, raw fruits and vegetables	1–3 days

Foodborne viruses			
Bovine spongiform encephalopathy	Infectious BSE prion material	Beef foodstuffs	Months to years
Gastroenteritis	Norwalk-like viruses	Fecally contaminated food, salads, sandwiches, ice, cookies, fruit, poorly cooked shellfish	24–48 hours
Viral hepatitis	Hepatitis A	Shellfish from contaminated waters, raw produce, uncooked foods	15–50 days, 30 days average
Foodborne toxins			
Botulism (in infants)	*Clostridium botulinum*	Honey, home-canned fruits and vegetables	3–30 days
Brucellosis	*Brucella abortus, B. melitensis, and B. suis*	Raw milk, goat cheese from unpasteurized milk, meats	7–21 days
Diarrhea	*Escherichia coli* O157:H7, and other Shigatoxin producing *E. coli*	Undercooked beef, unpasteurized milk and juice, *raw* fruits and vegetables	1–8 days
Paralytic shellfish poisoning	Dinoflagellates (neurotoxins)	Scallops, mussels, clams, cockles	30 minutes to 3 hours
Staphylococcal food poisoning	*Staphylococcus aureus*	Improperly refrigerated meats, potato and egg salads, pastries	1–6 hours

Table 6.2   Expert judgment–based estimates of incidence of foodborne illnesses and deaths, 2009

Food category	Percent	
	Total cases[a]	Total deaths[b]
Produce	29.4	11.9
Seafood	24.8	7.1
Poultry	15.8	16.9
Luncheon and other meats	7.1	17.2
Breads and bakery	4.2	0.6
Dairy	4.1	10.3
Eggs	3.5	7.2
Beverages	3.4	1.1
Beef	3.4	11.3
Pork	3.1	11.4
Game	1.1	5.2
Total	100	100

a. Total cases: ~13,000,000.
b. Total deaths: ~1,800.

Table 6.3   Ten rules for safe food preparation and consumption

1.	Choose food processed for safety.
2.	Cook food thoroughly.
3.	Eat cooked food immediately.
4.	Store cooked food immediately.
5.	Reheat cooked foods thoroughly.
6.	Avoid contact between raw and cooked foods.
7.	Wash hands repeatedly.
8.	Keep all kitchen surfaces meticulously clean.
9.	Protect foods from insects, rodents, and other animals.
10.	Use pure water.

exists in the intestines of chickens, dogs, and rodents, and it can be transmitted through food as well as drinking water (Chapter 7). Common sources include foods containing chicken, pork, and beef, with eggs and poultry being the primary sources of infection. In fact, about 12 percent of the chickens sold in U.S. supermarkets are estimated to contain this organism (*Consumer Reports*, 2007). Eggs can contain *Salmonella* on the outside and/or the inside. The latter is due to infections in the ovaries of the hens. Low-acid foods, such as meat pies, custard-filled bakery products, and improperly cooked sausages, are also common sources of outbreaks.

Control requires insisting that food handlers wash their hands carefully and frequently. Stores that market live poultry, most especially baby chicks, should provide hand-washing facilities. Obviously, baby chicks and other live poultry should not be given to young children, and all such animals should be kept separate from areas where food and drinks are prepared or consumed (CDC, 2009a). Several types of *Salmonella* infections are discussed below.

Salmonella typhimurium—The most recent major U.S. outbreaks involving this specific variety occurred in the United States in late 2008. They were traced to peanut butter crackers produced in Georgia, a recall was issued, and the public was warned. Peanut butter sold in jars was not contaminated. After reaching a peak in December, the numbers of new cases declined substantially during the spring of 2009, but illnesses were still being reported in early May. Between September 1, 2008, and April 20, 2009, about 725 people, ranging from infants to people 98 years of age and residing in 46 states, were infected (CDC, 2009b).

Shigella dysenteriae and *Shigella sonnei*—The first of these varieties is a common source of bacillic dysentery. Two-thirds of all cases, and most deaths, occur in children under 10 years of age. Illness in infants less than 6 months old, however, is unusual. Secondary attack rates in households can be as high as 40 percent. *S. sonnei* is a common cause of gastroenteritis. In both cases, the organisms are present in human feces, and transmission is favored by crowded conditions where personal contact is unavoidable. Food handlers can readily spread the infection. Flies can also transfer the organisms to nonrefrigerated food. Once in the food, they multiply. In the case of *S. dysenteriae*, ingestion of a large number of organisms is required for a person to become infected. In the case of S. sonnei, as few as 10–100 organisms can cause illness. As a result, person-to-person is a viable method of transmission for this organism. The incubation period in the case of *S. dysenteriae* is up to a week; in contrast, it requires only 2 to 4 days in the case of *S. sonnei* (Heymann, 2008).

About 450,000 *Shigella* infections were reported in the United States in 2008. Of these, most of which occurred during the summer, an estimated 78 percent were due to *S. sonnei* and 18 percent to *S. flexneri*, with *S. boydii* accounting for most of the remainder. Another variety, *S. dysenteriae*, is rare in this country. The highest incidence was among children younger than 5 years of age. Since more than 95 percent of the infections may be asymptomatic, the actual number may be 20 times higher than reported (i.e., more than 1 million (CDC, 2007).

VIRAL INFECTIONS

There are three principal varieties in this category.

Norwalk-like viruses annually cause an estimated 23 million episodes of gastroenteritis, 50,000 hospitalizations, and 350 deaths in the United States. As in cases of bacterial disease, they can be transmitted by fecal contamination and direct person-to-person contact. Typical sources include salads, sandwiches, fruit, and improperly cooked shellfish. These viruses are extremely contagious for several reasons, one being that the dose that will cause infection is low; the other being that patients continue to be infectious for up to 2 weeks after recovery. Exacerbating the situation is that they are resistant to chlorination as well as to temperature variations ranging from 0°C (32°F) to 60°C (140°F). Outbreaks are common in settings in which people are crowded and sanitation facilities are inadequate, such as at summer camps. There have also been outbreaks among passengers and crews on ships operated by several of the major cruise lines. One of the problems in its control is the lack of a simple and sensitive technique for detecting its presence (CDC, 2002).

Infectious hepatitis (hepatitis A) is a highly contagious disease caused by a virus whose symptoms are fever and general discomfort. In the developing countries of the world, this disorder occurs most frequently among school-age children. As a result, most adults are generally immune. In the developed countries, however, most cases occur among young adults. Common sources include sandwiches and salads that are not cooked or foods that are handled after cooking by infected handlers. Contaminated produce, such as lettuce and strawberries, and drinking water may also be sources of infection (Heymann, 2008).

Bovine spongiform encephalopathy (BSE), commonly known as *mad cow disease*, is a fatal neurodegenerative disease in cattle that causes a spongy degeneration of the brain and spinal cord. In cattle, it has an incubation period of about 4 years. According to the U.S. Department of Agriculture, all breeds are equally suscep-tible. In the United Kingdom, the most heavily impacted country, almost 180,000 cattle were infected and 4.4 million were slaughtered during efforts to eradicate the disease. An inquiry revealed that the epidemic was caused by cattle, normally herbivores, that had been fed the remains of other cattle that contained the infectious agent. The consensus is that infectious BSE prion material is not destroyed through normal cooking procedures and even contami-nated beef foodstuffs that are "well done" may remain infectious. The origin of the disease is not known. A similar disease in sheep is known as scrapie. By February 2009, it had killed 164 people in Britain and 42 elsewhere. Control depends on preventing the occurrence and spread of the disease in cattle and ensuring that none of the meat from infected animals is consumed by people (*Wikipedia*, 2009a).

TOXINS

In contrast to bacterial infections, in which the illnesses are caused directly by the organisms, other foodborne illnesses are caused by toxins produced by bacteria that are not, in themselves, harmful. The toxins may also be naturally present in food or produced by viruses and fungi that were intro-duced through improper handling. Regardless of their origin, the ingestion of the accompanying toxins can readily lead to illnesses and, in some cases, to death. Three of the more common bacteria that produce toxins are dis-cussed below.

Staphylococcus aureus. Under favorable conditions, *S. aureus* can pro-duce one or more enterotoxins that if ingested can abruptly (within 1 to 6 hours) lead to severe nausea, cramps, vomiting, and prostra-tion, often accompanied by diarrhea and sometimes with subnor-mal temperature and reduced blood pressure. Potential sources include infected cuts, boils, sores, postnasal drip, or sprays expelled during coughing or sneezing. They are also present in air, water, milk, and sewage, and an estimated 25 percent of the population are carriers. Meat (especially ham) and its products, poultry and its products, including dressing, and custards used for pastry fillings are common sources (Heymann, 2008). Handling cell phones can also be a source. Analyses in one health-care facility revealed that *S. aureus* was present on more than half of their phones (Ulger et al., 2009). These organisms have also been found on west coast beaches (Marchione, 2009).

Clostridium botulinum. This bacteria, which is present in the soil throughout the world, can be the source of an extremely deadly toxin. Conditions that promote their germination and growth include the absence of oxygen (that is, anaerobic conditions), low acidity (pH ~4.6), temperatures higher than 4°C (39°F), and high moisture content. Although rare and sporadic, foodborne botulism is a persistent cause of morbidity and mortality in the United States. It was a common problem for decades in Alaska, where several hundred Natives were infected through the consumption of beaver meat that was being fermented in plastic or glass containers. Most poisonings in the continental United States result from the consumption of fermented food (CDC, 2001). The toxins are extremely potent; a few nanograms (10^{-9} gram) can cause illness. Although the toxins can exist for long periods, they can be destroyed by boiling. Inactivation of the spores, however, requires higher temperatures (Heymann, 2008).

During 2006, the consumption of carrot juice led to four infections, with one death, in the state of Georgia. Based on subsequent investigations, the Food and Drug Administration (FDA) concluded that in order to protect consumers, carrot juice and other processed foods with no natural barriers to the germination of *C. botulinum* require additional chemical or thermal barriers. Accordingly, they modified their guidance on low-acid juices that require refrigeration. This included adding a validated juice-treatment method, such as acidification or appropriate thermal treatment, to reduce the risk of such contamination should any breaches in refrigeration occur (CDC, 2006b).

Escherichia coli. *E. coli* are increasingly being recognized throughout the world as potential sources of food poisoning. Outbreaks in Europe have revealed new and unexplained links between some bacteria and viruses that cause food poisoning. Outbreaks in the United States have been linked to bagged baby spinach, and *Salmonella* has been linked to imported cantaloupe melons and tomatoes. One of the most recent events involved the discovery of *E. coli* in cookie dough manufactured in a plant in Virginia (Associated Press, 2009b). Since such events can affect multiple countries simultaneously, some are not recognized and the scale of the problem has not been fully assessed. Such outbreaks have prompted some government officials to mandate that food products be tracked throughout their life cycles to assist investigators in follow-up investigations (Lynch, Tauxe, and Hedberg, 2009).

NATURALLY OCCURRING TOXINS
Foodstuffs and cooked food contain an array of naturally occurring toxins that would normally not be permitted as regulated additives. The common assumption that "natural" is safe, and "man-made" is suspect, is contrary to current scientific knowledge. In fact, a typical diet contains far more natural carcinogens than synthetic ones (American Council on Science and Health, 2004). Here are descriptions of two of the more common naturally occurring toxins:

Aflatoxins and other toxins. Peanut butter contains aflatoxins, one of which *(aflatoxin B)* has been shown to be acutely poisonous and carcinogenic in animals. In a similar manner, carrots contain *carotatoxin,* a fairly potent nerve poison, *myristicin,* a hallucinogen, and *isoflavones,* which have an estrogenic effect similar to female hormones. Intakes can readily lead to death. Of the many toxins found in plants and animals, only a few have been specifically associated with human illnesses. The most dramatic is paralytic shellfish poisoning, which is caused of a highly potent neurotoxin that is a metabolite of certain marine dinoflagelates. One of the most common of these is *Karenia brevis* (discussed below), which results in what are known as the "red tide."

Brevetoxins. Fortunately, only a few brevetoxins have been specifically associated with human illness. Nonetheless, poisonous concentrations of these toxins can accumulate in shellfish (mussels, clams, and occasionally scallops and oysters). Although the toxin appears to produce no effects in the shellfish, its consumption by humans can be deadly, producing, within 1 to 3 hours, numbness of the lips and fingertips, ascending paralysis, and finally, in cases of severe poisoning, death from respiratory paralysis (Heyman, 2008). Control can be accomplished by monitoring potentially affected waters and discontinuing seafood harvesting when dinoflagellates are present.

Sampling of the offshore waters can readily confirm the presence of *Karenia brevis* (CDC, 2008).

Not all foods that are suspected to be toxic are actually so. Tomatoes, which are native to the Americas, were cultivated by the Aztecs and Incas as early as 700 A.D. During the sixteenth century, however, they were avoided in England since they resembled toxic berries. In contrast, the French consumed them in abundance because they were considered powerful aphrodisiacs. Today, they are recognized for their taste and health benefits, one being that they contain the antioxidant, lycopene, as well as vitamins A and C, magnesium, phosphorus, potassium, riboflavin, sodium, and thiamine. They are also low in calories and readily digestible. In addition, recent studies show that people who consume seven or more servings of tomatoes per week reduced their risk of stomach cancer by 60 percent, and men who consume at least 10 servings per week are 45 percent less likely to develop prostate cancer (Reid, 2009). Methods for the microbiological examination of foods have been published by the American Public Health Association (Downes and Ito, 2001).

Inorganic and Organic Chemical Contaminants and Additives

Foods can contain a variety of inorganic and organic (i.e., those that contain carbon) chemicals. Some of these are purposefully added and others result from human actions. Heavy metals, such as lead, copper, tin, zinc, or cadmium, can leach from containers or utensils, particularly in cases in which acidic foods are being prepared or stored. Other chemicals can be introduced through accidental or inadvertent contamination with detergents or sanitizers. Vegetables, poultry, and livestock can be contaminated with pesticides, herbicides, fungicides, fertilizers, and veterinary drugs, due to improper conditions under which they are grown (Marshall and Dickson, 1998). As will be discussed later, they also contain antibiotics they are given to promote their growth (Webster, 2009). Although such contaminants are not a significant problem in the developed countries of the world, they can be in the developing countries.

Inorganic chemicals. Mercury discharged into rivers, lakes, and oceans in the form of inorganic salt or as the metallic element (which is not harmful to humans) can be converted by microbes to methyl mercury. In this form, it can pose a significant health risk. Large-scale poisonings by these compounds have caused deaths and cases of permanent damage to the central nervous system. In a classic episode in Japan in the early 1950s, industrial wastes containing mercury were discharged into Minamata Bay (*Wikipedia*, 2009b). More than 100 people who ate contaminated fish were poisoned, and 46 died. Even so, a recent survey (Scudder et al., 2009) showed that detectable quantities of mercury were present in every fish sampled in 261 U.S. streams across the country. These included relatively undeveloped watersheds in the Northeast and upper Midwest (Chapter 8).

A variety of other inorganic chemicals are, or can be, introduced into foods during processing. One of the most common is salt, a taste enhancer and preservative that, as discussed earlier, can cause a variety of health problems if consumed in excess. Also added to foods are sulfites and bisulfites, which, in aqueous solutions, form sulfurous acid, an antimicrobial agent. In addition, nitrites and nitrates serve as agents for curing and pickling meats and vegetables. These two substances can also gain access to some foods through uptake in agricultural crops that are produced using nitrogen fertilizers. One of the benefits of nitrates and nitrites is that they inhibit the growth of *Clostridium botulinum* in foods that are vacuum packed. One of the risks is that high concentrations of nitrate in baby food, much of which is converted into nitrites, can cause *methemoglobinemia* in infants. Also to be recognized is that during cooking, nitrites can react with secondary and tertiary amines to form nitrosamines, a potential carcinogen. Current formulations, however, significantly reduce this risk. To qualify for use, all such additives must be classified as GRAS, that is, they must be "generally recognized as safe" (FDA, 2009).

Organic chemicals. Multiple organic acids and their salts are used as preservatives in foods. These include benzoates, which inhibit the growth of yeasts and molds; sorbate salts, which inhibit the growth of yeasts and molds; and propionic acid and propionate salts, which are active against molds (CDC, 2006a; Federated Mills). Other organic chemicals gain access to food through the use of pesticides and herbicides on agricultural crops, examples being chlorinated hydrocarbons, polychlorinated biphenyls (PCBs), chlorinated dibenzo-p-dioxins, and chlorinated dibenzofurans. Tests in animals show that PCBs can cause reductions in immune system function, behavioral alterations, and impaired reproduction (ATSDR, 2001).

Organic contaminants can also be produced in foods, especially meat, through the cooking process. Browned or burned portions of meats that have been charbroiled, whether fried or smoked, contain heterocyclic aromatic amines—many of which have been shown to be highly mutagenic. Examples are benzo-a-pyrene and the polycyclic aromatic hydrocarbons, as well as numerous breakdown products of common dietary amino acids. Measures that have been suggested to avoid the production of these compounds include using alternative processes such as stewing, poaching, or boiling to cook meat and employing a microwave oven to cook fish and poultry. Related studies showed that bisphenols could be transferred to milk that was warmed in polycarbonate plastic baby bottles. Soon, thereafter, major food suppliers and retailers in this country voluntarily eliminated the use and sale of any products that were packaged in, or contained, this ingredient (Calafat et al., 2005). This is discussed in more detail in Chapter 7.

Antibiotic and Hormone Use in Animals

More than a half century ago, farmers began feeding antibiotics to chickens, pigs, and cattle to prevent the spread of infections and to reduce the amount of feed required to fatten them. During the intervening decades, studies have increasingly confirmed that such practices have contributed to an alarming increase in the resistance of many human bacterial pathogens to antibiotics. Nonetheless, this practice continues today in the United States. In contrast, the European Union took steps a decade ago to ban the use of antibiotics of human importance in farm animals for nontreatment purposes. Making this action even more significant was that *Staphylococcus aureus* is one of the most common causes of hospital- and community-acquired infections in the United States. Increases in resistance among drugs used on patients are estimated to cause perhaps as many as 70 percent of the 90,000 fatal hospital infections that occur in this country each year.

Yet, the U.S. Congress and the FDA continue to endorse such practices even though the Centers for Disease Control and Prevention (CDC) state that at least 17 classes of antimicrobials, including antibacterial antibiotics, antivirals, and antiparasitic drugs, are approved for farm-animal growth promotion in this country. These include many families of antibiotics, such as penicillin, tetracycline, and erythromycin, which are critical for treating human disease. In fact, it is estimated that 70 percent of antimicrobials being used in the United States are fed to chickens, pigs, and cattle for nonthera-peutic purposes. Fortunately, major U.S. poultry producers, stimulated by actions in other countries of the world and increasing public recognition of the serious negative impacts of antibiotic use, have taken the lead in ban-ning such uses in this country (Falkow and Kennedy, 2001). In the meantime, the FDA has granted approval to the domestic aquaculture industry to use antibiotics. This industry now consumes some 50,000 pounds of antibiotics annually, far more than those used in medicine (HSUS, 2009).

Care in Food Preservation and Handling

A variety of methods are available for safely preserving wholesome food, preventing contamination, and destroying organisms or toxins that may have gained access to or been produced within the food. One of the prereq-uisites for these methods to be successful is to seek to maintain the food in a condition that is not favorable for bacterial growth, namely, to avoid con-ditions that provide warmth, moisture, and a medium that is neither highly acid nor alkaline. Based on these and other considerations, the primary methods that have proved to be effective for preserving food are as follows.

Cooking. This renders food digestible and palatable. Although it also tends to kill many bacteria, this process alone will not preserve food. In fact, partial cooking may render protein foods (meat, eggs, milk, milk products) more susceptible to bacterial growth, permit-ting active increases in the number of harmful organisms or the toxins they may produce. Even when food is heated thoroughly and to a sufficiently high temperature to kill any microorganisms present, it must be eaten promptly or protected from subsequent spoilage.

Canning. This involves heating food sufficiently to kill any microor-ganisms present and then sealing it in a container to keep it sterile. Although acid foods—tomatoes and some fruits—need to be heated to the boiling point for only a few minutes, nonacid foods—corn and beans—must be heated to higher temperatures for a longer time to prevent undesirable changes in appearance and flavor, as well as, for example, to destroy the anaerobic microorganisms that produce the botulism toxin (CDC, 2006b).

Drying and dehydration. Air drying, one of the most economical and effective ways of preserving food, has been practiced for centuries. Today food can be dried in the sun or by artificial heating processes. Other methods include spray, freeze, vacuum, and hot-air drying. Once the food is reconstituted by the addition of water, bacterial activity resumes and it is essential that sanitary controls be applied.

Preservatives. As noted earlier, a variety of chemicals are purposefully added to foods either to inhibit the growth of microorganisms, kill them, or serve as flavor enhancers. These include salt, sugar, sodium nitrate and nitrites, salicylic acid, sodium benzoate, as well as propionates and sorbic acid. Each has associated risks and benefits. One additional method of preserving foods, especially meats, is smoking. This technique can also improve its flavor.

Refrigeration. Storing food at temperatures lower than 5°C (41°F) will retard the growth of pathogenic organisms and the more important spoilage organisms, but it does not prevent all changes. Proper air circulation and regular cleaning and sanitizing of chilled spaces are mandatory.

Freezing. Bacteria that cause food spoilage do not multiply at freezing temperatures. Once thawing begins, however, it becomes vulnerable to bacteria and the associated toxins they may produce. Refreezing will not make the food safe. Nor will freezing improve the original quality of the product. Thus, the selection of appropriate products for freezing is essential. One variation is "dehydrofreezing," in which the food is partially dehydrated (but still perishable) and then frozen. This process provides the space and weight savings of dehydration without depriving the food of its fresh color, flavor, and palatability.

Pasteurization. This is an excellent supporting technique for preserving food for a short time. Combined with refrigeration, it extends the useful shelf life of dairy products. Milk is generally heated to 63°C (145°F) for 30 minutes—or to 72°C (161°F) for 15 seconds—to kill the pathogenic organisms. Although some heat-resistant organisms will survive, subsequent refrigeration will preserve the milk for several weeks.

Irradiation. This involves exposing food to ionizing radiation at sufficiently high doses to kill a large fraction of any microorgan-

isms present. In this sense, it is analogous to pasteurization. It is
especially effective in destroying foodborne contaminants such as
Salmonella and *Escherichia coli,* and it also destroys *Trichinella* in
pork (Loaharanu, 2003). In some foods, however, this process
produces undesired changes in taste and palatability. Although
fears have been expressed about the formation of radiolytic
compounds, the types and quantities are no different than those in
foods preserved by other methods. Irradiation has been approved
in the United States for the preservation of pork, chicken, herbs
and spices, fresh fruits and vegetables, grains, and seeds used for
producing sprouts. In response to legislation, the USDA now offers
irradiated meat as part of its nationwide school lunch program
that provides daily meals to more than 25 million children. If
widely applied, this technique would significantly reduce foodborne illnesses.

Components of an Effective Sanitation Program

In addition to exercising care in processing, the prevention of foodborne
illnesses requires an effective sanitation program. Also essential are a safe
water supply, adequate garbage and refuse disposal, proper wastewater
and sewage disposal, and effective insect and rodent control. Other essential factors are discussed below.

> *Equipment and facilities.* Equipment used in the preparation or pro-
> cessing of food should be designed to facilitate cleaning. Transport-
> ing vehicles must be clean and should not carry other products.
> Refrigerated vehicles must be available for the transport of perish-
> able foods. All foods, particularly vegetables, can be stored above
> the floor, where they will remain dry and will not come in contact
> with powders and sprays used to control insects and rodents.

> *Personnel training and habits.* As noted earlier, personal hygiene is
> indispensable in the proper handling and preparation of food
> products. Antimicrobial cleaners should be used on the surfaces on
> which foods are prepared, and cleaning rags and sponges should
> be disinfected regularly or replaced. Food handlers must avoid
> contact between open wounds and foodstuffs, must wear clean
> outer garments, including a cap over the hair, and must avoid using
> tobacco products.

Table 6.4 Definitions of the federal safety standards and a description of the bases for the limits for the control of food contaminants and ingredients, United States

Ingredient	Definition	Safety standard or limit
Unavoidable contaminants	Inherent food substances that cannot be avoided	Adulterated if substance "may render food injurious to health"
GRAS substances	Substances "generally recognized as safe" by the scientific community	Must be "generally recognized as safe"
Food additives	Substances added for specific intended effects, including GRAS substances, color additives, new animal drugs, and pesticides	"Reasonable certainty of no harm"
Substances previously sanctioned	Substances explicitly approved for use by FDA or USDA prior to 1958	Adulterated if substance "may render food injurious to health"
Pesticides	Substances intended for preventing, destroying, repelling, or mitigating any pest, or intended for use as plant regulator	Tolerance based on whether substance is "safe for use," considering its benefits
New animal drugs	Substances intended for food-producing animals, excluding antibiotics	"Reasonable certainty of no harm"
Color additives	Dyes, pigments, or other substances capable of imparting color, excluding substances that have other intended functional effects	"Reasonable certainty of no harm"
Prohibited substances	Substances prohibited from use because they present a potential risk to public health or because the data are inadequate to demonstrate their safety in food	Must not be present in detectable amounts

Table 6.5 Federal agencies responsible for the safety of the U.S. food supply

Department of Health and Human Services:
Food and Drug Administration, which is responsible for the regulation of food labeling, safety of food and food additives, inspection of food processing plants, control of food contaminants, and establishment of food standards

Centers for Disease Control and Prevention, which analyze and report incidents of foodborne diseases

National Institutes of Health, which conduct research related to diet and health

Department of Agriculture:
Food Safety and Inspection Service, which is responsible for inspection and labeling of meat, poultry, and egg products, as well as grading of foods

Animal and Plant Health Inspection Service, which inspects food and animal products imported into the United States

Human Nutrition Information Service, which establishes food consumption standard tables for the nutritive value of food and provides educational materials related to food

Other agencies:
Environmental Protection Agency, which develops standards for the use of pesticides on food crops

National Marine Fisheries Service, within the Department of Commerce, which conducts inspections and establishes standards relative to the quality of seafood

Bureau of Alcohol, Tobacco, and Firearms, which regulates alcoholic beverages, and the Customs Service, which inspects food products imported into the United States (both within the Department of the Treasury)

Federal Trade Commission, which regulates food advertising

Standards and regulations. These should specify proper methods of processing, preparing, and selling food products; limitations on the types and quantities of chemicals that can be added to foods; restrictions on the quantities, types, and manner in which pesticides can be used on agricultural food crops; and proper labeling requirements for commercial food products. Listed in Table 6.4 are the principal federal agencies in the United States that have responsibilities related to food safety, and a brief description of their duties.

Enforcement and monitoring. U.S. agencies at the state and local levels have primary responsibility for the inspection of restaurants, retail food establishments, dairies, grain mills, and other food establishments. Their goals are to ensure the safe handling, proper labeling, and fair marketing of food products. Methods used to meet these responsibilities include inspection at the point of production or processing, examination of products at the retail or wholesale level of distribution, and licensing of establishments that manufacture or handle foods. Because it is impossible to inspect every food at every site of production, processing, and distribution, the incentives to comply with regulations depend primarily on the probability of detection and the penalties for noncompliance (which can include fines and legal proceedings). In addition, compelling economic and business factors encourage food handlers, processors, and distributors to comply with the regulations. No food processor wants to suffer the loss of customer confidence that can accompany a highly publicized foodborne disease outbreak.

Summarized in Table 6.5 are definitions of the safety standards and a description of the bases for the limits for the control of a range of food contaminants and ingredients in the United States. Excellent guidance on the proper preparation of foods to minimize the risks of foodborne illnesses and to ensure the safety of foods served in restaurants, grocery stores, and institutions such as nursing homes is given in the Food Code published by the FDA (2005). Providing guidance on food safety at the international level is the Codex Alimentarius Commission 2009).

Genetically Modified Food

Through genetic modification, it is possible to enable individual plant species to resist insect pests or fungi, tolerate a specific herbicide, have longer shelf life, provide increased nutritional value and higher yields (advances that would be of special benefit to populations in the developing countries), and reduce the amount of allergenic substances in foods, such as peanuts (Eubanks, 2002; NRC, 2008). Researchers are also identifying genes that can help plants tolerate arid conditions or grow in salty water. This would be of immense help in meeting the world's needs for food, as well as in conserving the ever-increasing shortage of fresh water (Chapter 7). In fact, it may be possible to identify the genes that will confer durable multipathogen resistance to plant diseases, a critical example being fungal infections in wheat, the second largest cereal crop in the world (Krattinger et al., 2009). New varieties of rice have also been developed that hold promise of solving hunger problems in many parts of the world (Pooladi, 2009.).

Nonetheless, the introduction of genetically modified (GM) plants also poses risks. These include cross-pollination and the escape of genes, leading to the creation of genetic pollution and "super weeds," the promotion of the development of insects resistant to natural toxins, and the possible introduction of allergens into foods (Metcalfe, 2003). For reasons of this nature, the European Union issued a report in 2002 indicating that GM canola is at "high risk" of cross-pollination with other canola crops. This coincides with other reports of cross-pollination between GM crops and similar varieties of other plants. Although various controls are applied in the United States during the testing of GM seeds, the USDA places no such requirements on the crops once they are approved (Kliebenstein and Rowe, 2009).

The General Outlook

There are multitudes of challenges facing those who are responsible for ensuring the safety of food. One of the most important, in terms of the United States is the fragmented nature of the national regulatory structure. These responsibilities are distributed among almost a dozen federal agencies, which, in turn, must interpret some 35 different laws. Consequently, there is no clear national coordination or oversight. While the FDA has jurisdiction over about 50,000 food processing and storage facilities, its budget permits its staff to visit less than one-third of these annually. In contrast, the USDA has some 7,500 inspectors, ten times the number in the FDA, to check 6,000 meat processors. As is obvious, the allocation of resources in this manner does not maximize the overall ability of the federal government to reduce the associated risks (Taylor and Hoffmann, 2001). Aware of these concerns, a food safety working group, established by the president, recommended stricter rules for the production of eggs, poultry, beef, leafy greens, melons, and tomatoes and that the FDA and the USDA make this a top priority (Associated Press, 2009c).

Another is the ever-increasing global nature of the production and distribution of food. This has led to a tripling of U.S. food imports during the last two decades. Nonetheless, applicable regulations in many other countries, such as the People's Republic of China, are far less restrictive than those in the United States. Due to the problems discussed above, the FDA is able to inspect only slightly more than 1 percent of all incoming shipments. Even so, during March, 2007, this led to the detention of nearly 850 shipments of grains, fish, vegetables, nuts, spices, oils, and other imported foods for issues ranging from filth to unsafe food coloring to contamination with pesticides to *Salmonella*. These ranged from baked goods from Canada, India, and the Philippines, to beans from Belgium, blackberries from Guatemala, frozen catfish from China, and jalapenos from Peru (Associated Press, 2007).

Such challenges, however, are not limited to those of national governments. Individuals throughout society have important roles to play. As participants in the chain from food production to consumption, consumers have obligations no less important than those of food processors. If they are to fulfill these obligations, they must learn to handle foods properly. If this is to be successful, it will depend on continuing programs of public education and the lessons learned implemented by each and every individual.

7

DRINKING WATER

ECORDS SHOW that the quest for pure water began in prehistoric times. Information on methods for treating water is available in Sanskrit medical lore, and pictures of apparatus to clarify water have been discovered on Egyptian walls dating back to the fifteenth century B.C.E. (*before the Common Era*). Treatment methods such as boiling, filtration through porous vessels, and even filtration through sand and gravel have been prescribed for thousands of years. In his writings on public hygiene, Hippocrates (approximately 460–354 B.C.E.) directed attention to the importance of water in the maintenance of health (Simon, 2001). The Romans demonstrated a similar awareness of the merits of pure water; witness the extensive aqueduct systems they developed, as well as their use of settling reservoirs to purify water, their rulings that unwholesome water should be used only for irrigation, and the passage of laws prohibiting the malicious polluting of waters (Frontinus, A.D. 97).

Water and Health

Water, the same as food, is essential to life. In fact, 66 percent of the human body and 75 percent of the brain are water. For this reason, a loss of as little as 1 to 2 percent will cause thirst or pain, the loss of 5 percent of body water can cause hallucinations, and a loss as large as 10 to 15 percent can be fatal. Although humans can readily survive a month without food, they can survive for a maximum of only about 2 weeks, depending on the circumstances, without water.

The first positive evidence that public water supplies could be a source of infection was based on epidemiological studies of cholera in the city of

London conducted by John Snow in 1854 (Chapter 3). His studies are particularly impressive because at that time, the germ theory of disease had not yet been established. A similar study by Robert Koch in Germany in 1892 provided evidence of the importance of drinking-water filtration as a mechanism for removing the bacteria that cause cholera. Subsequent experiments in the United States, relative to the control of typhoid fever, confirmed his observations and revealed the additional benefits of adding chemicals to coagulate the water prior to filtration.

One of the most important technological developments in the treatment of water during the twentieth century was the introduction in 1908 of chlorination. This provided a low-cost, reproducible method of ensuring that water being prepared for human consumption met acceptable bacteriological standards. The dramatic impact of this development, combined with the filtration of water, in reducing deaths from typhoid in Philadelphia, Pennsylvania, is shown in Figure 7.1 (Cochran and Cheney, 1982). Prior to that

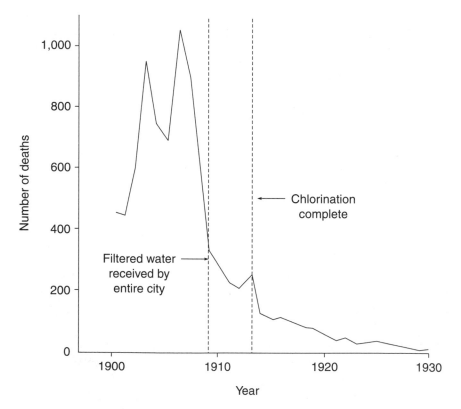

Figure 7.1   Deaths from typhoid fever, Philadelphia, PA, 1900–1930

time, typhoid fever had been a major contributor to illness and death in the United States. In fact, during the U.S. Civil War (1861–1865), about twice as many soldiers died from illnesses as did from injuries in battle. Although chlorine continues to be the most common disinfectant used for ensuring that drinking water complies with applicable bacteriological standards, the fact that it interacts with and produces unwanted by-products has led to the increasing application of other approaches. Methods for resolving this problem will be discussed later.

Primary Sources of Water

Broadly speaking, the basic source of all water on Earth is precipitation—rain, snow, and sleet. Only about 30 percent of this, however, falls on land areas—and that is not evenly distributed. Although the annual average precipitation in the United States ranges from 28 to 30 inches (71–76 centimeters), there are significant variations both geographically and seasonally within this country and worldwide. Compounding the situation is that about 70 percent of the precipitation that reaches land is evaporated or transpired (after uptake by vegetation) back into the atmosphere, 10 percent soaks into the ground and becomes groundwater, and 20 percent runs off into lakes, streams, and rivers, most of which ultimately flows into the oceans. The overall movement of water from precipitation through various pathways on Earth and back into the atmosphere is called the hydrologic cycle (Figure 7.2).

Availability of Freshwater

The total amount of water on Earth is estimated to be 321 million cubic miles (~1.338 million cubic kilometers; USGS, 2009). Of this, about 97 percent is present as saltwater in the oceans; only about 3 percent is present as freshwater (Figure 7.3). Nearly 70 percent of the world's freshwater is locked in glaciers and icebergs (USGS, 2009). Even so, the identified sources of freshwater are not continuously available; in fact, many geographical areas can be without water for extended periods. Climate change could also have significant impacts on a long-term basis (Chapter 18). The principal sources of drinking water are discussed below.

GROUNDWATER
Groundwater can be accessed through wells dug beneath the ground surface. Sources located in deeper reservoirs or aquifers can be accessed

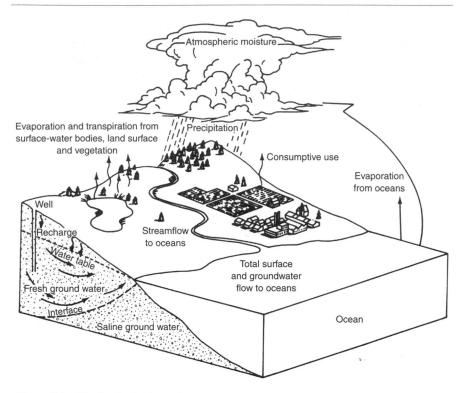

Figure 7.2   The hydrologic cycle

through wells that are driven or drilled (Figure 7.4). Springs, outcrops where the underground aquifer intersects the surface, are another source. The widespread use of groundwater is due to multiple reasons. It is commonly available at the point of need at relatively little cost, and reservoirs and long pipelines are not necessary. Unless it contains contaminants introduced by human activities, groundwater is normally free of suspended solids, bacteria, and other disease-causing organisms. Unfortunately, as of 1992, it was estimated that more than 10 percent of the U.S. community water supply wells and almost 5 percent of the rural domestic wells contained detectable concentrations of one or more contaminants, primarily agricultural pesticides (Glacier, 2006). About 1 percent contains one or more contaminants. Exacerbating the situation, major sources of groundwater are being contaminated. The Environmental Protection Agency (2001) has

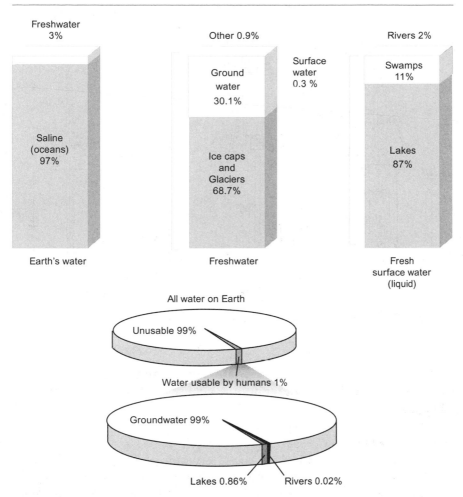

Freshwater
3%

Other 0.9%

Rivers 2%

Saline
(oceans)
97%

Ground
water
30.1%

Surface
water
0.3 %

Swamps
11%

Ice caps
and
Glaciers
68.7%

Lakes
87%

Earth's water

Freshwater

Fresh
surface water
(liquid)

All water on Earth

Unusable 99%

Water usable by humans 1%

Groundwater 99%

Lakes 0.86%   Rivers 0.02%

Figure 7.3 Distribution of Earth's water and relative amounts that are usable
and unusable

repeatedly stressed that once this occurs, it is next to impossible to restore
their quality. Another negative factor is that farmers and municipalities
worldwide continue to remove the water faster than the natural rate of recharge. Major portions of the Ogallala aquifer, which underlies the Great
Plains section of the United States, have already been depleted. The same
is true in India and the People's Republic of China. In fact, it is estimated

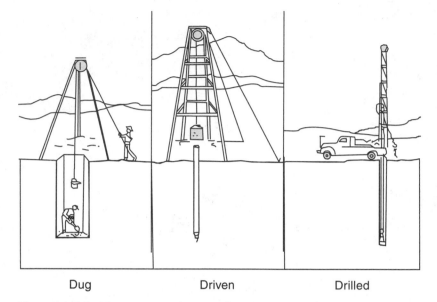

Dug   Driven   Drilled

Figure 7.4   The three principal types of wells—dug, driven, and drilled

that lack of water will reduce the production of grain in China and India by between 10 to 20 percent within the next several decades (Montaigne, 2002).

SURFACE WATER

Sources of surface water include lakes, reservoirs, and rivers. It may also come from protected watersheds. As is true for groundwater, questions have been raised on how these supplies should be managed and allocated. In the final analysis, it will be necessary for all users to learn to accept limitations and to share responsibility for these resources.

Most surface water sources require treatment prior to consumption. This can be avoided if certain methods and precautions are applied. One approach for meeting the needs of individual households is to collect rainfall from household roofs and store it in cisterns or barrels. Even so, without care, such sources are almost certain to have some degree of contamination. To reduce potential contamination, manually operated diversion valves can be installed to enable homeowners to delay collecting the water until sufficient precipitation has fallen to cleanse the roof. Additional protection can be provided by ensuring that the collection facilities are watertight and that in the case of cisterns, manholes or other ports of entry are leakproof. Experience shows that families in areas that have an annual average rainfall of 1 meter (~40 inches) can readily meet their needs through this approach. Rainfall can also be collected on a wider scale to provide drinking water to large municipalities. Cities employing this approach include New York, Boston, and Lisbon, where foresighted planners set aside large protected areas with natural and artificial lakes for collecting precipitation and runoff.

Water and Health

As was the case with food, there are multiple diseases that can be transmitted through contaminated drinking water. Due to the commonalities of the two sources, and because, as indicated earlier (Chapter 6), water is an ingredient of essentially all food products, that information will not be repeated here. In fact, because one out of every eight people lacks access to clean water, more than 3 million, most of them children, annually die from water-related health problems. If adequate water were available to enable them to wash their hands with soap, diarrheal diseases could be reduced by 45 percent (Rosenberg, 2010). In contrast, the relative number of deaths in the United States and other developed countries is far less. In fact, the average number of people affected in this country during each 2-year period from 1995 through 2004 was about 2,000 (Babin et al., 2008; Table 7.1).

Human Uses of Water

Water is essential for all human activities, ranging from drinking and cleaning to dust control. It is also a necessity for many industrial operations as well as for many recreational activities, as exemplified by fishing, boating, swimming, and waterskiing. Two other major uses are as transport vehicles of human waste (i.e., sewage disposal) and for irrigating agricultural crops. Some of these involve situations in which the water, after use, is discharged and later reclaimed for use. Others, as exemplified by irrigation, are consumptive in nature, in that the water ultimately evaporates into the atmosphere. The generation of electricity exemplifies both extremes: (1) nonconsumptive, except for normal evaporation in the use of hydropower to generate electricity, and (2) consumptive as in the use of cooling towers in electricity-generating power plants (Chapter 16), in which large quantities of water are necessary to condense the steam exiting the turbine generators. In so

Table 7.1 Outbreaks of illnesses due to contaminated drinking water (1995 through 2004), United States

Years	No. of states involved	No. of outbreaks	No. of people affected	No. of deaths	Size of outbreaks (range and median)	No. of outbreaks due to pathogens	No. of outbreaks due to chemicals	No. of outbreaks (cause not determined)
1995–1996	13	22	2,567	0	1–1,449 (22)	7	7	8
1997–1998	13	17	2,038	0	1–1,440 (10)	10	2	5
1999–2000	25	19	2,068	2	2–781 (13.5)	20	2	17
2001–2002	19	31	1,020	7	2–230 (6)	19	5	7
2003–2004	19	30	2,760	4	1–1,450 (7)	17	8	5

doing, much of it is evaporated. The alternative, to discharge the hot water into the environment, would warm lakes and other surface waters, with detrimental effects on aquatic life. In total, the United States uses about 350 billion gallons of water daily (Nationalatlas.gov. 2009). The primary uses are discussed below.

Personal use. This includes drinking, cooking, bathing, laundering, and excreta disposal. On a daily basis, flushing the toilet consumes 15–25 gallons (60–90 liters); bathing consumes another 15–20 gallons (60–80 liters). Total personal (domestic) water usage, however, depends on whether a home has a washing machine, dishwasher, or swimming pool, the extent to which water is employed to irrigate lawns, and other factors. Only about 1.9 liters per person (2 quarts) is actually consumed (used for drinking and cooking). The relative distribution of uses of water within the home is summarized in Figure 7.5.

Industrial use. The four largest industrial users are the manufacture of paper, refinement of petroleum products, and production of chemicals and of primary metals. These groups consume about 160 billion gallons of water per day. For example, it requires about 40,000 gallons (150,000 liters) to manufacture a new car and its four tires, and about 63,000 gallons (240,000 liters) to produce 1 ton of steel. As a result of improved plant efficiencies, increased recycling and conservation, more restrictions on discharges of liquid pollutants, and shifts to new industries and technologies that require less water, the consumption of water by industries in the United States has remained essentially constant for the past 25 years.

Irrigation of agricultural crops. An estimated 140 billion gallons of water (two-thirds of the total) are consumed daily in the United States for irrigating agricultural crops (Rosenberg, 2010). In fact, such use has increased by a factor of seven since 1900. Worldwide, this accounts for almost 60 percent of the withdrawals of freshwater. Based on current procedures, more than 50 glasses of water are required to grow the oranges needed to provide one glass of orange juice; 8 gallons (30 liters) to grow a single tomato; 9 gallons (35 liters) to process one can of fruit or vegetables (EPA, 2009); 120 gallons (450 liters) to produce one chicken egg (including the water required to grow a chick to the age at which she can lay eggs and to feed her subsequently); and 3,500 gallons (13,250 liters) to produce a steak

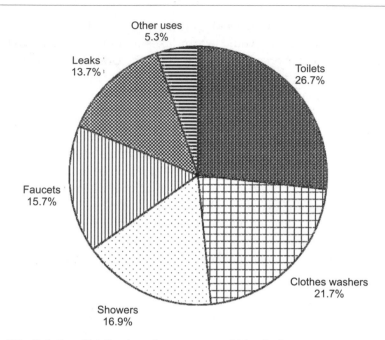

Figure 7.5   Relative distribution of water uses within the home

(Agricultural Water Conservation Clearing House, 2009; Canby, 1980; EPA, 2009). Compounding this problem is the increasing size of the global population.

Waste disposal. On average, each person in the United States. daily uses about 50 gallons (190 liters) of water for personal needs, including flushing toilets, bathing, washing clothes, and cleaning dishes, all of which becomes liquid waste. The water-carriage method of excreta disposal, an outgrowth of the development of the flush toilet, is particularly wasteful. It, for example, leads to the use of almost 250 gallons (950 liters) of purified water to transport 1 pound of fecal material to a sewage-treatment plant for disposal. Fortunately, major quantities of reclaimed water (i.e., treated sewage) are being used to irrigate greenbelts, parks, and other recreation areas.

Recreational, aesthetic, and transportation uses. Boating, sailing, swimming, waterskiing, and spray fountains, as well as the use of water for transportation in waterways and canals, are examples of this category. Of these, swimming pools can be a key source of public health problems. Pool water, for example, can contain microscopic organisms, even though it is visually clear. For this reason, the EPA (2003) has promulgated standards for recreational water. Otherwise, swimmers will be placed at risk of contracting diseases and infections, such as diarrhea and skin, ear, and upper-respiratory infections, particularly if the swimmer's head is submerged or the water is swallowed. Common sources of contamination include feces released from swimmers, especially infants, droppings from birds, and runoff from rain. The fact that there are some 400 million visits each year to swimming pools in the United States illustrates the potential magnitude of this problem (CDC, 2001).

Need for Water Conservation

As exemplified by the earlier discussion, it is mandatory that readily available methods for conservation and technologies for developing new sources of freshwater be implemented on a global basis. Although, with conservation, a person requires about 1,000 cubic meters (264,200 gallons) of water annually for purposes of drinking, hygiene, and growing food, the average use is many times that amount. This is due, as discussed earlier, to the vast quantities used in manufacturing industrial products and for cooling the waste heat produced by electricity-generating plants. Nonetheless, there are multiple ways in which water can be conserved. These include the application of economic and regulatory incentives, application of advanced technologies, and actions on an individual basis (Rogers, 2008). With the population of the Earth increasing by 83 million people annually, water demand will continue to increase unless we change how it is being used (Rosenberg, 2010; Royte, 2010).

Economic approaches. The average cost of drinking water remains low, about $2 per 1,000 gallons (3,800 liters), or one penny for forty 16-ounce glasses. That being the case, municipalities could increase the prices for water for both domestic and industrial users and thereby provide incentives to encourage conservation. In the meantime, this has led to the unnecessary exploitation of a precious natural resource, examples being the excessive irrigation of lawns, washing of cars, and failure to repair leaking faucets.

Advice and guidance to consumers. Homeowners can install low-flow shower heads and pressure activated flush systems for toilets (Table 7.2). Larger users, such as universities, businesses, and public facilities, can convert to the use of "dry" urinals for males. Public schools can also play a role through ongoing campaigns to provide information to their students on conservation measures. Print, radio, and television media should also be encouraged to do the same.

More efficient methods for irrigation. In many cases, less than half of the water being used reaches the crops. These losses could be dramatically reduced through converting from spray-irrigation

Table 7.2   Potential water savings by using water-efficient, instead of conventional, household systems

System	Water consumption		
	Liters	Gallons	Saving (%)
Toilets[a]			
Conventional	19	5	
Common low-flush	13	3.5	32
Air assisted	2	0.5	89
Clothes washers[a]			
Conventional	140	37	
Front loading	80	21	43
Dishwashers[a]			
Conventional	16	43	
Efficient	12	3	25
Showerheads[b]			
Conventional	19	5	
Common low-flow	11	3	42
Flow limiting	7	2	63
Air assisted	2	0.5	89
Faucets[b]			
Conventional	12	3	
Common low-flow	10	2.5	17
Flow limiting	6	1.5	50

a. Consumption per use.
b. Consumption per minute.

methods to drip-irrigation systems, which permit the water to seep slowly into the ground and into the root systems of the plants (Burt and Styles, 2007). The introduction of crops that could be irrigated with salt or brackish water might also lead to reductions in the use of freshwater resources. Another approach, which would reduce consumption by an estimated 10 percent, would be to line leaky irrigation canals with waterproof materials and to save water during the winter runoff, during nongrowing seasons, in large subsurface reservoirs that can be readily recharged. Such "water banks," which also avoid losses through evaporation, are in operation in Arizona, California, and other western states.

Matching the quality of the water to its intended use. In multiple cases, large volumes of drinking water are applied to tasks where supplies of lesser quality would suffice. These include irrigating lawns, fighting fires, washing cars, cleaning streets, and for recreational and aesthetic purposes. With increasing recognition of this fact, dual systems have been constructed in many arid areas, whereby separate plumbing systems deliver high-quality water for human consumption and water of lesser quality for other purposes. Another approach is to install "point-of-use" purifiers on household faucets and relevant home appliances to ensure that the water used in those cases is pure, while water that has not been treated can be used for other purposes.

Pathways of Exposure

There are several distinct avenues through which contaminated water can be a source of infection in humans. This is particularly true in the United States, where multiple recent disease outbreaks have been found to be associated with swimming and wading pools, water parks, fountains, hot tubs, and spas. In general, water can have effects on human health through four principal pathways.

Waterborne diseases. These, as was the case for food, result from the ingestion of water that contains the causative organisms, as in enteric diseases such as typhoid, cholera, and infective hepatitis. Prevention depends on avoiding the contamination of raw water sources by human and animal wastes or removing or destroying the contaminants prior to consumption.

Water-contact diseases. These can be transmitted through direct contact with organisms in the water. The most common example is *schistosomiasis,* which can be transmitted to people who swim or wade in water containing snails infected with the organism. The larvae, which leave the snail and enter the water, can readily penetrate the skin. Prevention requires the proper disposal of human excreta and deterring people from contact with infested waters. Another source of infection is the guinea worm, a parasite that will be discussed later.

Water-insect-related diseases. Examples are malaria, yellow fever, and West Nile virus, where water serves as a habitat for the disease transmitter, in this case the mosquito. Control requires eliminating their breeding areas, killing them, and/or preventing their contact with people.

Water-wash diseases. These result from lack of sufficient contaminant-free water for personal hygiene and washing. *Shigellosis,* trachoma, and conjunctivitis are among the diseases that may ensue.

Types and Impacts of Waterborne Diseases

The consumption of contaminated water accounts for an estimated 80 percent of all illnesses in the developing countries of the world. Groups at highest risk include the approximately 1 billion people lacking access to safe drinking water and the almost 2.5 billion without adequate sanitation facilities. The range of diseases include not only acute gastrointestinal illness (or gastroenteritis) but also *giardiasis* and *cryptosporidiosis.* By far, however, the most illnesses have been caused by the Norwalk virus (Surawicz, 2009). Multiple U.S. outbreaks are also associated with swimming and wading pools, water parks, fountains, hot tubs, and spas. Most of these are due to *Cryptosporidium parvum* (CDC, 1994).

One disease that has plagued populations for years, particularly in certain countries in Africa, is the large guinea worm nematode, *Dracunculus medinensis,* which can enter the body through the ingestion of contaminated water. Once there, the worms pierce the intestinal wall, grow to adulthood, and mate. Although the males die, the females make their way into the body, grow to a length of as much as 1 meter (3 feet), and end up near the surface of the skin, usually in the lower limbs, where they slowly emerge. This can lead to swelling and painful, burning blisters. To soothe the pain, sufferers tend to go into the water, where the blisters burst, allowing the worm to emerge and release a new generation of millions of larvae. If the worm, which is roughly the diameter of the wire in a paper clip, emerges and is broken, the portion remaining in the body can be infectious.

Through the efforts of former U.S. President Jimmy Carter, an international program was initiated in 1986 to eradicate this disease. In contrast to polio, which was conquered through the development of a vaccine, guinea worm disease is being eradicated through the simple approach of developing drinking straws equipped with filters that will remove the parasites from water. As part of this program, millions of these were provided to peoples in the affected countries with instructions on how to use them. As a result, a disease that once afflicted 3.5 million people in 20 nations in Africa and Asia has been reduced by more than 99 percent. In fact, only about 5,000 cases were reported in 2008, and these were limited to Sudan, Ghana, Mali, Ethiopia, Nigeria, and Niger (Staub, 2008).

Since the diseases transmitted by food and water are closely interrelated and the details on those common to these two avenues of intake were described in Chapter 6, that information will not be repeated. Suffice it to say, they include parasitic diseases, described above, as well as those of bacterial and viral origin.

Chemicals in Drinking Water

Chemicals in drinking water have been a concern for years. These range from arsenic and lead to those added through the prevention of dental caries (fluoride) and disinfectants (chlorine). Another problem arose in 2001 when, on the bases of epidemiological data, the limit for arsenic in drinking water was reduced from 50 to 10 parts per billion (ppb). Although most of the larger water districts in the United States already met the standard, approximately 4,000 smaller suppliers did not. Of the latter group, an estimated 97 percent served fewer than 10,000 people. Because of this, the cost per person to install the equipment to remove this contaminant was beyond their capabilities of complying. Exacerbating the situation, the inhabitants of many of the affected communities had lower incomes (Oates, 2002). This was ultimately resolved through the provision of assistance from federal and/or state sources to install the necessary remediation equipment.

A more serious situation occurred at about the same time, when Bangladesh, West India, and several other countries, based on recommendations of international agencies to reduce the potential intake of bacteriological contaminants, changed from obtaining their drinking water from surface sources to shallow wells. As a result, an estimated 35 million or more people unknowingly began consuming water that contained up to several thousand ppb of arsenic. Similar situations occurred in Vietnam and Thailand. As a result, millions of people were poisoned (Nordstrom, 2002). Effects at lower doses range from headaches to confusion, convulsions, and diarrhea. Higher doses can lead to effects on the lungs, kidneys, and liver, and ultimately to coma or death.

Drinking Water Standards, Implications, and Analyses

As noted earlier, the basic U.S. federal law pertaining to drinking water is the 1974 Safe Drinking Water Act, which was expanded and strengthened by amendments in 1986 and 1996 (Table 7.3). These require that drinking water and its sources, such as rivers, lakes, reservoirs, springs, and ground-water wells serving more than 25 individuals, be protected. In response, the EPA has developed a series of *primary* standards, designed to protect human health from both naturally occurring and human-made contaminants, and of *secondary* standards, designed to ensure that drinking water is aesthetically pleasing in terms of temperature, color, taste, and odor. The *primary* standards include maximum contaminant level goals (MCLGs) for selected inorganic contaminants (such as arsenic, barium, cadmium, chromium, lead, and mercury), volatile organic chemicals (including pesticides such as endrin, lindane, toxaphene, and methoxychlor, and certain of the chlorinated hydrocarbons such as the trihalomethanes), and radioactive materials (such as radium), as well as limits for the presence of coliform organisms (Table 7.4). The *secondary* standards include limits for iron, which can

Table 7.3   The primary U.S. legislation for the protection of drinking water

Safe Drinking Water Act and Amendments	1974 1986	Provides maximum limits for contaminants in public drinking water and techniques for their removal
	1996	Specifies requirements for specific analyses for emerging contaminants, such as *Cryptosporidium;* provides a revolving fund to assist state and local authorities improve their drinking-water systems

Table 7.4 National primary drinking-water standards, 2009

Contaminant	Sources	Limit
MICROBIOLOGICAL		MCLG[a]
Total coliforms	Human and animal waste	0
Giardia Lambia	Surface waters	0
Legionella	Indigenous to natural waters	0
Viruses	Human and animal fecal material	0
INORGANIC CHEMICALS		(mg L^{-1})
Antimony	Fire retardants, ceramics, electronics	0.006
Arsenic	Naturally occurring, pesticides, smelters	0.01
Barium	Pigments, epoxy sealants, spent coal	2.0
Cadmium	Corroded galvanized pipe, batteries, paint	0.005
Chromium	Mining, electroplating, pigments	0.1
Cyanide	Electroplating, plastics, mining, fertilizer	0.2
Fluoride	Naturally occurring, water additive	4.0
Lead	Leached from lead pipe, lead-based solder	0.0
Mercury	Crop runoff, paint manufacturing, fungicides	0.002
Nitrate	Animal waste, fertilizer, sewage	10.0
Nitrite	Animal waste, fertilizer, sewage	1.0
Selenium	Mining, smelting, coal/oil combustion	0.05
Thallium	Electronics, drugs, alloys, glass	0.0005
SYNTHETIC ORGANIC CHEMICALS		(mg L^{-1})
Chlordane	Insecticides for termites and crop insects	0.002
Dalapon	Grass control in agriculture and other areas	0.2
Endrin	Rodents, birds, pesticides on insects	0.002
Heptachlor	Insecticide in corn, alfalfa, vegetables	0.0004
Toxaphene	Insecticide used on cotton, corn, grain	0.003
VOLATILE ORGANIC CHEMICALS		
Benzene	Gasoline, solvent, pesticides	0.005
Carbon tetrachloride	Cleaning agent, coolant manufacturing	0.005
Ethylene dibromide		0.00005
Trichalomethanes	Byproduct of chlorine disinfection	0.1
Vinyl chloride	PVC pipe and solvents, synthetic rubber	0.002
RADIONUCLIDES		(Bq L^{-1})
Tritium (^{3}H)	Nuclear power plants	3.04×10^3
Total alpha activity	Radioactive waste, uranium deposits	5.55×10^{-1}
Total beta and photon activity	Radioactive waste, uranium deposits	$40\,\mu$Sv y^{-1}
Combined ^{226}Ra & ^{228}Ra	Radioactive waste, uranium deposits	1.85×10^{-1}

a. Maximum contaminant level goal.

discolor clothes during laundering; sulfates and dissolved solids, which can have the same effect as a laxative; and minerals that can, for example, interfere with the taste of beverages. They also include limits for the amount of suspended solids (turbidity) both for aesthetic reasons and because the efficacy of disinfection is related to the clarity of the water. To ensure that compliance with the *primary* standards can be achieved, the EPA has identified treatment processes that are capable of providing the required removals. Although no single treatment technique is effective for the removal of all inorganic chemicals, a combination of coagulation, sedimentation and filtration, or lime softening treatment (discussed later) has proved effective for removing many of them. One of the problems with contaminants such as pesticides and organic compounds is that water purification plant operators must anticipate which contaminant will be present and be ready to remove it. The use of multiple-purpose removal agents, such as activated carbon (i.e., a form of carbon that has been processed to make it extremely porous and thus to have a very large surface area available for adsorption of chemicals) is one approach for addressing these problems (Culp and Culp, 1974). As noted later, another very promising and rapidly developing approach is the use of membrane filtration technologies.

In the past, the measurement of biological contaminants on an individual basis was difficult and tedious. Since coliform organisms originate primarily in the intestinal tracts of warm-blooded animals, including humans, the accepted approach had been to use the presence of these organisms as an indication of fecal contamination. As a result of technological advances, this has changed. For example, test papers and/or strips are now available for diagnosing the presence of certain microorganisms on an individual basis. These include *Bacillus brevis* and *Escherichia coli*. A more sophisticated approach for such tests is the use of molecular probes, which can not only detect the presence of human feces but also determine whether an organism such as *Salmonella* is present. In a related manner, test strips and/or sticks are available to determine immediately the presence of a wide range of individual chemical elements (for example, aluminum, antimony, chlorine, chromium, cobalt, fluoride, iron, lead, nickel, potassium, and silver), as well as chemical compounds (for example, ammonia, boric acid, hydrogen sulfide, nitrate, sulfur dioxide). To encourage the use of up-to-date methods and acceptable procedures, the American Public Health Association, the American Water Works Association, and the Water Environmental Federation cooperatively prepare and publish, on a periodic basis, a book of standard methods for the sampling and analysis of a wide range of physical, chemical, and bacteriological contaminants in drinking water (Eaton et al., 2005).

Traditional Water Purification Processes

Preparing and distributing water for human use in the United States includes an estimated 160,000 public water systems that serve almost 310 million people. The associated annual operating costs exceed $3.5 billion (EPA, 2009). About half of the water comes from surface sources and half from ground sources. Most of the community systems are small. In fact, well over half provide water in towns with 500 people or fewer, and only about 250 are in towns with populations of 100,000 or more. An additional 15 million people (5 percent of the population) obtain their water from private wells. In addition, many industries have their own purification systems, and there are more than 100,000 "noncommunity" drinking-water suppliers—motels, remote restaurants, and similar establishments—that serve the traveling public.

The primary purposes of a water purification and treatment system are to collect water from a source of supply, purify it for drinking if necessary, and distribute it to consumers. Since about half of the groundwater supplies are distributed without treatment, the discussion that follows focuses on methods for purifying surface sources. The sequence in which the principal steps are applied is illustrated in Figure 7.6. As noted, the initial steps are to remove the suspended solids from the water. This is accomplished by raw water storage to permit natural processes (i.e., settling) to remove the

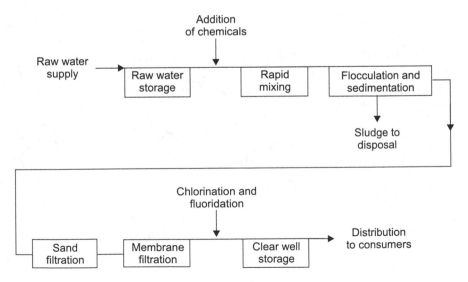

Figure 7.6   Principal steps in the water purification process

larger particles, the addition of chemicals to coagulate the particles, flocculation to facilitate the coalescence of the particles, followed by additional
sedimentation. Even so, many smaller particles will remain. The two processes most commonly applied to remove them are slow sand filtration or
rapid sand filtration.

SLOW SAND FILTRATION

In this relatively simple process, the raw water is passed slowly through a
sand bed 2–3 feet (60–90 centimeters) deep. These typically are carefully
constructed using graded layers of sand, with the coarsest layer, combined
with gravel, at the bottom and the finest sand at the top. Soon after a bed
becomes operative, a thin biological growth, the *zoogleal* layer, or *schmutz-
decke*, develops on top of and within the sand and dramatically improves
the removal of the particles as well as most bacteria and disease organisms,
including the cysts of *Giardia lamblia*. Because excess turbidity in the raw
water supply will rapidly plug the filter bed, preliminary settling is recommended. One of the primary limitations of this technique is the need for
relatively large land areas. Nonetheless, a filter bed area of 2,000 square
feet (185 square meters) will provide approximately 100,000 gallons (380,000
liters) of treated water per day. With proper care, slow sand filter beds can
be operated 30–200 days before the top layers of sand have to be scraped,
cleaned, or replaced. In other respects, such systems require minimal
maintenance (Leland and Damewood, 1990).

RAPID SAND FILTRATION

This is one of the most common processes. It is applied, however, after prior
steps (i.e., settling) have been taken to assist in removing the suspended
solids. This also removes color and reduces the concentrations of bacteria.
To accomplish this step, water is pumped or diverted from a river or stream
into a raw water storage basin. The next step is to add chemicals to the water
to create a coagulant. The chemical most commonly used in the United States
is aluminum sulfate ($Al_2(SO_4)_3 \cdot 14H_2O$), commonly referred to as alum. A
less frequently used chemical is ferric chloride ($FeCl_3$). The basic reactions
are almost identical:

$$Al^{+++} + 3HCO_3 \longrightarrow Al(OH)_3 + 3CO_2,$$

$$Fe^{+++} + 3HCO_3 \longrightarrow Fe(OH)_3 + 3CO_2.$$

The highly positively charged Al^{+++}, or the similarly effective Fe^{+++} ions, attract the negatively charged colloidal suspended particles in the water and, together with the $Al(OH)_3$ or $Fe(OH)_3$ form a gelatinous mass called floc. Rapid mixing is essential to provide maximum interaction between the positively charged metallic ions and the negatively charged colloidal particles. To ensure proper coagulation, the water is then slowly and gently stirred to enable the initial finely divided floc to agglomerate into larger particles and enhance settling. This process, called flocculation, is accomplished by moving large paddles slowly and gently through the water. As the water gradually moves along, the relatively large particles (including bacteria) are enmeshed in the floc, and ionic, colloidal, and suspended particles are adsorbed on their surface. This process, however, will not remove contaminants that are dissolved.

Although the settling process originally required a quiescent period of 2–4 hours, it is now being accomplished much faster through the use of high-rate settling tanks. These incorporate small-diameter tubes or parallel plates, set at an angle, through which the water passes. Because the solids travel a shorter distance before reaching a surface on which to deposit, and because this arrangement provides unique flow conditions, the required detention time is reduced to about 20 minutes. These additions also reduce the required size of the settling tank. Because the settled water will retain some traces of the floc, it is subsequently filtered. The filter beds are generally 2–3 feet (0.6–0.9 meter) deep and contain sand or crushed glass as the filter medium. Through a combination of adsorption, additional flocculation, sedimentation, and straining, the beds provide a final product of acceptable aesthetic quality. Another advantage of such beds is that they provide an effective method for removing particularly troublesome disease organisms, such as *Giardia lamblia* and *Cryptosporidium*. Because *Cryptosporidium parvum* is present in an estimated 65 percent or more of U.S. surface waters, the EPA requires that all large drinking water systems be monitored for this organism and that all surface waters, including those obtained from protected watersheds, be filtered prior to distribution for human consumption (Lee et al., 2002).

In time (12–72 hours), the filter beds become saturated with floc and must be cleaned by backwashing with purified drinking water. As might be anticipated, the accompanying solids present a formidable disposal problem. In fact, a typical water treatment plant will produce about 250 cubic feet of sludge—three large truckloads—per million gallons of water processed. As will be noted later, new coagulants and coagulant aids have been developed to reduce the quantities of sludge. Supplementing the above treatment processes in recent years has been membrane filtration. It has proved helpful as an additional, final step in removing multiple contaminants from drinking water (Adham et al., 2006). It must be applied, however, prior to the addition of fluoride (for the control of dental caries) or a disinfectant, such as chlorine; otherwise, the membrane filters would remove them (Figure 7.6).

FINAL STEPS IN PURIFICATION

Although sedimentation and filtration remove a significant portion of the microorganisms, these processes alone do not provide adequate protection. In the past, the most common disinfecting agent applied in the United States had been chlorine, which is usually added in sufficient quantity that a small residual will accompany the water entering the distribution system. Consumers were thereby protected in case bacterial contaminants later gain access to the supply. Unfortunately, the addition of chlorine to water that contains organic contaminants produces chlorinated hydrocarbons, which are known to be carcinogenic. One of the best methods to avoid this problem is to use another type of disinfectant, such as ozone or ultraviolet radiation. This is one of several reasons that increasing numbers of water treatment plants are switching to these types of disinfectants. This is discussed in more detail in the section that follows. In rural areas in developing countries, disinfection is also being accomplished by heating water using solar radiation. The characteristics of various disinfectants are summarized in Table 7.5.

After the preceding steps have been completed, the water is sent to storage and is ready for distribution to the consumers. In terms of the addition of fluoride, there are multiple factors to consider. It is a common additive to many toothpastes; conversely, certain household filters may remove some or all of it from drinking water. At the same time, people who depend on groundwater sources may not be provided an adequate amount, or they may be provided an excess. In fact, the consumption of groundwater with naturally relative high concentrations of fluoride has caused dental fluorosis in some cases (Wikipedia, 2009).

Additional steps that can be applied include the removal of iron and manganese, calcium and magnesium, organic compounds, and tastes and odors. Iron and manganese (which, as previously noted, can discolor clothes) are soluble in water only in the reduced state. If they are oxidized, for example, by aerating the water, they immediately become insoluble and precipitate. Although otherwise harmless to humans, calcium and magnesium give

Table 7.5 Characteristics of various disinfectants

Disinfectant	Characteristics
Chlorine	Widely used in the U.S; forms harmful by-products if water contains organic matter; maintains residual in distribution system; requires care in handling as a gas
Hypochlorite	Safer alternative to chlorine gas; can be purchased or produced on site by electrolysis of sodium chloride, but this process introduces chlorates and bromates as disinfection by-products and adds both sodium and chloride to treated water
Chlorine dioxide	Must be generated on site since it cannot be transported because of its potential explosiveness; is a strong oxidant that will kill *Cryptosporidium* while chlorine will not; does not provide a persistent residual in treated water; produces its own range of by-products that may be cause for concern
Chloramines	Normally used in conjunction with another disinfectant; does not effectively inactivate viruses or protozoa; does not produce chlorinated by-products; provides a persistent residual in distribution system
Ozone	Must be generated on site since it is highly reactive; produces no unwanted by-products and will inactivate viruses, bacteria, and protozoa, including *Cryptosporidium*; will also reduce tastes and odors and improve coagulation; does not provide a residual
Ultraviolet radiation	Will inactivate *Cryptosporidium*; effectiveness requires low turbidity; small size of units makes them suitable for installation in existing facilities; overall cost is about double that for chlorine; does not provide a residual

water the undesirable property of being "hard," that is, these chemicals make it difficult to develop a lather when a person uses soap during bathing or in washing dishes or clothes. These chemicals also leave a scum or ring in the bathtub. Although hardness is not generally a problem for supplies that are derived from surface waters, it frequently is in situations where supplies are derived from groundwater. The relative amounts of hardness in groundwater supplies in various regions of the conterminous United States are shown in Figure 7.7.

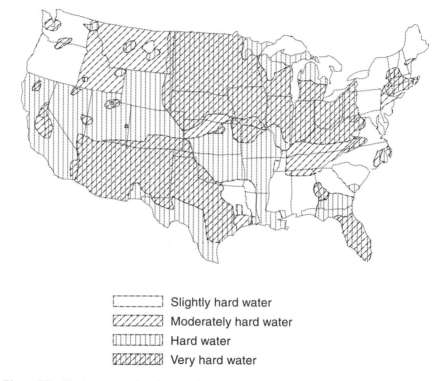

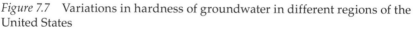

Slightly hard water
Moderately hard water
Hard water
Very hard water

Figure 7.7 Variations in hardness of groundwater in different regions of the United States

In the case of a large water-purification facility, the normal procedure for removing calcium is to add calcium hydroxide (lime) and sodium carbonate to the water to interact with the dissolved calcium to form insoluble calcium carbonate. This, in turn, precipitates and reduces the calcium concentration. This process also removes some of the magnesium. To remove hardness from small volumes of water, as in systems that serve an individual household, the process normally applied is ion exchange. Unfortunately, this replaces the calcium with sodium. Since, as noted earlier (Chapter 6), sodium is believed to cause high blood pressure in some people, care must be exercised in consuming water treated by this process. One approach is to connect the softener only to the hot-water line, thus restricting use of the treated water primarily to taking baths and washing clothes and dishes. Following this approach, the cold-water supply, normally used for drinking and cooking, will not contain added sodium.

Advances in Water Purification

In recent years, additional technologies have been developed and applied in the purification of water. Three of these are discussed below.

POLYMER COAGULANTS

The success of coagulation depends on how well the floc settles, supplemented by how effectively the remaining particles are removed by the filters. Because, as discussed earlier, suspended and colloidal solids in surface waters possess anionic (negative) charges, they are, in essence, prevented from coalescing into larger particles. One of the primary goals of coagulation is to neutralize these surface charges. Alum not only does this well but also reacts with the alkalinity in the water to form metal hydroxide precipitates that encapsulate the colloidal particles. A major advance was the development of polymer coagulants. Because these materials possess a cationic charge, they form a dense, rapidly settling floc and do not alter the pH or alkalinity of the water. They also are not as sensitive as aluminum sulfate to temperature and require only minor adjustments in dosage, even if the amounts of turbidity in the raw water vary over a relatively large range. In some cases, the amount of sludge has been reduced by more than half, and the length of time the filters can be used prior to backwashing has been increased by one-third or more (Laughlin, 2001).

ULTRAVIOLET RADIATION

It has been known for years that ultraviolet radiation was an effective germicidal agent. In fact, it has been used for years in Europe for disinfecting drinking water. Due to the previously cited problems with chlorinated by-products, combined with tighter restrictions on permissible limits for these contaminants in drinking water, the use of UV radiation is increasingly being applied in the United States. Its mechanism of action is that it penetrates the cell walls of microorganisms and affects their DNA in such a way that they cannot reproduce. Rather than being killed, the microorganisms are, in essence, inactivated. Another advantage is that UV radiation will inactivate both *Girardia lambia* cysts and *Cryptosporidium parvum* oocysts and is capable of effectively treating certain bacteria found to be unaffected by chlorine. Where desirable, UV can be used in combination with chlorine (added in reduced concentrations after the application of UV) to provide a residual disinfectant in the drinking water distribution system (Fleming, 2002).

Table 7.6 Effects of purification processes on specific characteristics of water

Process	Characteristic						
	Bacterial content	Color	Turbidity	Taste and odor	Hardness (calcium and magnesium)	Corrosive properties	Iron and manganese
Raw water storage	+	+	+	±	+	0	+
Aeration	0	0	0	+	0	+	+
Coagulation and sedimentation	+	+	+	0	0	−	+
Lime-soda softening	+	0	+	0	+	+	+
Sand filtration	+	+	+	0	0	0	+
Chlorination or ozonation	+	+	0	±	0	0	+
Carbon adsorption	−	+	+	+0	0	0	
Membrane filtration	+	+	+	+	+	+	+

Note: 0 (no effect); + (beneficial effect); − (negative effect); ± (sometimes beneficial, sometimes negative, effect).

MEMBRANE FILTRATION

In cases where chlorine is used as a disinfectant, one of the best methods for avoiding the production of unwanted by-products, as noted earlier, is to remove any organic compounds from the water prior to this step. The same is true for hydrogen sulfide and algae growths, both of which can produce bad tastes and odors when subjected to chlorination. Since membrane-filter technologies are effective in removing these types of contaminants, as well as cysts and viruses, they also are increasingly being applied in water purification systems. (Adham, et al., 2006).

The effects of various steps in the water purification process on specific characteristics of the raw water supply are tabulated in Table 7.6.

The General Outlook

Adequate consideration of the challenge of providing safe sources of drinking water on a global basis requires addressing a wide range of topics. These include the need for technological advances in watershed protection, drinking-water treatment and distribution system management, our understanding of the associations between water quality and its impact on the safety of food, and the influence of irrigation practices on water consumption and the quality of agricultural crops (CDC, 2009). There is also a need to improve methods for removing and/or disinfecting microbial contamination and for avoiding and/or removing the by-products of disinfectants such as chlorine.

If these efforts are to be effective, it is essential that a *systems approach* be applied. The "costs" of producing drinking water, for example, include those associated with managing the watershed from which the water is obtained, whether from surface or groundwater sources; the negative impacts of airborne and liquid wastes that degrade the quality of the water; and the enforcement of policies to ensure that consumers, and appliance manufacturers, effectively apply steps to limit their use of water. Excellent examples are the manner in which key household systems have been modified to reduce their use of water. Comparable attention should be directed to conserving the use of water outside the home, such as maximizing the use of native plants resistant to dry conditions.

These efforts should also include a nationwide program to educate individual members of the public in understanding how much water is necessary, in contrast to how much they use. Many of the western desert sections of the United States already face shortages. It is even a more challenging problem in the drier developing nations with large populations. Rivers such as the Nile, the Jordan, the Yangtze, and the Ganges are not only overtaxed, but their flow regularly ceases for extended periods. In a similar manner, the levels of the underground aquifers below New Delhi, Beijing, and many other burgeoning urban areas are also being reduced (Rogers, 2008).

Also to be recognized (as noted earlier) is that overuse of groundwater sources can lead to multiple problems, such as the intrusion of saltwater. This is exemplified by the movement of water from the Atlantic Ocean into well-water sources in some areas along the eastern coast of Florida. The result is that an estimated 15 percent of the world's population does not have access to safe drinking water. As sources continue to be depleted and/or contaminated, this percentage will undoubtedly increase. In addition, there are the overwhelming challenges of the increasing global population (Chapter 1), and the accompanying increasing needs for water not only for direct human consumption but also for hygiene, sanitation, food production, and industry. In the meantime, agricultural runoff, releases of atmospheric contaminants (Chapter 5), and releases of domestic and industrial liquid wastes (Chapter 8) will continue to reduce the quality of freshwater sources.

As so aptly described by Samuel Taylor Coleridge (Keach, 1997) in his poem "The Rime of the Ancient Mariner," there was "Water, water everywhere and not a drop to drink." Unless this reality is recognized, water supply and sanitation will continue to pose chronic environmental and public health challenges. In fact, although nations continue to compete for oil, natural gas, and uranium, water may very soon become our most limiting natural resource (Greenberg, 2009).

8

LIQUID WASTES

Liquid Wastes include cooking oil, fats, or grease, used oil and transmission fluids from motor vehicles, and runoff from a variety of sources, including livestock mass-production facilities, agricultural fields, and city streets and other paved areas following heavy rains. They also include wastes resulting from the processing of milk and other food products, the manufacture of biomedical products, tires, plastics, batteries, and household products, and those due to the construction and demolition of buildings.

One of the most common forms of liquid waste is human sewage, and efforts to dispose of it in a proper manner have been ongoing for thousands of years. Basic guidance on its disposal can be found in verses 12 and 13 of the twenty-third chapter of Deuteronomy, where God provided the following instructions to Moses: "You shall have a place outside the camp and you shall go out to it; and you shall have a stick with your weapons and when you sit down outside, you shall dig a hole with it, and turn back and cover up your excrement." An early and simple method for disposing of human excreta follows this guidance almost to the letter. That is exemplified by the pit privy, a hole in the ground with a small closed shelter and toilet built above it. Generally, the hole is approximately 3–4 feet in diameter and 5–6 feet deep. Privy designs range from the pit privy to those in which the excreta are collected in a bucket or tank for later removal and disposal. Double-vault pit privies are used by many people in the developing countries. Alternating the pits each year provides sufficient retention and decomposition to ensure that the most pathogenic organisms in the wastes are destroyed. Improved versions have a screen-covered vent pipe (Figure 8.1), which provides a natural pathway for removing odors and for trapping flies and other insects.

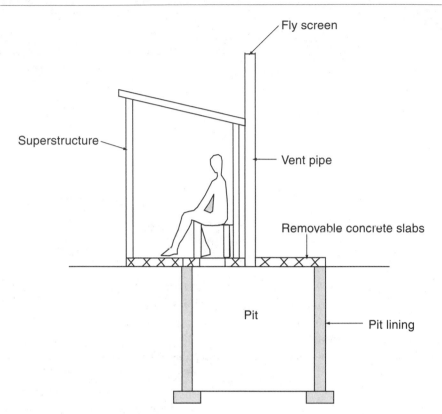

Figure 8.1   Pit privy with ventilation pipe

With the development of the flush toilet, sewage treatment and disposal entered a new era. As far back as the Mesopotamian Empire (3500–2500 B.C.E.), however, the toilets in some homes were connected to a stormwater drainage system that transported the wastes away. In larger homes in Babylon, toilets flushed by hand were connected to vertical shafts in the ground that were lined with perforated clay pipe that permitted the liquid to be absorbed in the surrounding soil (Wolfe, 1999). Although the royal palace of King Minos in Crete had toilets as early as 1700 B.C.E. for which water collected in rain-fed cisterns provided a continuous flow of cleansing water, it was not until 1596 that the modern flush toilet was invented. Even so, the valve controlling the inflow of water allowed considerable leakage. This problem was not solved until 1872, when Thomas Crapper, later knighted by the Queen of England, invented the first valveless water waste preventer.

The principles of his design continue to be applied today. Even so, as late as the 1880s, only one of every six people in U.S. cities had access to modern bathroom facilities.

Individual Household Disposal Systems

Subsequent widespread use of the flush toilet necessitated methods for disposing of the discharged wastes. Most larger municipalities constructed systems to transport the effluent to a sewer and then to some form of municipal treatment plant. Even today, however, roughly 90 million people in the United States (about 30 percent of the U.S. population), many of who live in small communities, are not served by sewers. They depend instead on some form of on-site subsurface sewage disposal system. The most common of these is the septic tank. Although more than 90 percent of those in the United States who live in larger communities are served by sewers, the national average is only about 75 percent (EPA, 2006).

SEPTIC TANK

A septic tank is generally constructed of concrete or plastic, with an inlet for sewage to enter and an outlet for it to leave (Figure 8.2). As sewage passes through the tank, solids settle to the bottom and are digested through the action of anaerobic bacteria that naturally develop. To help prevent the release of settled solids and floating material, the tanks have a divider at the bottom and a baffle at the top near the outlet. Under proper operating conditions,

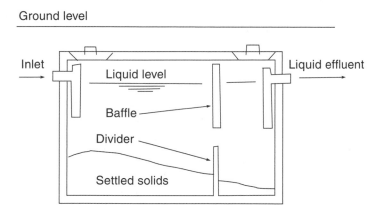

Figure 8.2   Cross section of a typical septic tank

the effluent is clear and is discharged into a drain field consisting of open-jointed or perforated pipe buried in the ground so that the liquid can seep into the soil. The purposes of the drain field are several. It disperses the septic-tank effluent over a wide area and promotes infiltration of the waste into the soil. Furthermore, natural bacterial populations in the soil continue the digestion of soluble organic materials in the septic-tank effluent. The soil also acts as a filter mechanism to adsorb pathogenic organisms remaining in the waste.

For proper performance, it is generally recommended that (1) the volume of the tank be at least 2,000 liters (500 gallons), (2) the soil in which the drain field is located be sufficiently porous to absorb the effluent, (3) the land area be adequate for absorption of the volume of flow anticipated, and (4) the tank be cleaned (solids removed) every 3 to 5 years. Otherwise, the solids will ultimately be carried out with the effluent and will seal the perforations in the pipes that distribute the sewage within the drain field. Before a septic tank is installed, field tests should be conducted to document that the adsorptive capacity of the soil and the land area available is sufficient to handle the anticipated flow (Fernando, 2008). In most jurisdictions, unless these conditions are met, a permit for the installation of a septic tank will not be granted. Even so, about one-quarter of the existing tanks malfunction either periodically or continually, leading to a situation in which the effluent will break through to the surface or find its way into a groundwater source.

In seeking to solve these problems, septic-tank systems have been modified to (1) include filters to avoid premature plugging of the disposal field with solids (Dix, 2001) and (2) incorporate a unit for equalizing the flow of wastewater being treated. Another change is the use of low-flush toilets, water-saving appliances, and reduced-flow showers to limit the volume of liquids being discharged into the tanks (Chapter 7).

OTHER TREATMENT SYSTEMS

In recent years, a variety of alternative treatment systems have been developed (Hetrick, 2001; Smith, 2009). These include units wherein the sewage is collected in a tank, mixed by a pump to break up the solids, and aerated. Under proper operating conditions, these are less prone than a septic tank to produce disagreeable odors. Because the effluent contains dissolved oxygen, the probability that the absorption field will be plugged by solids is also reduced. More sophisticated aerobic systems include features through which the effluent can be recycled and used for flushing the toilet again. A more recent development, based on a two-stage bio-filtration design using a single pump, has proved to be highly effective. In fact, such units can produce an effluent of comparable quality to that of centralized wastewater systems that employ external carbon dosing for tertiary denitrification (Smith, 2009).

Whether the treatment and disposal system involves a septic tank or an aerobic unit, its design, construction, and location of the absorption field are not the only factors that can affect its performance. Particularly troublesome is the discharge of household effluents containing antibacterial soaps, relatively high concentrations of bleach, pesticides, and strong disinfectants. These can kill the bacteria that stabilize the waste. Also troublesome are the previously mentioned grease, fats, oils, and food wastes (from garbage disposal units, for example). These can overload the system (Guy and Catanzaro, 2002). To avoid some of these problems, various types of biological, composting, incinerating, and oil-flushed toilets have been developed. One of the most popular of the composting toilets incorporates the waterless Clivus Multrum household excreta and garbage disposal system that was developed in Sweden. In this case, the toilet is located on the first floor of a house, with the composting portion in the basement (Figure 8.3). Advanced designs incorporate fans to promote aeration and heating elements

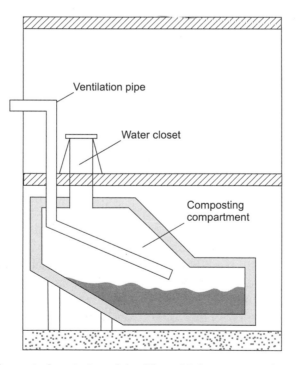

Figure 8.3 Schematic description of the Clivus Multrum water closet

to maintain optimum temperatures to hasten the composting process and to evaporate urine and other moisture. Operators of composting toilets also commonly add a small amount of absorbent carbon material (such as untreated sawdust, coconut coir, or peat moss) after each use to create air pockets for better aerobic processing, absorb liquid, and create an odor barrier. Advanced designs have been installed in multiple locations in Europe, and New Brunswick, Canada, and are being considered for installation in several U.S. outdoor recreational centers (Clivus Multrum, Inc., 2011).

Advent of Sewer Systems

Between the 1830s and 1850s, there was a series of epidemics of cholera and typhoid in London, Paris, Hamburg, and other European cities. These led to the previously discussed epidemiological studies conducted by John Snow (Chapter 3). Similar outbreaks occurred in the United States between 1832 and 1873, causing the deaths of thousands of people. Recognizing that the installation of septic tanks on an individual household basis was not feasible in metropolitan areas, the city of Hamburg, Germany, constructed the first comprehensive sewer system in 1843. This followed by almost 500 hundred years a much simpler system that had been built in Paris, France. Systems similar to the one in Hamburg were subsequently constructed in other cities in Europe, as well as in New York and Chicago. Even so, in all cases these systems served only as a vehicle for transporting the wastes for discharge into a nearby river or lake. Although some of the sewage in the smaller cities in Europe was used to irrigate nearby farmlands, this disposal method proved impractical and unsanitary for all but the smallest cities (Wolfe, 1999).

As would be anticipated, the water bodies into which these wastes were discharged soon became heavily polluted. Recognizing the need to treat such wastes, scientists in England, Europe, and the United States began developing mechanisms for using natural biological stabilization processes for treating such wastes. Playing a significant role in such activities was the Lawrence Experiment Station that was established by the Massachusetts State Board of Health in 1887. By 1890, the staff of this facility had published a report documenting the technical basis for the treatment of municipal wastewater. These and related activities led to the construction and operation of large-scale treatment facilities in the larger cities of the world (Wolfe, 1999). Details of the operation of such facilities are discussed later in this chapter.

Liquid Wastes: Nature and Classification

As noted earlier, the sources of liquid wastes extend far beyond those generated in individual households. Within a modern city such sources include commercial and office buildings, schools, restaurants, and hotels, as well as a wide range of industrial operations. Other sources, often neglected, include health-care products and pharmaceuticals. For example, antibiotics are used for the treatment of a host of infections, and although concern has been expressed about those excreted by the patients, a more important threat are the liquid wastes released by those who manufacture these products (Saravanane and Tamijevendane, 2009). Recognition of these and related problems has led to a series of laws that ensure that the environment is protected (Table 8.1). Not only is the nature of industrial wastes often significantly different from municipal sewage, but such wastes frequently contain toxic chemicals as well as heated water and various types of suspended materials. If discharged into rivers and lakes without treatment, they, as well as municipal sewage, can be major sources of pollution. If discharged onto the land, they can contaminate the soil and groundwater.

History of Water-Pollution Regulations

Recognizing the need for federal action, the U.S. Congress passed a number of laws that were designed to maintain the quality of U.S. rivers and streams. These included the River and Harbors Act of 1899 that barred the discharge or deposition of refuse in navigable waters without a permit. The first federal legislation, however, that was designed to address the conventional aspects of maintaining the quality of surface waters was not enacted until 1948, when Congress passed the initial Water Pollution Control Act. This provided funds for relevant federal research and associated investigations. Amendments in 1956 authorized the states to establish water-quality criteria and sponsored enforcement conferences to negotiate cleanup plans for bodies of water whose maintenance required the involvement of several states. More importantly, they provided federal support for the construction of new municipal sewage-treatment plants.

These initiatives were followed by the Water Quality Act of 1965, which was designed to assist the states in meeting the water-quality standards that had been established. Shortly thereafter, Congress passed the Water Quality Act of 1972 and the Clean Water Act of 1977 (U.S. Congress, 1972; 1976; 1977). The ultimate goal of these laws was to provide "fishable and swimmable"

Table 8.1 Major water-pollution laws, United States, 1972–1990

Law	Date	PL number	Purpose and/or scope
Federal Water Pollution Control Act	1972	92-100	Defined toxic pollutants as those that individually or in combination "upon ingestion, inhalation, or assimilation into organisms, either directly from the environment or indirectly by ingestion through food chains, will, on the basis of information available, cause death, disease, behavioral abnormalities, cancer, genetic mutations, physiological malfunctions"
Clean Water Act and Amendments	1977	95-217	Authorized the Fish and Wildlife Service to provide technical assistance to states in developing "best management practices" as part of its water-pollution control programs; the Department of Interior to complete the National Wetlands Inventory; and the Corps of Engineers to issue general permits on a state, regional, or national basis for any category of activities that are similar in nature, will cause minimal environmental impact when performed separately, as well as in combination
Water Quality Act	1987	100-4	Modified the National Pollution Discharge Elimination System by expanding the number of chemical constituents that must be controlled; significantly changed the thrust of enforcement by requiring increased monitoring and control of toxic constituents in wastewater and in discharges of polluted runoff from city streets, farmland, mining sites, and other nonpoint sources
Pollution Prevention Act	1990	101-508	Established a national policy to encourage the prevention of pollution at the source, with disposal to the environment acceptable only as a last resort; this has led to the substitution of less toxic substances for use in a wide range of industrial processes
Wet Weather Water Quality Act	2000	106-377	Directed specific attention to the problems of overflowing sewers, a type of event that has been occurring on an increasingly frequent basis in many U.S. cities, particularly during heavy rains

rivers and streams on a nationwide basis. As with the amendments to the Water Pollution Control Act, these laws have continued to provide federal funds for the construction of municipal sewage-treatment plants (Freeman, 2000). One of the provisions of the Clean Water Act that was fortuitous in terms of responding to acts of terrorism (Chapter 17) was a requirement that operators of facilities having the potential for major accidental liquid releases develop and maintain a plan describing the actions to be taken in case such an event occurs. Another law that has significantly reduced liquid-waste releases is the Pollution Prevention Act of 1990 (U.S. Congress, 1990; Table 8.1).

Classification of Wastes

A somewhat oversimplified approach classifies such wastes as *degradable* and *nondegradable*. Domestic sewage is the most common degradable waste; that is, it can be degraded or stabilized by bacteria. Many industrial wastes contain organic residuals that are also degradable. In fact, the quantities of degradable wastes released by industry vastly exceed those in domestic sewage. Primary sources include industrial facilities involved in food processing, meat packing, pulp and paper manufacture, petroleum refining, and chemical production. Notable examples of nondegradable constituents are those that contain inorganic substances, such as ordinary salt and the salts of heavy metals (for example, lead, mercury, and cadmium). A *third* group, which does not fall into either of these categories, consists of the so-called persistent chemicals that are best exemplified by the synthetic organic chemicals, such as DDT and the phenols that result from the distillation of petroleum and coal products. Although they can be altered by biological and chemical transformation, these processes are extremely slow.

Another approach is to classify wastes as those discharged by *point* and *nonpoint* sources. Examples of the former are wastes such as industrial and municipal discharges. Severe water pollution problems, however, are also caused by less obvious and more widespread sources of pollution, the so-called *nonpoint sources*. These include liquid runoff from agricultural lands, which accounts for 39 percent of the pollution being discharged into U.S. rivers (Figure 8.4). In localized cases, it can account for up to 80 percent of the degradation of such waters. The quantities of this type of waste are especially large during spring thaws (CEQ, 1993; 1998). For this reason, many of these sources have now been brought under regulatory control.

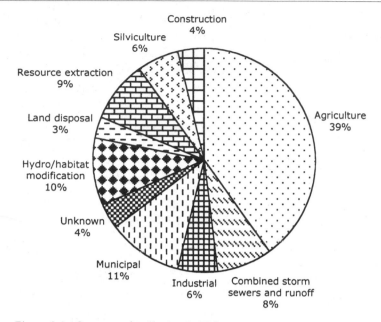

Figure 8.4   Sources of pollution in U.S. rivers

Other Major Sources of Liquid Wastes

Another source of increasing concern are the so-called concentrated animal feeding operations (CAFOs), in which upwards of 45 billion animals (i.e., chickens and hogs) worldwide, including 10 billion in the United States, are grown each year (EPA, 1998). By 2020, the global total is projected to increase as many as 100 billion. These are another example of a nonpoint source. A typical hog factory will have 12,000 or more animals that are fed and watered on a mass-production basis. At present, the most common method of handling the resulting wastes is to flush them into giant lagoons, some of which may have capacities in excess of 25 million gallons. Unfortunately, surface overflow and underground seepage from such lagoons have in many cases contaminated nearby surface and groundwater supplies. Odors and swarming flies often create noxious conditions for nearby residents. Excess spraying of wastes onto fields has led to runoff that pollutes surface waters. Exacerbating the problem is that pigs produce nearly twice as much waste as beef cattle and about three and a half times as much as chickens (Satchell, 1996). This is discussed in more detail in the section that follows.

Liquid wastes from urban areas are also increasingly recognized as a major nonpoint source of pollution. Because such areas are dominated by buildings ranging from high-rise offices to suburban single-family dwellings, plus multiple paved areas such as sidewalks and streets, a major portion of the rain that falls on such areas collects as runoff and flows into sewers. Prior to reaching the sewers, the water accumulates a host of organic and inorganic contaminants, including animal waste, infectious agents, pesticides, and fertilizers. Urban runoff is a special problem in cities in which the sewers were originally designed to handle only domestic wastes. When a decision is later made to direct runoff into the same sewers, they often overflow, the net result being that the runoff, combined with untreated domestic sewage, is released into the environment. Even if a separate storm-drain sewer system is installed, unless the collected runoff is properly treated, its release into the environment, for example, a nearby lake or stream, can create problems.

Impacts of Liquid Wastes

If liquid wastes contain toxic chemicals and/or pathogenic organisms and the receiving waters serve as raw sources for drinking-water supplies, the accompanying contaminants may have health impacts. If the wastes are applied to land areas for irrigation or other purposes, additional avenues of contact with humans may occur. As a result, there can be a variety of negative impacts.

DEPLETION OF OXYGEN

In cases where the wastes are released into a stream, bacteria within the water will attempt to stabilize the organic matter in the wastes. This process requires oxygen, which must be obtained from that dissolved in the water. Fortunately, there are several natural processes that continually replenish the oxygen. These include eddies and other turbulence, which serve as aerators, and the production of oxygen by green algae and various plants growing in the water. So long as these sources replenish the oxygen as rapidly as it is removed, aerobic conditions will be maintained and problems can be avoided. It is when the consumption of oxygen exceeds the supply that problems develop. Eventually, the concentration of dissolved oxygen (DO) may become insufficient to support fish and other forms of aquatic life. Interestingly, the more desirable varieties of fish will be the first to be affected, followed by the pollution-resistant lower orders, such as carp. If all the DO is consumed, anaerobic conditions will result. Instead of releasing carbon dioxide (which occurs under aerobic conditions), anaerobic decomposition produces methane or hydrogen sulfide. The stream or lake will, in turn, become dark and malodorous.

This is what happened in the spring of 2009, when retention lagoons in which the hog farmers in North Carolina had discharged their wastes overflowed during heavy rains and released significant amounts of waste. The accompanying lack of DO killed some 50 million fish (menhaden, or "pogy") where the Trent River emptied into the Neuse River on the eastern portion of the state (Book, 2009). Recognizing the increasing nature of these types of problems, the EPA had earlier changed its regulations pertaining to CAFOs to require that they (1) remove their requirements that all CAFOs apply for National Pollution Discharge Elimination System (NPDES) permits, (2) submit Nutrient Management Plans (NMPs) with their permit applications, and (3) require that the NMPs be reviewed by permitting authorities and the public (Griffiths, 2003). This problem was later addressed (Circuit Court, 2005).

Even so, the lack of oxygen is not the only example of the harmful effects of liquid wastes on aquatic life. Discharges of suspended solids, toxic chemicals, heavy metals, and other hazardous substances not only can also be harmful, but the receiving waters (even in cases where the wastes have been treated) and the fish and shellfish harvested from such waters can also be unsafe for human consumption. Analyses conducted by the U.S. Geological Survey showed that the range of pollutants in surface waters in the United States is widespread. Most of the samples, which were obtained from 139 sites in 30 states, were collected immediately downstream from suspected pollution sources, such as wastewater treatment plants, urban areas, or agricultural operations. Their analyses revealed the presence of a range of antibiotics, other prescription drugs, pesticides, and household chemicals, such as detergents and fragrances. The most commonly observed chemicals were steroids, caffeine, and components of insect repellent, disinfectants, and fire retardants, some of which may be a source of endocrine disruption. Although median concentrations were usually relatively low, maximum concentrations occasionally exceeded regulatory limits (Weinhold, 2002).

EUTROPHICATION

This is a process that occurs when excess nutrients are discharged into lakes. This, in turn, can lead to flourishing blooms of toxic blue-green algae. A person who takes a bath or shower in water containing these blooms can develop an allergic reaction resembling hay fever and asthma. Skin, eye, and

Table 8.2 National priorities for constituents of concern in liquid wastes

Priority	Pollutant group	Example
High	Nutrients	Nitrogen
	Pathogens	Enteric viruses
	Toxic organic chemicals	Polynuclear aromatic hydrocarbons
	Chemicals and biologicals	Urban storm water runoff
Intermediate	Selected trace metals	Lead
	Other hazardous materials	Oil, chlorine
	Plastics and floatables	Beach trash, oil, grease
Low	Organic matter	Municipal sewage
	Solids	Urban storm water runoff

ear irritations may also occur. In addition, the ingestion of water containing the blooms, either through drinking or swimming, may produce gastroenteritic or hepatoenteritic disorders (Pitois, Jackson, and Wood, 2001). Excess nutrients may also be one of the causes of the increased frequency of blooms of the so-called red tide, or paralytic shellfish poisoning, that occurs in coastal waters of many of the world's oceans. Another example of the impacts of excess nutrients is the so-called dead zone that has formed in the Gulf of Mexico off the coasts of Louisiana and Texas (Holden, 2002). Subsequent studies have shown that although the size of the most recent event is smaller than forecast, it still covers an estimated 3.000 square miles and, in contrast to the past when it was limited to water just above the ocean floor, was still not only severe where it did occur but also extended closer to the water surface than in most years (Rabalais, 2009). Concerned about these developments, the National Research Council has recommended that nutrients be considered a high-priority pollutant (Table 8.2).

Problems of Stormwater Flow

Although this problem has been addressed for many years, runoff from urban areas continues to be the primary contributor to the impairment of water quality nationwide (EPA, 2009a). It is another example of a nonpoint source. This is common in cities that have combined sanitary and stormwater sewers that overflow during periods of heavy rains. The impacts, however, are not restricted to the chemical and biological contaminants. The accompanying interactions, due to the tremendous increase in the velocity and quantity of the water flow, pose physical hazards on land-based structures as well as on aquatic habitats and stream functions. According to the EPA, more than 40,000, and perhaps as many as 75,000, such overflows occur annually in the United States and involve an estimated 750 systems in 32 states. Annual releases total 3 to 10 billion gallons of untreated wastewater that, as discussed earlier, contains solids, debris, and pathogenic pollutants. Causative factors include severe weather, mechanical failures, and sewers having insufficient flow capacities. In addition to posing serious water-quality problems and significant health threats, such releases represent a threat of contaminating drinking-water supplies and local watersheds. Possible ameliorative measures include standby electric generators and additional pump capacity that can be made available during emergency situations (WaterWorld, 2009). Given the shift of the global population to urban areas and the fact that this trend is expected to continue (Chapter 1), the magnitude of this problem is expected to continue to increase (NRC, 2008).

Assessing Water-Polluting Potential

Multiple methods are available for determining the concentrations of contaminants in liquid wastes. In the main, these are generic in nature (i.e., although they provide a broad measure of its polluting potential of the waste, they do not identify the individual contributors). One indicator is the concentration of suspended solids. Others are its nutrient content and the amount of chlorine required to disinfect the effluent after it has been treated by, for example, the activated sludge process (discussed later). The acidity or alkalinity of the waste can also be used as an indicator of its polluting potential, or "strength." Since the oxidation or stabilization of organic matter requires oxygen, the effective operation of a sewage-treatment plant makes it mandatory that an assessment be made of how much oxygen will be required to accomplish this task. A method for performing such an assessment will be described next. Such a method, however, will not necessarily provide information on the quantity of nutrients or toxic organic chemicals that are present. Since in many cases, these are also high-priority pollutants (Table 8.2), tests to evaluate their potential contribution to the impact of a waste may also be necessary.

BIOCHEMICAL OXYGEN DEMAND (BOD) TEST

The BOD test is the one most commonly used for assessing the amount of organic matter in domestic sewage or other nontoxic liquid wastes. This procedure is as follows. A sample of the waste is inoculated with bacteria and incubated in the laboratory for 5 days at 20°C (68°F). Since other incubation periods and temperatures can be applied, this one, to be specific, is described as the 5-day, 20°C, biochemical oxygen demand test (Eaton, et al., 2005). The required temperature is based on the fact that this is equivalent to that on a spring or fall day. An incubation period of 5 days is selected since, after that time, the demand is approximately 70 percent of that which would be exhibited if the sample were incubated until the bacteria have had sufficient time to stabilize 100 percent of the organic matter in it (Eaton et al., 2005).

Similar measurements of the BOD of the incoming waste, and at various stages within a sewage-treatment plant, provide an indication of the effectiveness not only of the individual treatment steps but also of the plant as a whole. Knowing the BOD of the effluent from the plant and the rate at which it is being discharged, coupled with the DO content and diluting volume provided by the receiving body, it is possible to estimate the extent to which the DO in, for example, a stream will be depleted. A related chemical test has been developed to assess the oxygen demand of toxic wastes that inhibit bacterial growth and therefore do not permit use of the BOD test. This test, which requires that the sample be chemically digested in the laboratory, yields a measure of the *chemical oxygen demand* of the waste and is called the COD test.

Treatment of Liquid Wastes

As implied by the previous discussion, methods for treating liquid wastes, particularly domestic sewage, are designed to stabilize or oxidize through biological processes the organic matter they contain. This can be most effectively achieved by providing conditions that will optimize the ability of natural biological processes to accomplish this task. This is therefore one of the primary goals in the design and operation of a sewage-treatment plant.

MUNICIPAL WASTES

Overall, the methods for the treatment of municipal sewage and other types of nontoxic liquid wastes include three stages: *primary, secondary, and tertiary.* Primary treatment is designed to permit the solids within the waste to settle and be removed. Secondary treatment is to enhance the application of the previously discussed biological processes for oxidizing the organic matter in the waste. Tertiary treatment involves a variety of processes tailored to the intended uses of the finished product. One of the more common tertiary or advanced methods for treating liquid wastes is very similar to the coagulation, settling, and filtration processes used in treating surface waters to make them acceptable for drinking (Chapter 7).

Each of these processes represents a progressive level of purification, and the required number of treatment stages depends on the degree of treatment necessary. With modifications, higher removals can be achieved. All municipal sewage-treatment processes begin with the primary stage. Under the 1972 amendments to the Clean Water Act, all wastewater-treatment plants in the United States must provide both primary and secondary treatment.

Primary treatment, as noted above, involves holding the sewage in a settling tank to permit the removal of solids by sedimentation. When performed on a separate basis, the settling process removes approximately half the suspended solids, providing a BOD reduction of 30–50 percent. Grease and light solids that float are removed from the settling tank by a scraper and are pumped along with the settled solids to a large closed tank called a digester, where they are held for anaerobic digestion. Digestion is most effective when the biosolids are heated to 32°C (90°F) or more. At 32°C, the biosolids are digested in about 24 days; at 54°C (130°F), they are digested in about 12 days. In colder climates, the methane gas produced in the process provides fuel for heating the digester and other applications within the treatment plant.

Secondary (or biological) treatment is generally accomplished through application of the *activated sludge process,* an aerobic process designed to stabilize the organic material in the sewage. This is accomplished by passing the waste through a large open tank where it is held for several hours and its oxygen content maintained by means of aerators (air diffusers) or mechanical agitators (paddles or brushes). Under these conditions, the microorganisms float as suspended particles. The effluent is then sent to a secondary settling tank, where the microorganisms settle out, and the settled effluent represents the treated product. The overall reduction in BOD is about 90 percent. Some of the microorganisms that have settled to the bottom of the secondary tank are pumped back into the aerated tank to maintain an adequate population of microbial growth. The rest of the growth is treated as biosolids and sent to a digester. It is common practice today to combine the primary and secondary stages on a flow-through basis (Figure 8.5).

Another method that combines *primary* and *secondary treatment,* is the *waste stabilization pond,* which has been used in other countries for many years. This method, however, was largely ignored in the United States until

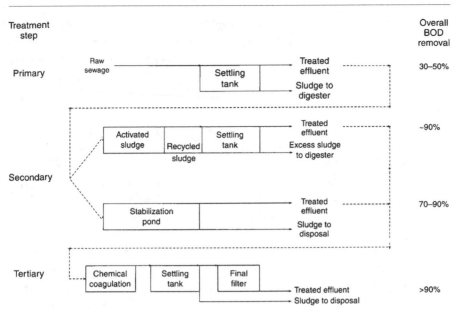

Figure 8.5 Primary, secondary, and tertiary stages in the treatment of municipal sewage

the 1950s (Gloyna, 1971). Nonetheless, because of their low construction cost, ease of operation, and minimal maintenance requirements, they are now in use, particularly in the warmer southern states, for the treatment of animal and pulp and paper wastes. As the name implies, the basic unit is an earthen pond having a depth of 1–2 meters (~3–7 feet), a width of 25 meters (~80 feet), and a length of 90 meters (~300 feet). The system is operated on a continuous flow-through basis. Most ponds operate biologically at two levels: the lower portion is anaerobic, and the upper portion is aerobic. In the border area, facultative bacteria (which can live under either aerobic or anaerobic conditions) are active. In some cases, wind-driven mixers are used to increase the amount of oxygen in the upper portions of the pond. Algae growth at the surface also helps ensure aerobic conditions. The primary precaution is not to locate such ponds in soils with fissures that would permit the waste to move through the ground without filtration, thereby contaminating nearby groundwater supplies. When a pond fills with biosolids, it must be cleaned and the cycle begun anew.

Tertiary treatment can involve any steps or processes that are applied after primary and secondary treatment. Examples are the process of disinfecting the treated sewage chemically (i.e., applying chlorine or UV radiation) or physically (i.e., applying microfiltration prior to its discharge into a bay, lagoon, river, stream, or lagoon or, for example, using it to irrigate a golf course, greenway, or park. If it is sufficiently treated, it might also be used for agricultural purposes (Wikipedia, 2009). Excess organic compounds are commonly removed by passing the treated waste through two granular carbon beds, each of which provides 30 minutes of contact time. Ozone may also be used to disinfect the waste as it passes from the first carbon bed to the second. Heavy metals and viruses can be removed by coagulating the waste, for example, with lime, followed by sedimentation. This process, however, creates large volumes of highly toxic sludge that must be disposed.

ADVANCED SYSTEMS

As in essentially all fields of environmental health, there have been significant advancements in methods for the treatment of municipal wastes. Several of these were developed in Europe. In the case of the activated sludge process, one such change is to add solid substrates, such as small polyethylene structures or vertical tubes, to the aeration tanks. These provide a convenient surface on which the bacteria can live, increasing the number of organisms by several orders of magnitude. This not only enhances the effectiveness of the treatment system but also enables plant operators to control, in a more rigorous manner, the age and numbers of bacteria in the aeration tanks (Ai et al., 2005).

Another approach involves changes in the methods for disinfecting the treated waste prior to discharging it into the environment. In the past, the common approach has been to add chlorine. As was the case for drinking water (Chapter 7), reactions of this disinfectant with organic compounds in the waste produces by by-products, such as chlorinated hydrocarbons. If added in excess and not followed by some form of dechlorination, the discharged wastewater will be harmful to aquatic organisms. The alternative is either to dechlorinate or use some other means of disinfection. For these reasons and because UV radiation is effective against bacteria, viruses, and parasites, it is now being applied in more than 2,000 wastewater-treatment installations in the United States (Sakamoto, 2000). Since turbidity can severely reduce the effectiveness of UV radiation, it is often necessary to filter the sewage prior to applying the UV process.

INDUSTRIAL WASTES

As might be anticipated, wastewaters from industrial operations contain a wide range of contaminants. For this reason, their treatment requires not only an expansion in the number of methods applied but also a change in their sophistication. Methods that can be applied, either singly or in combination, include physical, chemical, and biological processes.

Physical processes include those designed to remove suspended solids through filtration, centrifugation, or the previously described settling tanks; oils, greases, and emulsified organics through aeration, which causes such materials to float to the surface where they can be removed by skimming devices; and dissolved materials, such as organic chemicals, by passing the water through a semipermeable membrane (Furukawa, 1999) or through beds of activated carbon (Chapter 7).

Chemical processes include the addition of acids to neutralize wastes that are alkaline, bases to neutralize those that are acid, and chemicals to coagulate and precipitate suspended solids (as, for example, in tertiary treatment systems applied to the effluents from municipal sewage-treatment plants). Other methods include the use of ion-exchange resins to replace contaminants in the waste with innocuous chemicals.

Biological processes include the predigestion of brewery, winery, and meatpacking wastes under anaerobic conditions, often at elevated temperatures to accelerate the process, and the oxidation of certain types of industrial wastes, such as petroleum constituents, under aerobic conditions similar to those applied in the treatment of domestic sewage and wastes from animal farms.

Control of Nonpoint Sources

Due to the intermittent flow rate of nonpoint sources and the difficulties in designing facilities to treat them, primary efforts are being directed to the reduction of the volumes of such releases and the more harmful constituents they contain. The latter goal is being achieved in agricultural runoff through the optimization of pesticide application rates and timing. Other controls include the implementation of (1) the previously described more effective methods of distributing irrigation water, (2) conservation techniques such as reduced tillage, crop rotation, and winter cover crops, (3) the establishment of buffer zones, such as vegetative cover along stream banks, and (4) strategically placed grass strips and artificial wetlands to intercept or immobilize pollutants.

Even so, the problem of runoff in urban areas continues. Among the solutions being proposed is the application of methods for collecting and "harvesting" the rainwater, rather than sending it to the sewers. One approach is to cover parking lots with permeable surfaces of a honeycomb design that permit rainwater to drain into the soil. In addition to retaining the water, such an approach could also reduce expenditures for storm drains and sewers. The presence of contaminants in urban runoff can also be reduced by providing convenient disposal sites for used oil and household hazardous waste, collecting leaves and yard trimmings on a frequent basis, and using vacuum equipment for street cleaning. Other techniques are designed to slow runoff, allow more water to percolate into the ground, and filter out contaminants. In addition, weirs, movable dams, and detention areas can provide storage capacity in storm and combined sewer systems, thereby reducing the frequency and volume of combined sewer overflows. In this regard, the EPA (2009b) recommends that (1) stormwater discharges be considered *point sources*, requiring coverage under an NPDES permit, and (2) that *best management practices* be applied in their control.

Disposal of Treated Wastewater

Due to the previously cited problems in releasing treated municipal sewage into rivers, streams, and lakes, increasing attention is being directed to the disposal of treated wastewaters on land surfaces. Advantages of this approach are that it (1) returns nutrients to the soil, making them available to nourish agricultural crops, golf courses, parks, recreational areas, and forests, (2) provides a mechanism for reclaiming and preserving open spaces and existing wetlands, as well as for developing new wetlands which, in turn, provide habitats for wildlife, (3) creates an ideal environment that can provide nutrient sinks and buffering zones to protect streams and other areas, (4) provides a ready means, under proper conditions, for recharging groundwater sources, and (5) frequently results in reductions in wastewater treatment costs, thus saving funds for addressing other problems.

Another advantage is that a properly developed land disposal system can be operated for 20 years or more. Such a system can also serve as a viable and beneficial alternative to methods commonly employed for the secondary treatment of municipal sewage. The reuse of human waste in aquaculture, in particular, can produce significant benefits and achieve a variety of useful goals. In countries where nutrition requirements exceed food production, aquaculture can assist in closing the gap by using valuable nutrients that would otherwise be squandered. In countries where water quality must be improved, aquaculture can lessen the harmful impacts of excess pollution on water courses. In arid regions, it can make an important contribution to the conservation of scarce water resources (Edwards, 1992).

Disposal of Biosolids

The magnitude of the disposal of biosolids is enormous. The EPA estimates, for example, that U.S. municipal wastewater-treatment plants annually produce almost 8 million tons of such material. Additional biosolids are produced in the treatment of industrial wastes. Although it might seem logical to use such materials as a soil conditioner and fertilizer, questions related to aesthetic and public health concerns immediately arise. Among the most common are questions concerning the possible transmission of disease, especially when such materials are to be used to grow edible crops. Even the use of biosolids on lawns, parks, and golf courses has not been without expressions of concern. If not properly pretreated, any such applications can attract vermin. There have also been objections to the accompanying odors. Another concern is that biosolids tend to concentrate toxic heavy metals from the wastes being treated. If biosolids are incinerated, there is the problem of the release of toxic materials into the air.

Seeking to resolve these issues, the EPA has promulgated regulations that specify the type of treatment biosolids must receive prior to being sent to a landfill, applied to land as a fertilizer, or incinerated. In addition, the regulations are designed to reduce the volume of biosolids, stabilize the organic materials they contain (so as to reduce odors and the attraction of animals), and kill the full range of microorganisms (for example, certain bacteria, viruses, and parasites) they contain (Logan, 1999; Sims and Bentley, 2001). Concurrently, other concerns have developed. One is what restrictions or treatment should be applied to ensure that workers who handle such materials are protected.

Special Problems of Groundwater Contamination

Another concern is that discharges of liquid wastes from industrial, agricultural, and domestic sources can contaminate groundwater. If biosolids, previously discussed, are placed in a landfill, these types of problems can occur through the leaching of toxic materials and pathogens not only into groundwater but also into surface waters.

Many methods have been applied in seeking to remove and/or stabilize groundwater contaminants. These include the full range of previously discussed physical, chemical, and biological agents, applied both in situ and to the water after it has been removed from the ground. One of the earliest approaches is what is called "pump and treat." This, in essence, involves the installation of a series of wells to extract the water and treat it with chemicals, such as aluminum sulfate or ferric chloride, following in general the methodologies used to purify surface waters (Chapter 7). The treated water is then either used or pumped back into the aquifer. Such an approach, however, is frequently both ineffective and expensive, the primary reasons being that (1) the treated water, once pumped back into the ground, is immediately mixed with water that has not been treated, and (2) contaminants, which leach out from sediments within the aquifer, serve as a continuing source of pollution.

Dissatisfaction with the "pump and treat" approach has led to the development and application of a variety of alternatives. One is to inject chemical reactants into the aquifer either to (1) convert the contaminants into a nontoxic form, (2) precipitate and fix them in place, or (3) mobilize them so they can be effectively extracted and removed by the "pump and treat" approach. Such methods, however, are practical only if the reactants can be injected in a soluble form and the chemistry of the contaminants makes them suitable for reacting. Another approach is to excavate a portion of the aquifer and install subsurface permeable membranes or reactive barriers that will remove contaminants from the groundwater as it flows through them. As with the "pump and treat" approach, this method can be very expensive (Lovley, 2001).

The General Outlook

There are some 20,000 municipal wastewater-treatment facilities in the United States, having a total daily treatment capacity of more than 40 billion gallons and representing a capital investment of about $4 billion (CEQ, 1998). Due to years of neglect and changes required by the Clean Water Act (U.S. Congress, 1977), many of these facilities are in need of major repairs and/or upgrades. Estimates are that the costs associated with improving waste-collection systems, coupled with the need to manage problems such as the previously discussed sewer overflows, could exceed $300 billion over the next two decades. The urgency of meeting these needs is further demonstrated by the fact that surveys of streams, lakes, and estuaries in the United States show that about 40 percent contain water of a quality that is not adequate to support fishing and swimming. Agricultural runoff, alone, adversely affects some 70 percent of the impaired rivers and almost half of the impaired lakes (Gray, 1999; Table 8.3). Compounding these problems is the continuing introduction of new types of wastes, some of which will introduce

Table 8.3 Quality of U.S. waters, 2004

Type	Percent assessed	Good	Good, but threatened	Polluted and/or impaired	Primary pollutants	Primary sources
Rivers and streams	16	53	3	44	Silt, nutrients, and pathogens	Hydroelectric dams and runoff from wetlands, agriculture farms, and urban areas
Lakes, ponds, and reservoirs	39	35	1	64	Nutrients, metals, and silt	Hydroelectric dams and runoff from wetlands, agriculture farms, and urban areas
Bays and estuaries	29	70	1	30[a]	Pathogens, oxygen-depleting substances, and metals	Municipal point and nonpoint sources, and atmospheric deposition

a. Due to rounding, the sum of the three categories may not equal 100.

new complexities. One example is that more than 800 consumer products contain nanoparticles (Chapter 4). Since some of these particles contain metals that are toxic to bacteria, this could well be a future concern in terms of their potential effects on those that oxidize the organic matter in sewage (Bass, 2009).

In cases where it is not practical to eliminate the production of the waste, systems should be designed so that the wastewater can be recycled and reused. Ultimately, such an approach can lead to "closed-loop" systems that produce essentially no liquid discharges. In the automotive industry, not only is wastewater being treated, recycled, and reused, but the effluent from sewage treatment is also being treated and reused for irrigation. One of the recognized leaders in this field is the Toyota Motor Corporation. This company recycles more than 85 percent of their waste, and the remainder is used to produce energy. This has enabled them to achieve a goal of zero transfer of wastes to landfills (Toyota Motor Corporation, 2007). Also playing a role is the International Standards Organization (ISO), which specifies the requirements for environmental management systems and confirms their global relevance for organizations desiring to operate in an environmentally sustainable manner (ISO, 2007).

One of the basic sources of these and related problems is that, all too frequently, the management and control of liquid wastes have been addressed in isolation. An excellent example of applying a systems approach is the Pollution Prevention Act of 1990 (U.S. Congress, 1990). This not only reduced discharges of toxic industrial wastes to the environment and accompanying exposures to the public, but it also reduced exposures to workers engaged in affected industrial operations (Chapter 4). There is also a need to address liquid-waste problems from the standpoint of an entire watershed or drainage basin. An essential part of any such effort would be a careful review and evaluation of the interrelationships and relative impacts of each of the contributing polluting sources. Since funds to address all these problems simultaneously are not available, it is essential that a mechanism be established for setting priorities on which problems should be addressed first, backed up by adequate research to provide the scientific information on which to base the associated regulatory programs (Gray, 1999). In fact, such efforts should be expanded to address such problems on an international perspective in which the developed nations would not only share their expertise with the developing nations but also recognize that wastewater pollution can move from one country or region to another much the same as atmospheric pollutants. One example is the discharge of ballast water at dockside when oceangoing vessels take on cargo. Such water, which was taken on at one or more overseas ports, frequently contains multiple organisms and virus-like particles. The magnitude of this problem is illustrated by the fact that such activities result in the annual discharge into U.S. ports of an estimated 80 million tons (20 billion gallons) of water (Ruiz et al., 2000). Nonetheless, significant progress is being achieved. Based on the latest surveys of the nation's 2.6 million miles of rivers and streams, 55 percent were rated as good, 10 percent good but threatened, and 35 percent impaired. Comparable results were found in terms of the quality of our lakes and reservoirs (EPA, 2009a).

9

SOLID WASTES

ORIGINALLY, a major portion of this type of waste was so-called municipal solid waste (MSW). This contained everyday items such as product packaging, grass clippings, furniture, clothing, bottles and cans, food scraps, newspapers, appliances, consumer electronics, and batteries. These items came from homes, institutions such as prisons and schools, and commercial sources, such as restaurants and small businesses. Most such waste was not considered hazardous, and it was simply transported to the local land disposal facility, or "dump," where it was periodically set on fire to reduce its volume and to discourage the breeding of insects and rodents. As long as windblown debris and fires were contained, material was covered over and sealed daily (so that breeding and habitation by insects and rodents were controlled), and contamination of nearby groundwater supplies was avoided, the sanitary landfill was considered an acceptable method for disposal. At the same time, it had two major benefits: (1) it served as a site for weekend community gatherings, especially in the smaller communities, and (2) by doing so, it provided an opportunity for families to exchange items that would otherwise have been disposed. In fact, it was an example of recycling long before it was formerly recognized as such. Nonetheless, this approach often led to windblown debris and unsightly disposal facilities. During subsequent years, these and other deficiencies were recognized and improved measures were instituted: (1) wooden boxes were replaced by metal and high-integrity containers, (2) the installation of multiple barriers became common, (3) facilities were covered, and (4) records were maintained on the location and nature of specific wastes.

With the development of a "throw-away" society and an unprecedented demand for new products, the nature and characteristics of municipal solid

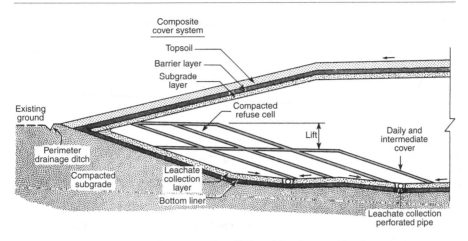

Figure 9.1 Cross section of a typical landfill and leachate-collection system

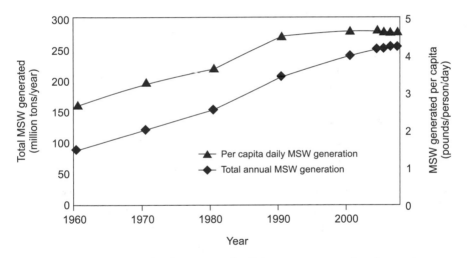

Figure 9.2 Trends in total U.S. municipal solid waste generated and quantity per capita, 1960–2009.

waste dramatically changed, and its volume significantly increased (EPA, 2009). Although such waste initially contained primarily what was classified as nonhazardous material, with the passage of time it contained paint, pesticides, and solvents, as well as construction and demolition debris that included oil and grease, lead, and toxic coatings on wood. Also present were many materials, such as plastics, that are not readily biodegradable. Because

Table 9.1   Trends in the disposition of U.S. municipal solid waste: 1960–2005 (as a
percentage of that generated); and the decreasing number of landfills: 1980–2005
(as a percent of those in operation in 1980)

Disposition	1960	1970	1980	1990	1995	2000[a]	2005[b]
Recycled/composted	6.4	6.6	9.6	16.2	26	30	32
Incinerated	30.6	20.8	9.0	15.5	16.8	16.7	15.9
Disposed in landfill	63	72.6	81.4	68.3	57.2	53.3	52.1
Total generated	100	100	100	100	100	100	100
Decrease in number of landfills[c]			100	27.5			15.5

a. Assumes 30 percent recovery.
b. Assumes 32 percent recovery.
c. There were 20,000 landfills in operation in 1980 and 3,091 in 2005. The estimated number in operation in 2008 was 1,234.

health authorities recognized the need for a more technically based method of disposal, land disposal facilities were gradually replaced by the sanitary landfill that has an accompanying leachate-collection system (Figure 9.1).

During the past half century, the situation has been exacerbated by the rapidly increasing quantity of waste per capita (Figure 9.2). There are, however, indications of progress in the percentage that is being recovered through recycling, sent to landfills, and composted. At the same time, there has been a significant reduction in the number of U.S. municipal sanitary landfills in operation. Of the approximately 20,000 such facilities that existed in 1980, only 1,234 remained in 2008 (Table 9.1). One reason for this is that government regulators, in seeking to make landfills safer, have made them more expensive to own and operate. Fortunately, this has led to the establishment of larger and better designed and operated disposal facilities on a regional basis.

Types and Classifications of Solid Wastes

Several terms are used for classifying solid wastes. In promulgating the Resource Conservation and Recovery Act (RCRA) of 1976, the U.S. Congress (1976) defined solid waste as *any garbage, refuse, sludge from a waste treatment plant, water supply treatment plant, or air pollution control facility and other discarded material, including solid, liquid, semisolid, or contained gaseous material resulting from industrial, commercial, mining, and agricultural operations and from community activities.* Specifically excluded were "solid or dissolved material in domestic sewage," as well as "industrial wastewater discharges regulated under the Clean Water Act" (EPA, 1986a; Table 9.2).

In a follow-up action, Congress defined *hazardous waste* as any *solid waste, or combination of solid wastes, which because of its quantity, concentration, or physical, chemical, or infectious characteristics may: (1) cause, or significantly contribute to an increase in mortality or an increase in serious irreversible, or incapacitating illness; or (2) pose a substantial present or potential hazard to human health or the environment when improperly treated, stored, transported, or disposed of, or otherwise managed.* To implement this definition, the EPA established two basic methods for designating hazardous wastes. Either they are *listed* or have been determined to have certain characteristics.

Wastes that are *listed* are those associated with various manufacturing and industrial processes and with certain commercial chemical products that have been specifically identified by the EPA as consistently posing a hazard to human health and the environment when discarded. Wastes are also characterized as hazardous if they exhibit certain properties, including ignitability, corrosivity, reactivity, or toxicity, based on test results or the knowledge of the waste generator. Although these characteristics are

Table 9.2   Principal programs and goals of the Resource Conservation and Recovery Act (1976)

Solid waste program (directed primarily at management and control of nonhazardous solid wastes)
 Primary goals:
 To encourage environmentally sound solid-waste management practices
 To maximize reuse of recoverable resources
 To foster resource conservation

Hazardous waste program ("cradle-to-grave" system for managing hazardous waste)
 Primary goals:
 To identify hazardous waste
 To regulate generators and transporters of hazardous waste
 To regulate owners and operators of facilities that treat, store, or dispose of
  hazardous waste

Underground storage tank program
 Primary goals:
 To provide performance standards for new tanks
 To prohibit installation of unprotected new tanks
 To provide regulations concerning leak detection, prevention, and corrective
  action

important, except for the matter of toxicity they relate only indirectly to the potential health impacts of a waste on the public and the environment. Various industries and the types of hazardous wastes they generate are listed in Table 9.3. Those generated by the computer industry are discussed later.

Table 9.3   Examples of hazardous waste generated by business and industry [a]

Waste generator	Typical wastes
Chemical manufacturers	Strong acids and bases Spent solvents Reactive wastes
Vehicle maintenance shops	Paint wastes containing heavy metals Ignitable wastes Used lead acid batteries Spent solvents
Printing industry	Heavy metal solutions Waste inks Spent solvents Spent electroplating wastes Ink sludges containing heavy metals
Leather products manufacturers	Waste toluene and benzene
Paper industry	Paint wastes containing heavy metals Ignitable solvents Strong acids and bases
Construction industry	Ignitable paint wastes Spent solvents Strong acids and bases
Cleaning agents and cosmetics manufacturers	Heavy metal dusts Ignitable wastes Flammable solvents Strong acids and bases
Furniture and wood manufacturers and refinishing	Ignitable wastes Spent solvents
Metal manufacturing	Paint wastes containing heavy metals Strong acids and bases Cyanide wastes Sludges containing heavy metals

a. Does not include computer manufacturers.

Table 9.4   Types, regulatory body, and characteristics of commercially generated waste

Type of waste[a]	Regulatory body	Typical content
Nonhazardous	State and local governments	Refuse, garbage, sludge, municipal trash
Hazardous	EPA or authorized states	Solvents, acids, heavy metals, pesticide residues, chemical sludges, incinerator ash, plating solutions
Radioactive	USNRC or Agreement States	High- and low-level radioactive waste, naturally occurring and accelerator-produced materials
Mixed	USNRC or Agreement States	Combinations of the above

a. About 99 percent of the total U.S. waste is nonhazardous and about 1 percent is hazardous. The volume of radioactive and mixed waste combined is less than a few ten thousandths of one percent of the total.

The next category addressed was *nonhazardous* waste. In this case, rather than defining what it is, Congress included in the RCRA certain categories of waste that were excluded from the definition of *hazardous waste*. The most common group excluded was household waste. Also excluded were agricultural wastes used as fertilizer, mining overburden returned to the mine site, and certain wastes produced in the combustion of coal. As a general rule, the waste produced by homeowners is designated as MSN, or municipal nonhazardous waste, and the chemical waste produced by industry is classified as hazardous. Another type of waste, *mixed waste,* was addressed in the Federal Facility Compliance Act of 1992 (U.S. Congress, 1992). This is defined as a waste that contains *both hazardous chemicals and radioactive materials.* The typical contents of hazardous and nonhazardous, mixed, and radioactive wastes, and the regulatory agencies responsible for their disposal, are summarized in Table 9.4.

Wastes of Special Interest

There are multiple wastes within the preceding classifications that represent major challenges in their management and disposal. This is vividly recognized by anyone who has observed a junkyard containing masses of rusting cars and trucks. The sheer magnitude of this problem can also be illustrated by the 280 million automobile tires that are annually discarded in the United States, as well as similar numbers of home appliances, such as refrigerators, stoves, washing machines, and clothes dryers.

By far, one of the most important of these is the variety of hazardous wastes generated by the use of computers and similar electronic devices.

ELECTRONIC PRODUCTS

The rapid increase in electronic waste due to the use of computers, cell phones, and other personal electronic devices is indicative of how urgent the need for improved methods for addressing the solid-waste disposal problem is (Morgan, 2006). Several factors compound this problem: (1) the small size and short useful life span of such devices, (2) a lack of understanding of their adverse impacts on the environment and public health, (3) the fact that these impacts occur throughout the life cycle of such products, extending from the acquisition of the raw materials to their manufactures and disposal, (4) the sheer magnitude of the problem, and (5) the absence of recycling policies. Unfortunately, this has resulted in China becoming a dumping ground for such products once they are discarded. The accompanying environmental and public health impacts in that country have been devastating. Tests for lead among children living in areas in that country where workers are removing the toxic materials from e-waste have mean concentrations of lead in their blood of $15.3 \mu g$ dl^{-1}, contrasted to a recommended remedial action concentration of $10 \mu g$ dl^{-1} (Huo et al, 2007). The problems, however, are not restricted to China. Studies show that the concentrations of polybrominated ether (a flame retardant) in the eggs of the peregrine falcons in California have reached concentrations up to 4.1 ppm lipid weight. This has raised the specter of endangered species (Holden et al., 2009).

On average, each U.S. household has at least four small (<10 lb) and three large (>10 lb) electronic waste products in storage (i.e., so-called *e-waste*). Overall, these and related products represent more than 1.36 million tons of toxic wastes. Although globally the United States is the highest producer of such waste, this country has no legally enforceable policies that require comprehensible recycling of e-waste or the elimination of hazardous substances from electronic products (Ogunseitan et al., 2009). To provide perspective, it is estimated that 100,000 computers are discarded daily in the United States and that 75 percent of all the personal computers sold are now surplus (Harris, 2008). Table 9.5 lists the toxic materials commonly found in desktop computers. Other discarded electronic products include

Table 9.5 Toxic materials in desktop computers

Material	Components	Chronic health effects
Arsenic	Doping agents in transistors, printed wiring boards	Skin sores, hypertension, peripheral vascular disease, skin and bladder cancer (ingestion), lung cancer (via inhalation and ingestion)
Beryllium	Printed wiring boards, connectors	Lung damage, allergic reactions, chronic beryllium disease, reasonably anticipated to be a human carcinogen
Cadmium	Batteries, blue-green phosphor emitters, cathode ray tubes, printed wiring boards	Pulmonary damage, kidney disease, bone fragility, reasonably anticipated to be a human carcinogen
Chromium	Housings, hardeners	Exposure occurs primarily via inhalation; may affect respiratory, cardiovascular, muscular, endocrine, gastrointestinal, hematological, hepatic, renal, dermal, and ocular systems[a]
Cobalt	Batteries	Respiratory irritation, reduced pulmonary function, asthma, pneumonia, and lung cancer (inhalation)
Gallium	Semiconductors, printed wiring boards	Evidence of carcinogenesis based on laboratory studies of animals
Lead	Radiation shielding, printed wiring boards	Damaging to children, includes blood anemia, kidney damage, colic (severe stomachache), muscle weakness, and brain damage; may lead to death[b]
Mercury	Batteries, switches, printed wiring boards	Chronic kidney and brain damage, including changes in hearing and vision, and difficulties with memory[c]
Nickel	Cathode ray tubes, printed wiring boards structural components	Allergic skin reactions, chronic bronchitis, lung cancer, and nasal sinus[d]
Polyvinyl chloride, bromine,	Plastic coated internal wiring, and external cables	Cancers of the liver, brain, and lung, and probably the lymphohematopoietic system, cellular interactions, synergism, fetal death, and malformations[e]

a. ASTDR. 2000. *Toxicological Profile for Chromium.* Agency for Toxic Substances and Disease Registry. Atlanta, GA.
b. ASTDR. 1999. *Toxicological Profile for Lead.* Agency for Toxic Substances and Disease Registry. Atlanta, GA.
c. ASTDR. 1997. *Toxicological Profile for Mercury.* Agency for Toxic Substances and Disease Registry. Atlanta, GA.
d. ATSDR. 2005. *Toxicological Profile for Nickel.* Agency for Toxic Substances and Disease Registry. Atlanta, GA.
e. Wagner, J. K. 1983. *Toxicity of Vinyl Chloride and Poly (Vinyl Chloride). A Critical Review. Environmental Health Perspectives* 82, 61–66.

batteries, of which more than 3 billion are sold annually, and television sets. Many of these products, especially the cathode-ray tubes in the older television sets, contain toxic materials such as lead, mercury, chromium, and polyvinyl chlorides. The older picture tubes destined for disposal in the United States during the next decade could contain as much as a billion pounds of lead. Depending on the various materials in computer wastes, the potential toxicological effects include brain damage, kidney disease, and cancer.

Fortunately, due to efforts of manufacturers, many of these problems are being resolved. The toxic materials that remain in modern television sets

Table 9.6   Techniques for minimizing the production of hazardous wastes

Inventory management and improved operations
Inventory and trace all raw materials
Emphasize use of nontoxic production materials
Provide waste minimization or reduction training for employees
Improve receiving, storage, and handling of materials

Modification of equipment
Install equipment that produces minimal or no waste
Modify equipment to enhance recovery or recycling options
Redesign equipment or production lines to produce less waste
Improve operating efficiency of equipment
Maintain strict preventive maintenance program

Production process changes
Substitute nonhazardous for hazardous raw materials
Segregate wastes by type for recovery
Eliminate sources of leaks or spills
Separate hazardous from nonhazardous and radioactive from nonradioactive wastes
Redesign or reformulate end products to be less hazardous
Optimize reactions and raw material use

Recycling and reuse
Install closed-loop systems
Recycle on-site for reuse
Recycle off-site for reuse
Exchange wastes

Treatment to reduce toxicity and volume
Evaporation
Incineration
Compaction
Chemical conversion

and computers are limited to substances such as polyvinyl chloride (PVC), bromine, and phthalates. Although new laptops still contain bromine, the amounts vary and are continuing to be reduced. Nonetheless, almost half of the plastic-coated internal wiring and external cables contains PVC. Although the power cords provided with most laptops continue to contain pththalates, other toxic substances, such as lead, cadmium, and mercury, have been eliminated (Greenpeace, 2007). Techniques for minimizing the production of hazardous wastes are summarized in Table 9.6.

Through these efforts, the amount of waste being generated per person is being reduced, and the percentage being recycled is being increased. The net effect is a reduction of almost 30 percent during the past 20 years in the percentage of municipal waste that is being sent to landfills for disposal (Padgett, 2001).

WASTE TRACKING

Another approach is to use computers to track individual items of municipal waste in real time. This can provide information not only on what is being discarded but also where it ends up. Researchers at the Massachusetts Institute of Technology visited the homes of hundreds of volunteers in Seattle, Washington, and tagged about 3,000 pieces of the almost 800,000 tons of waste that is discarded in that city annually. These included pizza boxes, plastic bags, empty cans and bottles, used printer cartridges, Styrofoam cups, slippers, and scrap metal. The resulting information revealed whether each item was recycled, composted, or placed in a municipal landfill, and how long was required for it to reach its destination (Le, 2009).

National and International Waste Transport

Although some homeowners prefer to combine their waste, it is much better, especially in terms of the environmental impacts, to separate the various items into specific categories. Only then, can they be readily disposed in the proper manner, for example, metals, glass, and plastics that can be recycled, organic materials that can be composted, and paper that can be incinerated. Although, as noted earlier, the number of landfill disposal sites has been decreasing, this has led to the establishment of larger and better designed and operated disposal facilities on a regional basis. Even though it appears that the existing disposal capacity is adequate, this has led to several major changes. One of the more important is a large increase in the transfer of municipal refuse from one state to another. In fact, tens of millions of tons of refuse are being handled in this manner and it involves all but 3 of the 50 states (Wolpin, 2002).

Another developing problem is an increase in the exporting of solid wastes from the developed to the developing countries for disposal. Some are being arranged through "silent trades," which are negotiated in secret. Others, however, had been arranged under contracts signed with governments of the importing countries. Unfortunately, the benefits of such contracts were assessed by officials in the receiving countries solely in terms of the economic income. One such arrangement led to the disposal of several thousand tons of hazardous wastes at inland and coastal sites in Lebanon in 1987. As would be anticipated, this action produced a critical environmental situation that was exacerbated by the fact that the human, technical, and financial resources to manage the associated environmental and public health impacts were not available.

Recognizing the need for action, in 1989 the United Nations Environmental Programme convened a meeting in Basel to review these matters and develop recommendations for avoiding the repetition of such events. Three primary recommendations were an outgrowth of these deliberations: (1) before any such waste is shipped, appropriate officials within the recipient country must be notified and indicate their consent, (2) officials in the countries through which the waste would transit must similarly be notified, and (3) officials in the transient and importing countries must provide their written consent to the arrangements. Participants in the convention also stipulated that they considered the participation in "silent trades" to be a criminal act (Jurdi, 2002a). These recommendations were expanded in scope at a follow-up conference, held in Geneva in 1994. One recommendation was that all transboundary movements of hazardous waste from OECD to non-OECD countries be banned (Jurdi, 2002b).

Health and Environmental Impacts

Numerous epidemiologic studies have been conducted to evaluate whether the health of people living near hazardous waste disposal sites is being adversely affected, particularly through an increase in cancer rates. Most such studies have been inconclusive. Even in the case of the highly publicized Love Canal episode, subsequent health reviews and analyses revealed no evidence of higher cancer incidence there than in the rest of New York State (Golaine, 1991). Although other studies have shown apparent associations between living near a hazardous waste disposal site and increased risks of certain types of cancer, as well as birth defects, investigators are careful to point out that, because of limitations in the data, it is too early to reach any definitive conclusions. For several reasons (Chapter 3), such an outcome is not unexpected. First of all, the earliest recognizable effects of low-level chemical exposures (headache, malaise, minor skin irritation, and respiratory tract complaints) tend to depend on many conditions. In addition, many of the illnesses (such as cancer) that might be anticipated have latency periods of 10–40 years. Under these conditions, it is difficult to establish patterns of exposure and equally difficult to gather data on an exposed population sufficiently large to verify a definitive relationship (Chapter 3).

While the impacts of solid waste on human health are important, attention needs also to be directed to the impacts on animals in other segments of the environment. Certain particular types of solid waste, namely, plastic fishing gear, six-pack beverage yokes, plastic sandwich bags, and Styrofoam cups, discarded into the ocean, entrap and kill an estimated more than 1 million seabirds and 100,000 marine mammals every year. In fact, plastics may be as much a source of mortality to marine mammals as oil spills, heavy metals, or other toxic materials combined (Shea, 1988). Recent surveys in the far reaches of the Pacific Ocean reveal the presence of extensive areas that contain up to a million pieces of plastic per square mile. Some of these result from materials illegally jettisoned from ships; other portions occur as a result of accidental spills. Because plastic is lighter than seawater, it floats on the surface for years, gradually breaking up into smaller and smaller particles that end up in filter-feeding animals, such as jellyfish. Due to their nature, plastics adsorb toxic chemicals and become part of the food web when eaten by turtles. In a similar manner, birds take in larger pieces of plastic when they mistake them for fish. Although an International Convention for the Prevention of Pollution from Ships bans the dumping of plastics at sea, the agreement is not enforced on the open ocean. Newer biodegradable plastics that have been developed offer hope for ultimately solving the problem (Hayden, 2002).

Case Study—Chromated Copper Arsenate Wood

In the early 1970s, wood treated with chromated copper arsenate (CCA) began to be widely used in the United States for the construction of structures to be used in outdoor land, aquatic, and marine environments. The treatment process, which involves applying the chemical under pressure so that it enters the pore spaces of the wood, was designed to prevent fungal and microbial decay. By the late 1990s, such wood was being used in almost 80 percent of the U.S. preserved-wood market. In fact, by the early part of the twenty-first century, it was estimated that nearly 450 million cubic feet (13 million cubic meters) had been sold in this country. Because fungal and microbial decay are especially troublesome in Florida, the use of CCA-treated wood there was quite extensive. In fact, it was used to construct the boardwalks and decks in essentially all of the 150 state parks.

Studies initiated in the 1990s showed that arsenic from the treated wood was leaching into the soil beneath such structures. As a result, the concentrations in some soils in Florida were in excess of federal and state limits. This led to concerns that children playing on such equipment might ingest arsenic through, for example, licking their hands. Tests showed even higher rates of leaching from similarly treated wood that was used to construct docks and marinas (Tom, 2001). Although tests of workers who were regularly exposed to the raw materials used in the wood found levels of arsenic that were deemed to be insignificant, many members of the public continued to be concerned. Based on evaluations of the associated risks, including those to children, the EPA concluded that it was not necessary for homeowners to remove their backyard decks and picnic tables or to dismantle swing sets and jungle gyms. Questions continued to be raised, however, and the accompanying concerns were subsequently heightened by studies that showed that the level of arsenic on the surfaces of such products may persist for as much as two decades (Lavelle, 2002).

As a result of these concerns and those of state regulatory agencies, several major retailers voluntarily agreed to discontinue selling such wood, and four U.S. manufacturers agreed to withdraw the chemical from the treatment of wood for residential use by December 31, 2003 (Lavelle, 2002). Today, all CCA-treated wood sold in the U.S. will be required to be accompanied by detailed safety handling information. Its use, however, will be restricted to certain industrial applications, such as pier marine pilings, highway barriers, and plywood used in the roofs of homes. Concurrently, the Consumer Product Safety Commission has agreed to request public comments on petitions that could lead to an outright ban on the use of CCA-treated wood (Tom, 2001). Exacerbating the situation, when such facilities are dismantled, they are frequently disposed in so-called construction and demolition landfills. In Florida, such landfills are not required to have liners. As might be expected, analyses of nearby groundwater aquifers, revealed arsenic concentrations more than double the EPA limit for drinking water (O'Connell, 2003). In other cases, it was found that the CCA-treated wood from construction and demolition projects was being burned to generate electricity, with the ash being applied to agricultural fields, such as those used to grow sugar cane. In still other cases, the wood was being ground into a mulch that was applied to the soil. Tests showed that some of the ash contained arsenic concentrations of several hundred parts per million. Chromium concentrations were also high (Tom, 2001).

Trends in Waste Management

For many years, agencies and organizations responsible for protecting the environment accepted the wastes that were generated and sought to develop satisfactory methods for their treatment and/or disposal. This is referred today as the "end of pipe" approach. With the coming of the previously mentioned throw-away society and the rapid expansion of industrial activities, environmentalists and the U.S. Congress soon realized that the generation of waste was becoming overwhelming and that new approaches had to be developed. As discussed earlier (Chapter 8), the U.S. Congress addressed these problems in the Pollution Prevention Act of 1990, which stated that it was the policy of the United States that, whenever feasible, pollutants that cannot be prevented should be recycled, and those that cannot be recycled be handled and disposed in an environmentally safe manner (U.S. Congress, 1990).

Today, the generally accepted philosophy is that waste management and disposal should not, and cannot, be regarded as "free-standing" practices that require their own justification. They must be made an integral part of the processes that generate them. In accord with this view, the challenges and potential difficulties of waste management and disposal must be addressed at the time the decision is made to initiate a given process or operation. In response, the EPA announced that it was committed to a policy that places the highest priority on waste minimization (EPA, 1993b). Under this approach, the EPA requires that generators of hazardous waste certify on their shipping manifests that they have a waste-minimization program in place. In a similar action, the U.S. Department of Energy adopted a policy such that if it is determined that a proposed activity will generate wastes that have no available option for disposal, it should not be approved (USDOE, 1999).

Enhancing the Disposal of Wastes—Recycling

There are two primary methods for accomplishing this objective. The first is to reuse them within the process (i.e., recycling lead storage batteries) or to transfer the waste to another industry that can use it as input to its production process (Debham, 2009). If such efforts are to be successful, it is mandatory that they receive the unequivocal support of all levels of corporate management. One way to achieve this is to make recycling a part of the in-house culture. The development of such a culture, however, should not be restricted to industrial organizations. It should become a part of the culture of individual households. The success of such efforts is demonstrated by the fact that millions of people in the United States routinely sort their trash, fill recycling bins, demand to be able to purchase products made of recycled materials, and avoid products with wasteful packaging. More than 80 percent of U.S. cities have curbside recycling programs, and more than 60 percent have programs for collecting recyclables from multifamily buildings. It can also occur at the national level, as exemplified by Executive Order 12873, issued by U.S. President Bill Clinton in 1993, that required all federal agencies to purchase only recycled copy paper. The U.S. Congress took similar action in 1996 through passage of the Mercury-Containing and Rechargeable Battery Management Act. This Act required operators of stores that sell batteries to accept them back at the end of their lives for possible recycling of their toxic components.

Organizations at the state and local level can also play a key role. More than 40 states have established recycling goals. One of the most ambitious is Rhode Island, which is seeking to recycle 70 percent of its garbage. In a similar manner, the state of Oregon has mandated that its Department of Environmental Quality conduct an annual comprehensive survey of progress in the management of solid waste. At the local level, nearly 4,000 communities now levy user charges, often called "pay-as-you-throw," or unit-based pricing, on municipal solid waste (Portney and Stavins, 2000).

Treatment of Hazardous Waste

Treatment is defined as *any method, technique, or process, including neutralization, that is designed to change the physical, chemical, or biological character or composition of a hazardous waste so as to neutralize it, recover energy or material resources from it, render it nonhazardous or less hazardous, or to make it safer to transport, store, or dispose of, more amenable to recovery or storage, or smaller in volume* (EPA, 1993b). It may be thermal (i.e., incineration), chemical, or biological (especially for wastes containing organic materials). Where methods for neutralizing or rendering a waste nonhazardous are either not available or are ineffective, immobilization (stabilization) can often be effective, especially for inorganic hazardous wastes.

The general goal is to convert hazardous waste into a solid form for disposal. Treatment may be initiated at any stage prior to or following solidification, for example, in tanks, surface impoundments, incinerators, or land treatment facilities. Because many of these processes are waste specific, the EPA has not attempted to develop detailed regulations for any particular type of process or equipment; instead, it has established general requirements to ensure safe containment (EPA, 1986a, 1986b). In general, four processes are being used to treat solid wastes.

INCINERATION

Incineration deserves special mention because it is one of several processes available both for reducing the volume of solid and hazardous waste and for destroying certain toxic chemicals it contains. The use of plastics in packaging, however, has created a corresponding increase in the amount of PVC in solid waste. When burned, such plastics produce hydrochloric acid. This extremely corrosive compound can destroy incinerator components such as metal heat exchangers and flue-gas scrubbers and can threaten human health if released into the atmosphere. Hydrochloric acid can also be produced in incinerators by the combustion of foods and wastes that contain chloride salts. Compounding these problems, incomplete combustion of some organic materials in the presence of chlorides can produce dioxins, a toxic group of compounds.

These and other potential threats to human health have led to stringent regulations on emissions from incinerator facilities, particularly in light of the realization of the health effects of extremely small airborne particulates (Chapter 5). Although modern technology will provide almost any degree of cleanup required, the economic costs can be high. One response has been to construct and operate centrally located incinerators to serve a group of waste producers. In many communities that have a limited capacity for direct disposal of solid waste in landfills (for environmental, political, economic, and other reasons), incineration has become the principal method of intermediate treatment. One reason is that the resulting ash is generally in a physical and chemical form that is more readily disposed than the original waste and it is biologically and structurally more stable. In addition, many of the compounds it contains are insoluble, so their long-term leaching by rain and groundwater is minimized.

HEAT TREATMENT

Heat, applied at moderate temperatures, is effective in treating soils, particularly those that are contaminated with volatile solvents such as creosote and diesel and gasoline fuels. This approach has been used for years to enhance the removal of oil from the ground. The heat can be applied either through submerged electric heaters or steam-injection wells. Electric heating is especially effective in clay soils, which are not very permeable and thus tend to have higher moisture content. The presence of water not only enhances the conductance of the electricity but also produces steam, which expands and dries the clay matrix and, in turn, either volatilizes or immobilizes the contaminants. In the latter case, the contaminants can either be removed or destroyed in place. An interesting by-product of this approach is that, in some cases, the presence of heat has attracted thermophilic bacteria that have assisted in stabilizing the contaminants. This approach also has the benefit of not requiring that the contaminated soil be removed and transported elsewhere. Since 70 percent of Superfund sites (discussed later) are contaminated with solvents, heat treatment methodologies may have widespread application (Black, 2002).

SOLIDIFICATION AND/OR STABILIZATION

Solidification and/or stabilization of solid wastes can be accomplished using several techniques. Plasma power is one of the newer technologies that is being applied. The temperatures that this technology is capable of producing (in excess of 7,000°C) can melt or vaporize contaminated soil and a full range of typical wastes and garbage, producing a glass- or sand-like residue. Through this process, hazardous and toxic chemicals and biological agents are reduced to their elemental components. This technology, which was developed to provide heat shields to protect spacecraft during reentry, is being used in Japan for the treatment of municipal solid waste and automobile shredder residue. One plant, which has a capacity of 20 tons per day, went into operation in 2002. The hot gases that are produced, which consist primarily of hydrogen and carbon monoxide, are sent to a secondary combustion system where they are mixed with water to form steam. The steam can then be used to run a turbine and generate electricity. Most of the sand-like residue is mixed with cement to form bricks that are used in pavements. A similar plant has been constructed in France, and plans are under way to construct one in the state of Georgia that will have the capacity to process upwards of 100 tons of tires per day. The steel in the tires will be drawn off as ingots. Since tires also contain sulfur, the off-gases will be treated to remove this contaminant (Link-Wills, 2002).

CHEMICAL TREATMENT

One of the common applications of this technique is in the treatment of corrosive solids, such as lime or cement kiln dust. These can be neutralized using either chemicals or acidic wastes from other operations within a plant. Specially formulated solutions are being used to leach organic or inorganic contaminants from soils either onsite or offsite. Through this process, some compounds can be chemically converted to related but much less mobile or less toxic versions; for example, chromium VI can be converted to less toxic chromium III. In a similar manner, some chlorinated organics, such as poly-chlorinated biphenyls, can be degraded in soils or other solids using various sodium-based reagents.

BIOLOGICAL TREATMENT

One new development is the bioreactor landfill, one of its many benefits being that it accelerates the stabilization of the waste. This is accomplished by the addition of moisture to the waste to create an environment favorable to the microorganisms responsive for waste decomposition. This is entirely different from the traditional approach of managing solid-waste landfills in a manner that discourages decomposition by minimizing moisture intrusion. Furthermore, the enhancement of biological decomposition leads to the potential recovery of space within the disposal facility, onsite treatment and storage of the leachate, potential reductions in long-term and post-closure costs and liability, and, in the case of anaerobic landfills, more reliable production of gas for energy recovery (Hughes et al., 1996; Editorial Staff, 2009).

Waste Disposal

Disposal, by definition, means *the discharge, deposition, injection, dumping, spilling, leaking, or placing of any solid or liquid waste onto, or into, the land or water*. It has crucial ramifications because such an approach may permit the waste and/ or its constituents to enter the terrestrial environment, be emitted into the air, or be discharged into surface waters. The potential contamination of ground-water is also a concern. The range of options for the management and disposal of solid and liquid wastes are summarized below (EPA, 1986b):

Landfills. Disposal facilities in which the waste is placed into or onto
the land. In most cases, the wastes are isolated in discrete cells
within trenches. To prevent leakage, landfills must be lined and
have systems to collect any leachate or surface runoff. In this
regard, it is important to distinguish between landfills and *surface*

impoundments. The latter are typically considered storage units; they are not an effective method for disposal. Hazardous wastes need more secure disposal facilities (Figure 9.3).

Waste piles. Noncontainerized accumulations of insoluble, solid, nonflowing, hazardous waste. Although, in some cases, waste piles serve as a site for final disposal, most provide temporary storage pending transfer of the waste to its final disposal site.

Land treatment. A disposal process in which solid waste, such as sludge from municipal sewage treatment plants, is applied onto or incorporated into the soil surface (Chapter 8). Under proper conditions, microbes occurring naturally in the soil break down or immobilize the hazardous constituents.

Underground injection wells. These are steel- and concrete-encased shafts placed deep in the ground into which wastes are injected under pressure. Although used in the past on a regular basis, this method is being applied in the United States only in the case of oil and gas wells that are exempted from hazardous waste regulations.

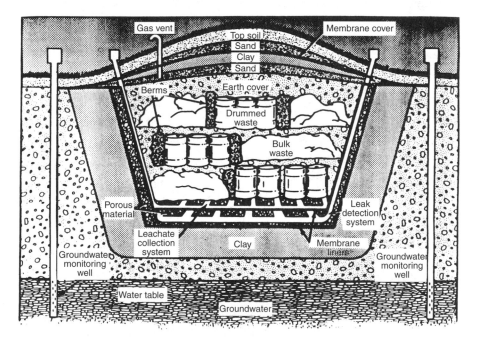

Figure 9.3   Land burial facility for hazardous waste

Updating the Restrictions on Municipal Waste Disposal

More than 200,000 sites in the United States are being or were formerly used as sites for the disposal of municipal wastes. Although the wastes were classified as nonhazardous, about 35,000 of these sites are known to have received hazardous chemicals and other materials from small-quantity industrial generators. In addition, as noted earlier, a certain amount of the waste from most households contains hazardous materials. Through the 1984 amendments to RCRA, Congress mandated that the EPA develop new criteria to provide better protection of the public from the potential health risks associated with these facilities (EPA, 1986b). In response, the EPA requires disposal practices for municipal waste that closely parallel those for industrial (hazardous) waste. These requirements, which apply to all aspects of the siting, design, construction, operation, and monitoring of such facilities, are summarized below (EPA, 1993a).

Location. Landfills must not be located on a floodplain. They must also not be built on wetlands unless the proposed operator can show that the landfill will not lead to pollution. In addition, they cannot be located in areas subject to landslides, mudslides, sinkholes, or major disruptive events such as earthquakes. Likewise, they cannot be located near airports, where birds that are frequently attracted to such facilities might constitute a danger to aircraft.

Design. Landfills must be designed to avoid contaminating groundwater. Ancillary requirements include lining the bottom of the landfill with clay and covering it with an impervious synthetic-material liner, coupled with a system to collect and treat any leachate (liquids) that may collect within the liner.

Operation. The disposed waste must be covered daily with dirt to prevent the spread of diseases by rats, flies, mosquitoes, birds, and other wildlife. In addition, access to the landfill must be restricted to prevent illegal dumping and other unauthorized activities, the site must be protected by ditches and levees to prevent storm-water flooding, and any runoff that occurs must be collected and controlled.

Monitoring. Generally, landfill owners or operators must install monitoring systems to detect groundwater contamination. Monitoring for changes, such as subsidence, that may be indicative of possible problems is encouraged. If contamination is observed, the concentrations must be reduced to ensure compliance with federal limits for drinking water. Methane gas, generated through decomposition of the waste, must also be monitored and controlled, if necessary.

Closure and postclosure care. Upon ceasing operation, landfills must be closed in a way that will prevent subsequent problems. The final cover must be designed to keep liquid away from the buried waste, and, for 30 years after closure, the operator must continue to maintain the cover, monitor the groundwater to be sure the landfill is not leaking, and monitor and collect any gases that are generated.

Updating the Classification and Controls for Superfund and Associated Sites

Thousands of U.S. waste disposal sites established during the past 30 to 50 years were improperly designed or operated and have leaked or have the potential to leak hazardous waste into the environment. Recognizing the severity of this problem and the urgent need for cleanup of these sites, Congress passed the Comprehensive Environmental Response, Compensation, and Liability Act (U.S. Congress, 1980; Table 9.7).

SUPERFUND SITES
CERCLA, more commonly referred to as the Superfund Act, authorized the EPA to investigate various waste disposal sites and to identify them as potential Superfund sites. By 1994, more than 40,000 such sites had been

Table 9.7 Principal federal laws related to the management and disposal of solid waste, United States

Year	Law	Public Law
1976	Resource Conservation and Recovery Act	84-580
1976	Toxic Substances Control Act	94-469
1980	Comprehensive Environmental Response, Compensation, and Liability Act (Superfund Act) (CERCLA)	96-510
1986	Superfund Amendments and Reauthorization Act (SARA)	99-499
1992	Federal Facility Compliance Act	102-386

so identified. Those sites with the highest levels of contamination and deemed to present the most serious threats to health are placed on what is called the National Priorities List (NPL). By September 1995, almost 1,400 sites had been so designated. For these, the EPA identifies the potentially responsible parties and gives them an opportunity to implement cleanup. If they fail to do so, the EPA arranges for the cleanup, using Superfund money, and then seeks to recover the costs from the responsible parties. As of 1995, work was under way at more than 90 percent of the NPL sites, final cleanup activities were in progress at about 35 percent, and such activities had been completed at another 25 percent (CEQ, 1997).

As part of what might be called a "streamlining" effort, the EPA concluded in late 1995 that approximately 24,500 of the potential Superfund sites were of such low priority that they could be removed from the list. Through that effort, the number of sites that remained in the Superfund inventory was reduced to about 15,500. The sites that were removed are now covered by the Brownfields Development Initiative (discussed later). The net impact of this change is that slightly more than 1,300 sites remain on the NPL or have been proposed for listing (CEQ, 1997). About 200 of these sites are former municipal landfills; many others were contaminated by operations of the U.S. Departments of Defense and Energy.

One approach that can be used for the cleanup of these sites is to excavate the contaminated material and transport it to a new burial site. Because in many cases the quantities involved are enormous, various methods for on-site treatment are being developed. These involve one or more of the previously described physical and chemical processes. Which treatment option is selected depends on the types of contaminants and the relevant properties of the soil—for example, its clay and humus content. In the case of soils containing organic chemicals, the most proven separation technology is volatilization, using vapor extraction and/or forcing air through the soil. Many sites are also being remediated by solidification/stabilization. Another technique is to render the contaminated soil inert by mixing it with additives, such as cement (Fox, 1996).

Another approach being extensively evaluated is biological treatment, discussed earlier. This offers two distinct advantages: (1) it is inexpensive, and (2) it has the unique potential for rendering hazardous constituents nontoxic. In contrast, however, to the more conventional applications of such processes (i.e., the treatment of domestic sewage), biological treatment of contaminated soils is an immature field that, although offering high expectations, is confronted with numerous scientific and engineering challenges

(Hughes, 1996). Treatment can be pursued through the introduction of new organisms or by the attenuation by organisms already present.

THE BROWNFIELD REDEVELOPMENT INITIATIVE

In addition to the previously cited low-priority facilities that were removed from the list of Superfund sites in 1995, there are up to 600,000 abandoned, idled, or underused industrial and commercial facilities in the United States, many of which have low levels of contamination but are in need of cleanup and restoration. In many cases, these sites are located in economically depressed areas. The Brownfields Redevelopment Initiative is designed to stimulate their cleanup and reuse, the goal being to revitalize the properties and restore their usefulness. Recognizing the benefits of this program, many state and local governments have agreed to provide economic incentives to private-sector companies that redevelop such areas. Such incentives include grants, tax exemptions or abatements, low-interest loans, waiver of impact and permit fees, expedited development approvals, and marketing and promotional assistance (Verbit, 2001). The EPA is also providing strong support. Its staff has agreed that if an enforceable arrangement can be entered into by a responsible state/local agency and a willing developer, its primary role will be to observe and ensure that progress is being made.

Buoyed by the success of the program, Congress passed the Small Business Liability Relief and Brownfields Revitalization Act, which became law in 2002. This increased the incentives provided for the cleanup and reuse of brownfields (Isler and Lee, 2002). As a result, long-neglected parcels of land in many areas of the country, particularly in cases where the degree of contamination is low to moderate, are now being converted into valuable new property. Common uses are to convert the land into parks and/or sites for industrial buildings. For those sites in which the extent of the contamination is very low, only minimal cleanup may be required. Where the concentrations of contaminants are relatively high, it may be necessary (as in the case of some Superfund sites) to excavate the contaminated soil.

Initially, the public was almost universally opposed to the construction of houses and schools or creating open spaces on reclaimed brownfield sites. In the late 1990s, however, this view changed for at least two reasons: (1) such sites provided a readily available supply of land on which to build new housing units which were in high demand, and (2) regulators were careful to ensure that the degree of cleanup was acceptable to the local community.

GREYFIELDS

Although not as well known as brownfields, the redevelopment of grey-fields represents another major effort under way to revitalize certain areas within cities. Greyfields are exemplified by failing malls and strip malls that do little if any business. A key characteristic of such areas is that they represent large tracts of land accompanied by both empty retail space and parking lots. Outclassed by newer, more modern malls and shopping centers, these districts have simply not generated sufficient revenue to sustain their use. The land they occupy, however, can be very useful and economically profitable to local communities. The concept of mixed-use redevelopment has demonstrated that greyfields can be converted into vibrant city centers that will be both profitable and sustainable (Chen, 2002).

Management of Radioactive Wastes

As is true of hazardous chemical wastes, the management and disposal of radioactive wastes have received extensive attention. Federal organizations involved include the U.S. Congress, the EPA, the U.S. Nuclear Regulatory Commission (USNRC), and the U.S. Department of Energy (USDOE). In general, the U.S. Congress passes relevant legislation, the EPA sets applicable environmental standards, and the USNRC develops regulations to implement the standards. Also exercising roles are the Agreement States (discussed below) and the independent Nuclear Waste Technical Review Board, established by the Nuclear Waste Policy Amendments Act (U.S. Congress, 1987), which has the responsibility to provide independent scientific and technical oversight of the program for managing and disposing of high-level radioactive waste and spent nuclear fuel. The principal laws for the regulation of the management and disposal of radioactive wastes in the United States are listed in Table 9.8.

LOW-LEVEL RADIOACTIVE WASTES

Low-level radioactive waste (LLRW) primarily consists of a wide range of items that have become contaminated with radioactive materials. More than 90 percent are generated through use of radioactive materials in hospitals and related institutions, educational and research establishments, private and governmental laboratories, commercial nuclear power plants, and nuclear fuel-cycle facilities. Specific examples are radioactively contaminated medical tubes, swabs, and syringes, carcasses and tissues from laboratory animals, and contaminated equipment, tools, clothing, shoe

Table 9.8   Principal federal laws related to the management and disposal of
radioactive waste, United States

Year	Law	Public Law number
1954	Atomic Energy Act	85-703
1978	Uranium Mill Tailings Radiation Control Act	95-604
1980	Low-Level Radioactive Waste Policy Act	96-573
1983	Nuclear Waste Policy Act of 1982	97-425
1986	Low-Level Radioactive Waste Policy Amendments Act of 1985	99-240
1987	Nuclear Waste Policy Amendments Act (Created the Nuclear Waste Technical Review Board)	100-203
1992	Energy Policy Act of 1992	102-486

covers, cleaning rags, and mops. A significant source, in the case of nuclear facilities, is the waste generated in decommissioning activities conducted by DOE and the U.S. Navy and related activities. For regulatory purposes, LLRW has been subdivided into Classes A, B, and C on the basis of the types and quantities of radioactive materials they contain. Of the three, Class A is the least hazardous, and Class C is the most hazardous.

Low-level radioactive waste disposal (LLRW). With increased understanding of the necessary precautions, the disposal of LLRW has followed a progression of improvements, beginning with simple landfills and ultimately leading to earth-mounded concrete bunkers (Figure 9.4). The National Council on Radiation Protection and Measurements (NCRP, 2003) has played a leading role in providing guidance on such activities. All LLRW disposal sites in the United States are licensed under the Low-Level Radioactive Waste Policy Amendments Act of 1985, which established the so-called Agreement States. This permits the USNRC to transfer regulatory authority for the licensure and inspection of radioactive material licensees and LLW disposal facilities to those states that meet stipulated requirements. The disposal of high-level radioactive wastes, however, remains the domain of the federal government. Specific regulatory actions that are transferred to the Agreement States include the authority to (1) form regional compacts for the establishment of disposal sites and (2) limit which states can send wastes to each site. As of 2009, 37 states had met the requirements to become Agree-

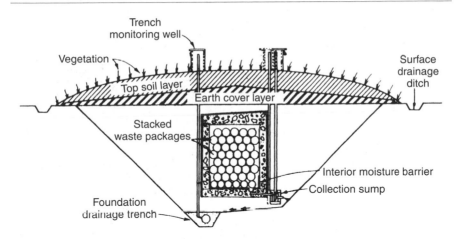

Figure 9.4 Earth-mounded concrete bunker for disposal of low-level radioactive waste

Table 9.9 Low-level radioactive waste compacts, United States 2009

Compact	States included
Atlantic	Connecticut, New Jersey, South Carolina[a]
Northwest	Alaska, Idaho, Montana, Oregon, Utah,[b] Washington,[c] Wyoming
Rocky Mountain	Colorado, Nevada, New Mexico

a. The state of South Carolina has licensed Barnwell to receive all classes of LLW from the Atlantic Compact.

b. The state of Utah has licensed Energy Solutions to receive Class A waste only, but it is open to all regions of the United States.

c. The state of Washington has licensed Hanford to receive all classes of LLW from both the Northwest and Rocky Mountain Compacts.

ment States, 3 had signed "letters of intent" to do so, and 10 remained non–Agreement States (USNRC, 2009; Table 9.9). Three LLRW disposal facilities are operating in the United States, all of which use shallow land disposal and dispose of the wastes, depending on their nature, with or without concrete vaults. These are: (1) the Barnwell facility in the state of South Carolina, (2) the Hanford facility in the state of Washington, and (3) the Energy Solutions facility in the state of Utah. The first two of these accept all three

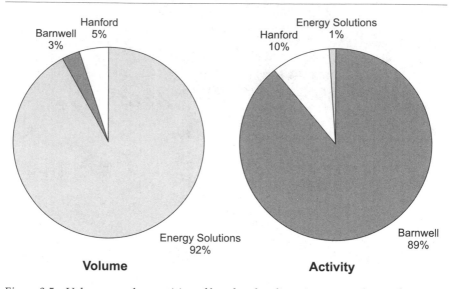

Figure 9.5 Volumes and quantities of low-level radioactive waste shipped to U.S. disposal facilities, 2008

classes of LLW; the third accepts Class A only. An analysis of this information shows that 37 states have no option for disposing their Class B and C wastes.

A quantitative perspective on the differences in Class A wastes, as compared to Class B and Class C, is shown in Figure 9.5, which lists the volumes and quantities of radioactive materials in the Barnwell, Energy Solutions, and Hanford disposal facilities. As noted, Energy Solutions contains a large volume of waste with relatively very low activity. In contrast, Barnwell contains a low volume of waste with relatively very high activity. This comparison also illustrates why small quantities of Class A wastes can be stored on-site until they have decayed, in which case they are then disposed with ordinary trash, or until a sufficient amount has accumulated to warrant shipment to a LLRW disposal site.

URANIUM MILL TAILINGS

Uranium mill tailings contain low-level radioactive waste produced primarily as a result of activities pertaining to national defense. They include the materials (so-called tailings) that remain after the uranium metal has been separated from the original ore. On a relative basis, the volumes (measuring in the millions of cubic yards) are significantly larger than those of the low-level waste generated by the commercial sector. For this reason, mill tailings are handled "in place," that is, they are stabilized and provided with a cover to protect them from wind and water erosion. The Army Corps of Engineers has jurisdiction over their disposal.

TRANSURANIC WASTES

Transuranic wastes (TRU) contain, as the prefix, "trans," implies, elements heavier than uranium that have half-lives and concentrations in excess of certain stipulated limits. Examples are ^{238}U, ^{237}Np, ^{239}Pu, and ^{241}Am, all of which emit alpha radiation. As such, they represent little, if any, threat as a source of external exposure. Nonetheless, because these radionuclides tend to be long lived and had long been considered toxic if taken into the body, they are being disposed in the Waste Isolation Pilot Plant (WIPP), a deep underground repository that was constructed by the USDOE in a salt dome in southern New Mexico. This facility, which is regulated by the EPA, completed its first decade of operation in 2009 (Moody and Sharif, 2009). At that time, 58,000 cubic meters (76,000 cubic yards) of waste, more than one-third of the existing inventory, had been stored (Moody and Sharif, 2009). In this regard, it has been known for decades that ^{238}U is primarily a chemical hazard, not a radiological one, the reason being that its half-life (4.5 billion years) makes it almost equivalent to being stable.

HIGH-LEVEL RADIOACTIVE WASTES

High-level radioactive wastes in the United States include spent (depleted) fuel removed from commercial nuclear power plants and fission product wastes that were produced in the process of purifying plutonium for nuclear weapons. A large percentage of the waste has been vitrified. Both of these forms of waste await disposal. Although the ^{239}Pu in the spent fuel can be recovered through chemical processing and used as fuel in commercial nuclear power plants, there is a ban on such activities in the United States. Nonetheless, a sufficient surplus of ^{239}Pu from dismantled nuclear weapons is available to meet this demand.

It is generally agreed that the best method for the disposal of high-level wastes is in a deep geological repository. Many countries are moving ahead with plans based on this approach. Sweden and Finland, for example, are making progress in their plans for selecting a site and initiating construction of an underground repository. In the former case, two possible sites are under consideration, with the specific site to be confirmed. The schedule calls for the proposed repository, which will be located in bedrock 500 meters (~1,650 feet) beneath the surface, to be ready for commercial operation in 2015. In the case of Finland, the goal is to have the proposed repository, which is to be located on the island of Olkiluoto, ready to accept high-level waste in 2020. Other European countries that are conducting research and exploring possible repository sites include Switzerland and the Czech Republic (Sperber, 2002).

In some ways, it would appear that the Department of Energy, which has the responsibility for the management and disposal of high-level radioactive waste in the United States, is far ahead of the world in its progress. In fact, a repository, located relatively deep in Yucca Mountain, Nevada, has been built in accordance with the standards developed by the Environmental Protection Agency (EPA, 2008). Nonetheless, due to associated controversies, primarily political in nature (Moeller, 2010, 2011), work was terminated on this facility by the Administration in 2009, "while developing an alternative." A *Blue Ribbon Panel* was subsequently appointed to "provide

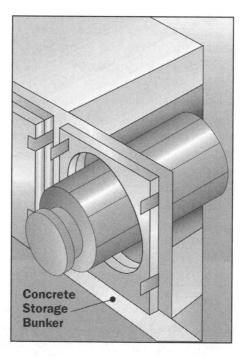

Figure 9.6   Concrete storage bunker for on-site storage of spent fuel

advice and make recommendations on issues including alternatives for the storage, processing, and disposal of civilian and defense spent fuel and nuclear waste" (Chu, 2009). In March 2010, however, the Department of Energy filed a motion with the USNRC to withdraw their application for the repository.

Prior to that decision, detailed supporting analyses had documented that none of the radionuclides in the waste proposed for disposal would represent a risk to the health of either the environment or the public (Moeller, 2009). This, which is discussed in more detail in Chapter 15, is based on new developments in our understanding of the health risks of the more abundant radionuclides in the waste. In the meantime, each U.S. nuclear power plant removes 25–30 tons of spent fuel annually, yielding a nationwide total of about 2,200 tons (2,000 metric tons). As of 2009, the accumulated total was 60,000 metric tons (66,000 tons). When initially removed, the spent fuel is stored at the nuclear power plants submerged in large basins of water. After sufficient cooling, as a result of radioactive decay, it is stored above ground encased in steel and placed inside concrete bunkers (Figure 9.6; USNRC, 2009).

The General Outlook

The generally accepted philosophy today is that waste management and disposal must be recognized as an integral part of any type of industrial operation. Effective implementation of this philosophy requires that the challenges and potential difficulties of waste management and disposal be considered at the time the decision is made to initiate the industrial operation that will generate them. This is particularly true for the multitudes of consumer products that enter the solid-waste stream on a daily basis. In this regard, one might readily ask whether an industry that claims to be environmentally responsible should not be held accountable if the products it sells are purposefully designed to fail (even under routine use) and/or rapidly become obsolete. Prominent examples are cell phones, television sets, computers, and motor vehicles. Why must a person purchase a new car simply because the old one is no longer in style? Perhaps the entire world is in need of a change in culture.

10

ANIMALS, INSECTS, AND RELATED PESTS

MORE THAN HALF of all the estimated 1,700 infectious diseases that have plagued humankind, past and present, originated in animals. In fact, of the slightly more than 150 so-called emerging diseases, almost 75 percent are zoonotic in nature (Normile, 2009)—that is, those that can be transmitted from animals to humans and vice versa. One relatively recent example is acquired immune deficiency syndrome (AIDS), in which the human immunodeficiency virus (HIV) moved from monkeys into chimpanzees and, later, from chimpanzees into humans. The more capable a virus is to propagate in humans, the more likely it is to cause a pandemic. Recognizing that international travel promotes the transmission of all such diseases, officials at the Centers for Disease Control and Prevention (CDC) have developed unified global programs for increased monitoring to provide early identification of such infections. Without such programs, the knowledge necessary to cope with future unanticipated disease episodes will not be available (Wolfe, 2009).

Recent applications of such knowledge are exemplified by the procedures developed for the annual selection of the vaccines that need to be developed to protect against newly emerging forms of certain diseases. Prominent examples include Influenza A and H1N1 ("swine") flu, which has been shown to have a seasonal global circulation (Russell et al., 2008). These procedures, for example, include identifying both the strain of influenza that will most likely pose a threat and the country in which it will originate (Dougherty, 2009).

Magnitude of the Challenge

Scientists estimate that there are more than 3 million insect species on Earth. Of these, nearly 1 million have been identified, including more than 100,000 species of butterflies and moths, more than 100,000 of ants, bees, and wasps, and almost 300,000 of beetles. At any one time, the total number of insects is believed to be about 1 million trillion (Wilson, 2002). Even so, about 4,000 new varieties are discovered every year. Some, such as the honeybee and the silkworm, bring financial benefits; in fact, honeybees are responsible for the pollination of some $15 billion worth of agricultural products in the United States each year (Editorial Staff, 2009). Other insects, such as the butterfly and lightning bug, are aesthetically pleasing. Still others, such as flies, mosquitoes, boll weevils, corn borers, termites, and locusts, are destructive and may even be dangerous to humans. The mosquito, in particular, is the transmitter of a wide range of disease agents, continuing to cause millions of deaths annually through the transmission of diseases such as dengue and yellow fever, encephalitis, West Nile fever, filariasis, and malaria (Spielman and D'Antonio, 2001).

Nature and Control of Zoonotic Diseases

The types of infections and countries of origin of diseases that can be transmitted from animals to humans are summarized in Table 10.1. Related information on those that can be transmitted from humans to animals is summarized in Table 10.2. Among the more common animals involved are rodents, which have been a public health problem for centuries. Most famous is the role of rats in the multiple epidemics of bubonic plague, collectively known as the Black Death, that swept Europe in the fourteenth century. One of the earliest began in 1347 in Genoa, when ships arriving from Black Sea ports left behind rats carrying infected fleas (Duplaix, 1988). The subsequent spread of bubonic plague depopulated some 200,000 towns and in 3 years killed 25 million people, 25 percent of the population of Europe. This remains the worst calamity in human history. In 1665, a similar epidemic in London killed 100,000 people; in the 1890s it struck in San Francisco, and today it continues to exist in Africa, North and South America, and Asia. A major outbreak in India in 1994 resulted in at least 1,000 cases and almost 100 deaths. Globally, up to 3,000 cases of bubonic and pneumonic plague occur each year, including from 10 to 15 sporadic cases in the United States. The causative agent is the bacterium *Yersinia pestis*.

Table 10.1   Sources, types of infections, and countries of origin of diseases
transmitted from animals to humans

Disease	Source	Type of infection	Country of origin
AIDS	Chimpanzees	Virus	Central Africa
Chagas disease	Wild & domestic mammals	Protozoa	Tropical America
Dengue fever	Primates	Bacteria	Africa
Hepatitis B	Apes	Virus	Central Africa
Influenza A	Birds	Virus	Worldwide
Plague	Rodents	Bacteria	Europe
Rabies	Bats and dogs	Virus	Europe
Sleeping sickness	Wild & domestic ruminants[a]	Parasite	East & West Africa
Vivax malaria	Monkeys (genus *Macaca*)	Virus	Asia
Yellow fever	Primates	Bacteria	Africa

a. Antelope, bison, buffalo, camels, cattle, and deer.

Table 10.2   Diseases that are transmitted from humans to animals

Disease	Type of infection	Infected animals	Country of origin
Tuberculosis	Bacteria	Cattle	Central Africa
Measles	Bacteria	Mountain gorillas	Western Africa
Yellow fever	Virus	Monkeys	South Africa
Poliomyelitis	Virus	Chimpanzees	Central Africa

RODENTS

Four species of rodents continue to be of environmental and public health
concern in the United States: the Norway rat, the roof rat, the house mouse,
and bats. Species of importance in other parts of the world include the
Polynesian rat (*Rattus exulans*), which has spread from its native Southeast
Asia to New Zealand and Hawaii, and the lesser bandicoot (*Bandicota ne-
sokia*), which is predominant in southern Asia, especially India, and whose
body can be more than a foot long. The main economic impact of rodents is
their widespread destruction of food, particularly grains (Canby, 1977).

Understanding the characteristics of the rodents that pose problems is essential to their control.

The Norway rat (*Rattus norvegicus*) is characterized by its relatively large size and short tail. Norway rats frequent the lower parts of buildings and inhabit woodpiles, rubbish, and debris. They also burrow under floors, concrete slabs, and footings and live around residences, warehouses, chicken yards, and in sewers. They nest in the ground and have a range of 100–150 feet.

The roof rat (*Rattus rattus*) is characterized by its smaller size and longer tail. Roof rats live in grain mills, dense growth in willows, and old residential neighborhoods. They are excellent climbers, frequently occupying shrubbery, trees, and upper parts of buildings. They usually nest in buildings and have a range similar to that of the Norway rat.

The house mouse (*Mus musculus*) is characterized by its small size, including small feet and eyes, and long tail. Mice live in buildings and in fields, and their range is limited, 3–10 meters (10–30 feet).

Bats are characterized by having modified forelimbs that are covered with membranous skin that extends to the hind limbs and serve as wings. They are about 3.5 inches long, have a wingspread of 14 inches, are nocturnal in nature, frequently live in caves, and fly primarily during the twilight.

Additional details on the distinguishing characteristics of the Norway rat, the roof rat, and the house mouse are visually illustrated in Figure 10.1.

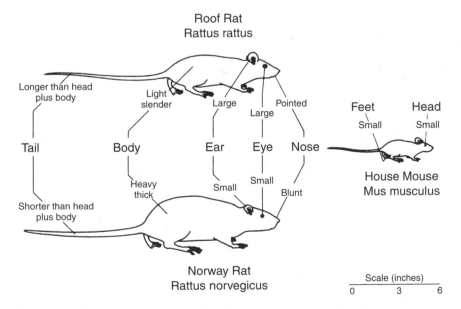

Figure 10.1   Distinguishing characteristics of three species of rodents

RATS

Among the fastest reproducing mammals, rats have a gestation period of 21–23 days, and they can reproduce every 60–90 days. A typical litter ranges in size from five to nine, and they have a life span of 9–12 months. Rats are very intelligent and survive in a hostile human environment by means of complex social mechanisms.

Characteristics and Impacts. Although their vision is poor, rats have a keen sense of smell; they like the same food as people and prefer it fresh. Yet they can also eat decayed food and consume contaminated water with no apparent ill effects. Rats have a well-developed sense of touch via their nose, whiskers, and hair. They have an excellent sense of balance, can fall three stories without injury, and are accomplished swimmers.

Rats also have a strict social structure. Although those in neighboring colonies will tolerate one another, those in nonadjacent colonies are openly hostile. They are rarely seen in the daytime. Cowardly, they will seldom pick a fight with people. They follow established paths, which are readily identified by the presence of droppings (feces), grease marks (where the rats have rubbed), urine stains (which can be located using ultraviolet light), and their characteristic odor. Rats are nonetheless extremely adaptable and can survive under adverse conditions (Canby, 1977). Scientists returning to Pacific islands that were virtually destroyed during nuclear weapons tests in the early 1950s found flourishing colonies of rats. The same species that lives in burrows in the United States and in attics in Europe can live in palm trees in the South Pacific. Other species, finding shortages of food on land, have learned to dive into lakes and ponds to catch fish.

In addition, rats can have significant economic impacts. One can annually eat 10 bushels of grain or 40 pounds of food. This results in the destruction of 20 percent of world's crops. In locations such as India, they compete with humans for food. In the United States, rats are estimated to cause about $1 billion in losses per year. This includes fires caused by rodents chewing on electrical wires. In fact, about 25 percent of the fires in rural areas are caused by rodents (Canby, 1977).

Control. The control of rats is complex because it is so closely tied to human behavior and to large-scale social and economic factors. Effective control programs are possible, however; they consist of three basic elements.

Eliminating food sources. Rats cannot live and reproduce without food. Making food unavailable to them, however, requires strict control of garbage and refuse, which in turn requires comprehensive public education. Garbage should be stored in metal cans with tight-fitting lids and collected twice a week; otherwise, storage containers will be filled, and residents will switch to plastic bags or cardboard boxes, which rats can easily tear open.

Rat-proofing. Basic to the long-term control of rodents is that buildings must be designed and constructed not to have any openings large enough for rats to enter or leave. In addition, all existing buildings should be surveyed to confirm that they are similarly rat-proof. Where deficiencies are found, all openings should be sealed (a young rat can squeeze through a 1/2-inch opening), and a concrete floor or underground shields installed outside the building walls to prevent rats from burrowing underneath. Buildings that cannot be rat-proofed should be demolished.

Traps, fumigants, or poisons. Once all buildings have been rat-proofed and food has been made unavailable, the rat population can be reduced by trapping, fumigation, or poisoning programs both inside buildings and in adjacent outdoor areas. Traps eliminate the need for poisons, but the rats that are caught must be collected and buried or incinerated. In the process, any fleas that survive may transfer to people. Fumigants consist of gases, such as calcium cyanide and methyl bromide. They provide a quick kill but require care to prevent exposures of humans during use. Poisons are generally placed in food (baits). Examples include warfarin, a slow-acting anticoagulant rodenticide; red squill, a bitter-tasting red powder that causes heart paralysis; zinc phosphide, a fast-acting black powder with a garlic odor that reacts with acid in the rat's stomach to produce phosphine gas; and norbormide, a fast-acting poison that causes shock and impairment of blood circulation.

Effective use of poisons is difficult. Rats have developed efficient feeding strategies that enable the members of a colony to avoid the baits and to adjust to sudden changes in the food supply. Both wild rats and those grown in the laboratory tend to avoid any contact with novel objects in their environment. Typically, they avoid a new food for several days, and they may never sample it if their existing diet is nutritionally adequate. Eventually, small sublethal quantities may be ingested. If the feeding animals become sick, the entire colony thereafter avoids the new food. Other approaches being developed or considered for rat control include single-dose chemosterilants that can sterilize both male and female rats and new rodenticides that are rat specific and thus not hazardous to nontarget animals.

Disease transmission. Rats affect the quality of human life in many ways. In addition to bubonic plague, rats can transmit typhus fever through infected fleas, *salmonellosis* through food contaminated by their urine, and rat-bite fever through a spirochete in their blood. In poor housing conditions, infants, paraplegics, and people under the influence of alcohol or drugs, are especially vulnerable to rat bites. On babies, the targets for mutilation are often the nose, ears, lips, fingers, and toes. Unfortunately, their impacts do not end there. For many inner-city residents, the presence of rats is a vivid and gruesome symptom of community environmental degradation, a token of the larger pattern of social and economic breakdown and disorder in the real world of the urban poor. The appalling quality of life in such conditions often becomes clear to others only during urban renewal, when rats from buildings that are being torn down or renovated move into adjacent neighborhoods.

MICE

Mice can be either a direct source of diseases or a host for insects that transmit them.

Characteristics. The house mouse can live in any structure to which it can gain entrance. This includes any that have a hole 1/4-inch in diameter or larger.

Disease transmission. Although the effects of mice are primarily of a nuisance nature, some varieties play a major role, either directly or indirectly, in the transmission of disease. One of the more serious is Lyme disease, for which the white-footed mouse serves as one of the hosts for the ticks that transmit the disease. Overall, almost 250,000 cases of this disease occurred in the United States between 1992 and 2006, and the number each year is increasing (CDC, 2008). In some cases, rats may also serve as a host for the ticks that transmit Lyme disease (Matuschka et al., 1993). Other diseases in which mice play a dominant role are those caused by the hantaviruses, which for many years have caused episodes of pulmonary disease and killed thousands of people in East Asia. An outbreak in 1993 that occurred in the southwestern United States was due to the transmission of the Sin Nombre virus by the deer mouse (*Peromyscus maniculatus*). It can be transmitted to humans either through direct contact with infected rodents, with rodent droppings or nests, or through inhalation of aerosolized virus particles from mouse urine and feces. Another is the Whitewater Arroyo viral disease, which killed several people in California during 1999 and 2000 (Enserink, 2000).

Control. Due to their living habits, the control of mice can be difficult. In the case of the undomesticated varieties that live outdoors essentially all of the time, control is virtually impossible. Those that are domesticated live outside during the summer but tend to move into buildings with the onset of cold weather and heavy rains. During the times they are indoors, however, various forms of baits can be effective. In terms of the Sin Nombre virus, the best approaches are to limit the exposure of humans to rodents and their excreta. This can be accomplished by eliminating available food sources, limiting possible nesting sites, sealing holes and other possible entrances, and placing "snaptraps" to catch and/or kill them. Since brooms and vacuum cleaners can spread the viruses, they should not be used to clean contaminated areas.

BATS

Characteristics. Although bats are commonly viewed as pests, they provide many benefits to humankind. One of the most important is their control of mosquitoes and insects that annoy people and harm agricultural crops. One cave, for example, can hold 100,000 bats, which can consume 100,000 pounds of insects annually. In so doing, they are the predators of insects that annually cost U.S. farmers and foresters billions of dollars in crop losses. In fact, bats are as important as birds in regulating such insects (Williams-Guillem, Perfecto, and Vandermeer, 2008). Since the seeds in their droppings are spread beyond the forest edge, bats help reforest neighboring areas. In a similar manner, they also have helped restore habitat ravaged by logging (Holden, 2008). Finally, bats are major pollinators of plantains and avocados, and they are the sole pollinators of the agave plant (Brown, 2009).

Control. It would be easy to control bats by poisoning them in their caves. Nonetheless, as described above, their beneficial effects (most especially in the United States) far outweigh their negative impacts. For that reason, as well as because of recent declines in bat populations due to the scourge known as "white nose syndrome," which leads to a secondary infection that weakens their immune system (Zimmerman, 2009), controls are not being applied.

Nature and Control of Insects

Insects are highly specialized. Houseflies, for example, have hundreds of eyes mounted in such a way as to provide them with wide-range vision coupled with unusual visual powers. Some insects can detect sex attractants more than 15 miles away. One of the unique characteristics of certain insects is their ability to protect themselves from cold weather. Those with dark colors survive by absorbing sunlight; others gain heat by basking on dark surfaces or have heavy layers of hair or scales that retard heat loss. Some survive subfreezing temperatures by lowering the freezing point of their body fluids by producing compounds that function in a manner similar to the antifreeze used in automobile engine cooling systems (Conniff, 1977).

Insects and related pests infect multitudes of people with diverse agents of disease, causing the deaths of about 2.7 million globally each year, predominantly children in sub-Saharan Africa. The economic impacts in that region of the world are vast, amounting to half a billion dollars annually (Satchell, 2000). It is estimated that malaria, transmitted by *Anopheles* mosquitoes, causes as many as several billion cases of illness and 1.3 to 3.0 million deaths annually, mostly in Africa (Breman and Mills, 2004; Breman and Holloway, 2007). In addition, an estimated 120 million people are infected due to lymphatic filariasis, a parasitic disease caused by microscopic worms transmitted by mosquitoes. It is endemic in tropical regions of Asia, Africa, Central and South America, and the Pacific Island nations, putting 1 billion people at risk globally, and it is thought to have affected humankind for about 4,000 years (Molyneux, Hotez, and Fenwich, 2005).

Insects also have an enormous economic impact on agricultural production. They attack all stages of plant life, eating seeds, seedlings, roots, stems, leaves, flowers, and fruit; after the harvest, they eat the stored product. Flies and other insects can reduce the yield of milk from dairy cows and eggs from chickens and can cause cattle to lose weight. In many parts of the world, the persistent biting of mosquitoes, black flies, and other bloodsucking insects seriously impairs the productive capacity of workers, sometimes even bringing their activities to a standstill. Details on the more important diseases they transmit are provided in the sections that follow.

MOSQUITOES

Essentially every person has heard the buzzing of mosquitoes and suffered their bites. Their characteristics, especially through interactions with humans, often provide the key to their control (Spielman and D'Antonio, 2001).

Nature and Characteristics. There are more than 2,500 species of mosquitoes (Shaw, 2001). Though seemingly frail, they show remarkable abilities in flight: those that fly during the day navigate by polarized light from the sun; those that fly at night navigate by the stars. Their wings move faster than those of a hummingbird—an estimated 250–600 strokes per second (Conniff, 1977)—and produce the familiar whine that is their mating call.

Only female mosquitoes bite people. Some bite only in daylight; others bite only at dusk or at night. They subsequently land on water and lay their eggs in a raft-shaped mass smaller than a grain of rice. The larvae hatch 2 days later and swim and feed in the water, breathing through a tube at the surface. After 12 days, they give rise to the pupa stage. Two days later, the young mosquito emerges (Figure 10.2). Fortunately, less than 5 percent of the eggs become mature adults, and each fall the initial frost kills most of the adult mosquitoes. Nonetheless, mosquitoes can thrive almost anywhere, from the heat of the arid wastes to the frigid Arctic (Shaw, 2001).

Disease Transmission. As noted earlier, mosquitoes are most well known for transmitting malaria. Its vector is the *Anopheles* mosquito (*Anopheles quadrimaculatus*), which breeds in swamps. Another prominent mosquito, the yellow fever mosquito (*Aedes aegypti*), breeds predominantly in artificial containers (cans, bottles, old tires) and is the urban vector of yellow fever and hemorrhagic dengue fever (HDF). The latter, which is the most serious form of dengue fever, is caused by one of four different viruses. It is endemic in more than 100 countries in Africa, the Americas, the Eastern Mediterranean, Southeast Asia, and the West Pacific. Although prior to 1970 only nine countries had reported epidemics of DHF, by 1995 the number had increased four-fold, and by 2009, cases were being reported in 42 countries of the Americas. This increase can be directly attributed to the fact that funding for a major regional effort to eradicate *Aedes aegypti* mosquitoes led by the Pan American Health Organization during the 1950s and 1960s was

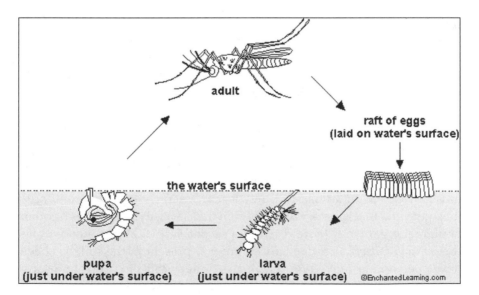

Figure 10.2   The life cycle of the mosquito

reduced in most of the South American countries and discontinued in the United States in 1970. As a result, by 2002 *Aedes aegypti* mosquitoes had reemerged, and cases of DHF were, once again, reported in Nicaragua in 1994 and in Costa Rica in 1995. DHF also reappeared in Venezuela, Colombia, Brazil, French Guiana, Suriname, and Puerto Rico. By 2003, 24 countries in this region had reported confirmed DHF cases. Although prior to 1970 only nine countries had reported epidemics of the hemorrhagic form, by 1995 the number had increased four-fold, and, in 2008, 850,000 cases were reported, including more than 38,000 of HDF, with an accompanying 584 deaths (Roses, 2009). Dengue fever is now endemic in most of these countries (CDC, 2009).

Control. The control of mosquitoes and mosquito-transmitted infections involves three basic steps: (1) reducing the mosquito population by eliminating their breeding habitats by draining land areas in the case of *Anopheles quadrimaculatus,* (2) applying insecticides or other agents to kill the adult mosquitoes or their larvae, and (3) preventing mosquitoes from biting people and providing medical treatment to individuals who have been, or are subject to being, infected.

As with rodents, the effective control of mosquitoes requires extensive knowledge of their characteristics, including their life cycles, breeding habits, and their role as vectors of disease. Even with this knowledge, mosquito control remains complex due to the large number of species involved and their widely different breeding places, biting habits, and flight ranges. Shoreline towns may be troubled by salt-marsh mosquitoes, while freshwater mosquitoes harass inland towns. Control is also complicated by the rapid ability of the mosquito to develop new behavioral patterns in response to insecticide control programs, such as the shift from indoor to outdoor blood feeding and resting locations.

Whereas earlier programs to kill mosquitoes provided temporary relief at best, the discovery and exploitation of *Bacillus thuringiensis israeliensis* (BTI), a natural enemy of mosquito larvae, appears to be changing this situation dramatically. This bacterium was discovered in the gut of dead mosquito larvae in an oasis in the Negev Desert. It appears to kill only mosquito and black fly larvae; it has shown no toxicity to humans or other nontarget organisms. More recently, a second larvicide that incorporates the *Bacillus sphaericus* bacterium, commonly present in the soil, has been developed and approved for use in the United States to control mosquitoes such as the *Culex,* which breed in municipal wastewater lagoons and storm-water basins.

Chemical insecticides continue to be widely used to control adult mosquitoes. In the past, the most commonly used insecticide was DDT [1,1,1-trichloro-2,2-bis(p-chlorophenyl)ethane]. Unfortunately, as was the case with other organochlorine compounds such as aldrin and dieldrin (ATSDR, 2002a), DDT proved to be persistent. This observation, combined with the fact that it is bioaccumulated within the environment, led the EPA to ban its use in the United States in 1972. A similar ban was placed in effect in Europe. These actions were widely challenged (Satchell, 2000). Later, the use of certain other organochlorine compounds (such as aldrin and dieldrin) was also prohibited. As a result, malathion, a far less persistent insecticide, is now the most commonly used insecticide for the control of mosquitoes in this country. While the use of DDT remains controversial, it offers many advantages in the battle against malaria. Two are that it is cheap and its effectiveness is intertwined with the behavioral characteristics of the mosquito that transmits malaria. *Anopheles* mosquitoes typically bite people indoors at night. In so doing, they may triple or quadruple their weight, making flight difficult. For this reason, those that have fed immediately fly to a nearby wall to rest and to excrete the excess fluid from their bodies. If the wall has previously been sprayed with DDT, it will be absorbed into their waxy body coating, they will be killed, and the transmission cycle will be interrupted. Since infectious mosquitoes are those that are being killed, the net effect is far more beneficial than reducing their absolute numbers (Shaw, 2001).

Following the bans on DDT, cases of malaria immediately began to increase on a global basis. Concurrently, the Agency for Toxic Substances and Disease Registry (ATSDR) concluded, on the basis of an exhaustive review of the literature, that there was no evidence that exposure to DDT at concentrations present in the environment causes birth defects or other developmental effects in people (ATSDR, 2002b). In a similar manner, the World Health Organization (WHO) has repeatedly stated its opposition to banning its use, particularly in view of the large number of deaths caused by malaria in the developing countries. Nonetheless, the bans in the United States and Europe continue.

Other controls include the elimination of breeding zones by digging drainage canals, restricting construction and other practices that create stagnant water, changing the salinity of existing waters, and raising and lowering the water level in lakes, such as those created by dams, to disrupt the life cycle of the mosquito. On a personal basis, people can stay indoors except on breezy or hot afternoons, install screens on doors, windows, and porches, wear protective clothing, and apply mosquito repellents to the skin. The most effective repellents are those that contain diethyltoluamide (DEET). To avoid harm to children, only products containing less than 35 percent DEET should be used.

More recently, scientists at the University of Oxford have discovered a method for modifying adult male mosquitoes genetically so that the female offspring of the female mosquitoes with which they mate are not able to fly. Since, as noted earlier, only female mosquitoes bite people, they are not able to fly and mate. Should it be possible to do this on a mass scale, it could prove to be a method that will help block the transmission of malaria and other mosquito-related diseases (Associated Press, 2010).

FLIES AND RELATED PESTS

Various kinds of flies can also serve as major transmitters of disease agents. One example is river blindness (*onchocerciasis*), a parasitic disease caused by the bite of small black flies that breed in rivers. Although this disease occurs primarily in Africa, it has also threatened hundreds of thousands of people in Mexico, Guatemala, Venezuela, Colombia, Ecuador, and Brazil. Globally, as of 2002, there were an estimated 17–18 million victims of river blindness, of whom at least 300,000 had lost their vision. Because people, due to the fear of being bitten, are no longer willing to cultivate fertile bottom-land near rivers, this disease has had detrimental impacts on agricultural production. Fortunately, as a result of international cooperation, including vigorous support from the Carter Center (2009), national public health leaders, and donations from a prominent U.S. drug company, this situation is rapidly changing. In 1987, the drug company agreed to donate the drug *invermectin* free of charge to countries where *onchocerciasis* was endemic. Since this drug kills the larvae but not the adult worms of *Onchocerca volvulus*, the parasite that causes the disease, continuing annual or biannual treatments are required. Today, annual treatments are being provided for all eligible communities. In 2008, these efforts provided protection for 60 million people in 26 African countries in which more than 99 percent of the cases occur. Today, only a few infections remain within the human population (Diawara et al., 2009), and additional infections and transmissions have ceased (Medical News TODAY, 2009).

Principal Species. Three members of the fly family will be discussed. One is the housefly (*Musca domestica*), which is present in many of the temperate zones and may be a carrier of the agents for several diseases. The others are the screwworm fly (*Cochliomyia hominivorax*), which can have a devastating impact on livestock, and the Mediterranean fruitfly (*Ceretitis capitata*), which can have a similar impact on citrus and other fruit.

The housefly. Gray and about a quarter-inch long, the housefly breeds in a variety of decaying animal and vegetable matter, and its larval stage is the maggot. In rural areas, horse, pig, cow, or chicken manure frequently serves as a breeding habitat; human excreta can also be involved where proper disposal methods are not observed. The larval stage lasts 4–8 days; the pupa stage, 3–6 days (Figure 10.3). In warm weather, the average time from the laying of eggs to the emergence of the adult is 10–16 days. Flies live 2–8 weeks in midsummer; in cooler weather, up to 10 weeks. Although flies have been reported to travel several miles in 1 day, most flies present in a given area probably originated nearby.

Although their role in transmitting disease is difficult to document, houseflies pick up and transmit a wide range of pathogens (including viruses, bacteria, protozoa, and the eggs and cysts of worms), both externally (on their mouth parts, body and leg hairs, and the sticky pads of their feet) and internally (in their intestinal tract). As a rule, the pathogens present in the larvae are not transmitted to the adult fly, and most pathogens do not multiply in them. Fortunately, the germs on the surface of a fly often survive only a few hours, especially if exposed to the sun. In contrast, pathogens can live in their intestinal tract and be transmitted to humans when the fly vomits or defecates (Kristiansen and Skovmand, 1985). In order to eat, the housefly regurgitates a fluid that dissolves its food. Part of this effluent

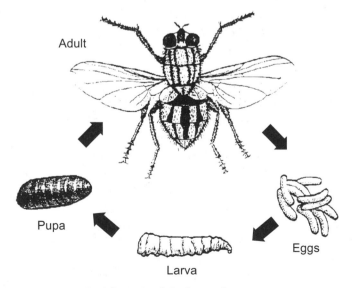

Figure 10.3   The life cycle of the housefly

may remain behind on the food when the fly departs and may contain pathogenic organisms. For these reasons, flies may play a role in the transmission of diarrhea, dysentery, conjunctivitis, typhoid, cholera, and larvae infestations. Of these, bacillic dysentery (*shigellosis*) is probably the most important.

The screwworm fly. The adult has a metallic blue body and three vertical black stripes on its back between its wings. In addition to being about twice the size of the housefly, it lays its eggs in fresh wounds of warm-blooded animals. Any accidental or surgical wound, or the navel of a newborn animal, can serve as the site for initial invasion by screwworm maggots hatching from the eggs laid by the screwworm fly. In warm areas populated by screwworm flies, few newborn calves, lambs, kids, pigs, or the young of larger game species escape attack. The maggots hatch in 12–24 hours and begin feeding on the flesh head down, soon invading the sound tissue. They become full grown in about 5 days, drop out of the wound, burrow into the ground, and change to the pupa (resting) stage. During warm weather, the adult flies emerge from the pupa case after about 8 days, live for 2 to 3 weeks, and range for many miles. The larvae that feed in the wound cause a straw-colored and often bloody discharge that attracts more flies, resulting in multiple infestations by hundreds of thousands of maggots of all sizes. Death of the infected animal is inevitable unless it is found at an early stage and treated. Obviously, screwworm flies can have devastating effects on livestock growers. Losses from infestations along the Atlantic seaboard in 1958 were estimated at $20 million (Richardson, Ellison, and Averhoff, 1982). Losses from a major outbreak would be many times that amount.

During the past half century, a novel method has been developed for the control of the screwworm fly. It is based on the fact that the female mates only once even if the mating does not produce viable offspring (Galvin, 1996). Capitalizing on this characteristic, the International Atomic Energy Agency (IAEA) initiated a program in the late 1950s through which hundreds of thousands of males were sterilized at a facility in Vienna, Austria, by exposing them to ionizing radiation. Their release proved to be dramatically effective. In fact, by 1960 the screwworm had been eliminated in the United States, as well as in parts of Mexico, Central America, and Libya. This technique was subsequently successfully applied throughout the western hemisphere, as well as the melon fly in Japan, as a backup in the control of the tsetse fly on the Tanzanian island of Zanzibar, and in several of the sub-Saharan African countries (IAEA, 1999).

The Mediterranean fruitfly. Also known as the medfly, this fly is slightly smaller than the common housefly, has yellowish orange spots on its wings, and thrives in warm climates. Scientists believe that it originated in West Africa. By 1850, it had spread throughout the Mediterranean region. It appeared in Australia in the late 1800s and in Brazil and Hawaii in the early 1900s. In 1929, it was discovered in Florida.

A medfly typically lays her eggs in a ripe, preferably acidic fruit by drilling tiny holes in the skin or rind while the fruit is still on the tree. Choice targets are oranges, grapefruit, peaches, nectarines, plums, apples, and quinces. In 2–20 days the eggs hatch into larvae, which eat their way through the fruit, causing it to drop to the ground. The larvae later burrow into the ground, where they pupate. Adult flies emerge after some 10–50 days and repeat the cycle.

Although quarantines of fruit and other measures have brought the medfly under control, infestations recurred in Florida and Texas between 1930 and 1979, and again in the 1980s. The medfly also appeared in California in 1975, 1980, 1987, and 1990. Because the export of fruit is prohibited from any areas where the medfly has been detected, the economic impact is tremendous and could approach a billion dollars each year in crop damage in California alone.

Control. Several approaches can be used to control flies, such as the medfly, which do not possess the vulnerability of the screwworm fly. The specific technique depends on the habits of the species in question. Although installing screens in buildings helps reduce contact between houseflies and people, it does nothing to reduce the fly population. Attaining that objective requires other approaches, one of the most important of which is a careful and effectively applied sanitation program. Keeping garbage and excreta covered and disposing of them promptly and properly will eliminate major breeding grounds. Timely disposal of garbage, especially decaying fruit, and prompt removal and disposal of infested fruit that has fallen from trees have proved effective in controlling the medfly in Israel and Italy.

Chemical insecticides are widely used for killing flies inside buildings. The two basic approaches are to wipe or spray the insecticides on indoor surfaces or to hang tapes impregnated with insecticides from indoor ceilings. In all cases, however, care must be taken to avoid contaminating foodstuffs. Outdoor control includes the application of larvicides to breeding areas and the use of bait stations and sprays. As was the case with mosquitoes, the principal insecticides initially used for the control of flies were the organochlorine compounds, such as DDT, dieldrin, and chlordane. As a result of the previously described controversies surrounding the use of DDT, there has been a similar shift in these efforts to the use of less persistent organophosphorus compounds (such as malathion), the carbamates (such as Sevin), and the pyrethroids (such as permethrin).

In some cases, unique approaches have been developed in the use of insecticides for the control of specific fly species that transmit certain diseases. In the case of the tsetse fly, one method that has proved effective is to impregnate cow-sized rectangular sheets of cloth with synthetic pyrethroids, an insecticide, plus a mixture of chemicals, such as acetone and octonel, that are exhaled by cattle as they breathe, plus phenols that are present in cattle urine. In essence, the sheets of cloth are designed to represent a "fake" or "artificial" cow. Once the flies land on the impregnated cloth and come into contact with the insecticide, they are killed. More than 60,000 such cloth cows have now been deployed on ranches in Zimbabwe, and the technique is also being applied successfully in parts of Zambia, South Africa, and the Ivory Coast. This has led to the virtual elimination of human deaths, and nagana infections in cattle were reduced from some 10,000 in 1984 to about 50 per year today. An additional benefit is that this approach is much less expensive than the procedures employed in the past (Lecrubier, 2002).

FIRE ANTS

Although fire ants are one of the major pests in the United States today, they are a nonnative species. They were reportedly introduced from South America in 1918 in soil used as ballast on ships that docked in Mobile, Alabama. They subsequently migrated northward and today are present in most all of the southern states in the eastern United States. Once entrenched, they dominate and regularly constitute up to 99 percent of the total ant population. In the course of their activities, they damage electrical equipment, air conditioners, and farms, as well as domestic lawns and gardens. They also bite and sting people, releasing a venom comparable to that of a bee. In addition, they have significant ecological ramifications. In the Galapagos Islands, they eat the hatchlings of tortoises and have also attacked the eyes and cloacae of the adult reptiles. In the Solomon Islands, they have reportedly occupied the areas where incubator birds lay their eggs, and their stings have reportedly blinded dogs. In Gabon in West Africa, they have reportedly had the same effects on cats and elephants (Hayashi, 1999).

Control. Although their presence can be limited, to some extent, by treating the individual mounds with over-the-counter insecticides, reinfestation rapidly occurs since individual mounds are often interconnected via underground tunnels. One promising approach is the introduction of the pinhead-sized phorid fly (also called the humpback fly), which kills fire ants through its reproductive process. This fly lays its eggs in a fire ant, the emerging larvae makes its way into the ant's head and then proceeds to eat out the inside of the head. Eventually, the head falls off and a new fly emerges. As the new generation of fire ants emerges, the cycle resumes. Since the fire ant appears to be the only host that provides this service, evaluations indicate that the introduction of the phorid fly should not have any effect on any other part of the environment. Based on this favorable information, phorid flies were first introduced into Alabama in 1998 as part of a regional project designed to control fire-ant populations throughout the South. Surveys have shown that phorid flies have migrated and produced well beyond their points of release. Plans also call for releasing a different kind of phorid fly, known as *Pseudacteon curvatus*, which appears to be more cold tolerant and better suited to the hybrid fire ants common in this region of the state. Although it is recognized that phorid flies will never succeed in eradicating fire ants, the application of this approach, combined with other methods, should enable specialists to restrict their populations to manageable levels (Graham, 2009).

TERMITES

Although best known for their destructive impacts on wooden homes and similar structures, experts estimate that termites have been on Earth for more than 180 million years. Also less recognized is the fact that they have similar effects on ecosystems. Nonetheless, they do provide benefits. For example, they recycle wood and other plant materials, and their tunneling efforts aerate the soil, which, in turn, improves soil composition and fertility and helps the ground to absorb water.

Control. Scientists have recently developed a bait that termites carry back to their nests that destroy the colonies. In addition, scientists are seeking to exploit a primary social characteristic of termites, and related insects, namely, the origin of the division between reproduction and work among individual members of a colony. Nowhere is this separation more pronounced than in these types of insects, where workers rarely produce offspring even though they are often capable of doing so. Through a combination of behavioral and RNA interference, the gene required for the reproductive division of labor between the queen and the workers has been identified. If plans materialize, it may be possible that the termite social organization can be influenced. This, in turn, could lead to the application of a novel strategy for insect control through the use of chemical genetic factors that cause anarchy within their societies (Korb et al., 2009).

TICKS

Ticks are a good example of noninsect pests that can be important vectors of disease. For many years, the primary disease of concern relative to this vector was Rocky Mountain spotted fever, but the recent upsurge in this disease and, most especially, Lyme disease, has caused renewed interest in these pests (Marques, 2008).

Characteristics and Disease Transmission. Ticks are leathery-bodied, eight-legged arthropods with mouthparts that enable them to penetrate and hold fast in the skin and withdraw blood from animals. The female mates while attached to a host and usually feeds for 8–12 days. Since as young nymphs they are only as large as a poppy seed, they often are not detected. For these and other reasons, ticks play a major role in the transmission of Lyme disease in the eastern United States. The first cases were reported in Connecticut in 1975; it is now known to be present in more than 40 states. In fact, Lyme disease is the most common arthropod-borne disease in this country, with almost 18,000 cases having been reported in 2000. This represents more than a doubling of the number of annually reported cases since 1990. A contributing factor to this increase was the shift of major farming activities from eastern to western regions of the United States, the accompanying abandonment of many of the farms in the Northeast, and the subsequent regrowth of trees that provide an ideal habitat for both deer and their ticks. Most cases now occur in the northeastern, mid-Atlantic, and north-central regions of this country, and the highest numbers occur during June and July, reflecting the May and June peak months of the host-seeking activities of the infective ticks (Matuschka et al., 1997).

Control. As with mosquitoes and malaria, the control of ticks and Lyme disease can be complex. On a long-term basis, there is a need to avoid changes that increase outdoor environments that facilitate transmission of this disease. On a shorter-term basis, the disease can be limited by reducing tick populations through area control and vegetation management with insecticides, avoiding tick-infested areas, applying chemical repellents (for example, DEET), and promptly removing any ticks that become attached (CDC, 2002). Prompt removal is essential because the Lyme disease spirochete, *Borrelia burgdorferi*, is not likely to infect the patient before the vector ticks begin to engorge (Matuschka and Spielman, 1993). Also important is the control of ticks on pets and in buildings.

As is the case with other such pests, effective control of ticks requires a fundamental understanding of their biology. Recognizing that one of the hosts for the ticks that transmit this disease is the white-footed mouse, scientists developed a system for distributing cotton balls impregnated with permethrin in areas foraged by these animals. When mice collect the cotton to use as a liner for their nests, the ticks they carry are killed (Spielman and Kimsey, 1997). A list of insects and non-insect pests, and their public

Table 10.3 Public health impacts of various insect and non-insect pests

Vector	Impact
Flies	Bacillic dysentery, typhoid, cholera, yaws, trachoma, annoyance
Mosquitoes	Malaria, yellow fever, dengue, encephalitis, West Nile virus, filariasis, bites, annoyance
Lice	Epidemic typhus, louse-borne relapsing fever, trench fever, bites, annoyance
Fleas	Plague, endemic typhus, bites, annoyance
Mites	Scabies, rickettsial pox, scrub typhus, bites, allergic reactions, annoyance
Ticks	Ehrlichosis, Lyme disease, tick paralysis, tick-borne relapsing fever, Rocky Mountain spotted fever, tularemia, bites, annoyance
Bedbugs, kissing bugs	Chagas disease, bites, annoyance

Table 10.4 Global impacts of tropical disease infections

Disease	Insect Vector	Number of countries affected	Number of people infected (millions)[a]	Total population at risk (millions)[a]
Chagas disease	Triatomine (kissing bugs)	21	16–18	90
Leishmaniasis	Sandflies	80	12	350
Lymphatic filariasis	Mosquitoes	76	120	900
Malaria	Mosquitoes	>100	300–500[b]	>2,400
Onchocerciasis (river blindness)	Black flies	37	17–18	100

a. Numbers are approximate.
b. Number of cases occurring annually.

health impacts, is summarized in Table 10.3. A summary of the global impacts of tropical infections is presented in Table 10.4. As in all cases of insect control, attention is increasingly being directed to the implementation of an integrated approach (Molyneux, Holtz, and Fenwick, 2005).

The General Outlook

Globally, malaria annually continues to kill 1 million people, mostly children under 5 years of age, especially in Africa. Although mosquitoes serve as the vector, humans serve as the reservoir. Reducing the incidence of malaria anywhere helps protect people everywhere. An equally important benefit of such efforts is that, in multitudes of cases, the people who live in the developing countries are poor because they are sick. They simply do not have the energy to work and earn income. By reducing their burden of disease, not only would the developed countries be reducing disease and suffering but they would also reap a multiple return on their investment through increased trade and economic gains. The WHO (2009) has emphasized this point on multiple occasions.

Also to be recognized is that, in most instances, the increase in infectious diseases can be directly tied to one or more environmental changes that have facilitated contact between the vectors of disease and their human hosts. The origins of such changes include the unprecedented rate of growth in the global population and the rapid rate of urbanization. Lyme disease is an excellent example of the latter. These and other environmental trends demonstrate the dynamic relationships that link an ever-changing landscape, the vectors that exploit these instabilities, and the pathogens that may thereby affect human health (Telford, Pollack, and Spielman, 1991). Similarly, the continued proliferation of rodents is almost totally a result of urbanization, the deterioration of many of our inner cities, and the lack of proper garbage and refuse disposal. Some scientists predict even more dramatic changes if the predicted climate changes materialize (Chapter 18).

If these challenges are to be addressed in an effective manner, it will require the concerted efforts of all nations. That such efforts can be successful is illustrated by the outcome of the program coordinated by the WHO that led to the conquering of river blindness in Africa. Another is an experiment in Europe, in which scientists are seeking to use mosquitoes as flying "needles" to deliver a "vaccine" of live mosquito parasites through their bites (Campbell, 2009). Still another is the success of the controls that led to a 99.7 percent reduction in guinea worm disease in Africa (Chapter 7). Still another is the $10 billion global effort initiated by the Bill & Melinda Gates Foundation to ensure that existing vaccines for the prevention of a host of diseases ranging from chicken pox, hepatitis A and B, influenza, measles, mumps, polio, pneumonia, smallpox, tetanus, and whooping cough reach those who need them and to support research to develop improved versions of these vaccines as well as those for preventing other diseases (Bill & Melinda Gates Foundation, 2010).

11

INJURY PREVENTION
AND CONTROL

FOR PURPOSES of evaluation, injuries are generally classified as unintentional (e.g., vehicular accidents) or intentional (i.e., acts of violence). According to the National Safety Council (NSC, 2009), the largest sources of both disabling injuries and deaths are motor vehicle accidents. These caused an estimated 2.3 million disabling injuries and 43,100 deaths in the United States in 2007. Although comparable data on injuries are not available, the latest records indicate that the estimated number of deaths in 2009 had been reduced to 33,808, compared to 37,423 in 1950 (Associated Press, 2010). The second-largest source was poisoning, which caused an estimated 22,500 deaths in 2007. These were primarily among 47-year olds, followed by those aged 40–41 and 43–50. The third leading number of deaths (21,500) occurred as a result of falls, primarily among persons aged 73 and older. These three sources alone caused about three-quarters of all unintentional deaths. The distribution of the sources of all disabling injuries and deaths during 2007 is summarized in Figures 11.1 and 11.2, respectively. Related details are summarized in Table 11.1 (NSC, 2009).

In 2006, 42.4 million visits to U.S. hospital emergency departments were injury related. This represented an annual rate of 14.4 such visits per 100 persons. An additional 6 million were treated in hospital outpatient departments, and tens of millions more were treated in the offices of physicians. About 3 million of these patients were hospitalized (NSC, 2009). Globally, it is estimated that unintentional injuries annually account for about one-third of all hospitalizations and about 3 million deaths. In many countries,

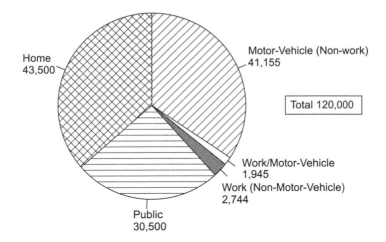

Figure 11.1 Relative sources of deaths, United States, 2007

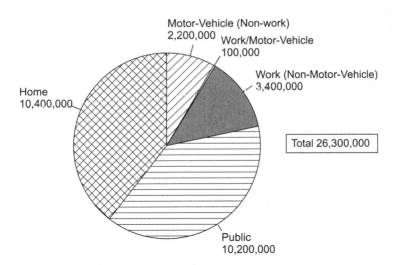

Figure 11.2 Relative sources of disabling injuries, United States, 2007

the problems of injury and injury control have assumed an importance equal to that of infectious diseases (Chapter 10). Concurrently, more than 50,000 deaths occur in the United States each year through intentional injuries, two-thirds from suicides, and one-third from acts of violence.

Table 11.1 Unintentional deaths and disabling injuries by class, United States, 2007

Classes	Deaths	Deaths per 100,000 people	Change (%) from 2006	Disabling injuries[a]
(1) Motor-vehicle (total)	43,100	14.3	−0.3	2,300,000
Public-non-work	40,955			2,200,000
Work	1,945			100,000
Home	200			<10,000
(2) Work (total)	4,689	1.6	−8	3,500,000
Non-motor-vehicle	2,744			3,400,000
Motor-vehicle	1,945			100,000
(3) Home (total)	43,500	14.4	−11	10,400,000
Non-motor-vehicle	43,300			10,400,000
Motor-vehicle	200			<10,000
(4) Public (total)	30,500	10.1	26	10,200,000
Total for all classes[b]	120,000	39.8	−0.4	26,300,000

a. Defined as disabling beyond the day of injury. Since these are not reported nationally, the estimates are an approximation.

b. Deaths and injuries for the four classes add to more than the total for all classes due to rounding and the fact some deaths and injuries are reported in more than one class.

Addressing the Problem

According to Julian Waller (1994), injury control as a public health endeavor began in Germany in 1780, when Johann Peter Frank urged that injury and its prevention be addressed not only by individuals but also by nationwide public health programs. In the mid-1900s, several state and local health departments in the United States initiated modest data-collection efforts and child safety, burn prevention, and other programs. Unfortunately, the effects of these programs on behavior, morbidity, or mortality were never fully evaluated.

The first major breakthrough occurred in the early 1940s when Hugh De Haven, an engineer at Cornell University, published a paper that initiated a conceptual revolution in injury control (De Haven, 1942). He showed how people had successfully survived falls of 15–45 meters (50–150 feet), in some cases with only minor injuries, through the dispersion of kinetic energy in amounts as high as 200 times the force of gravity. Through this process, he demonstrated that it is possible to disconnect the linkage between accidents and the resultant injuries. His studies, in turn, led to the development and introduction of automobile seat belts by Nils Bohlin, a Volvo engineer, in 1945.

A decade and a half later, J. J. Gibson (1961) observed that injury events have only five agents, namely, the five forms of physical energy: kinetic or mechanical energy, chemical energy, thermal energy, electricity, and radiation. William Haddon expanded this concept through the initiation of a movement to incorporate sounder scientific and public health concepts into the development of accident and injury prevention programs. Instead of relying primarily on attempts to change human behavior, he applied an environmental approach. In so doing, he followed the lead of De Haven by concentrating on the prevention of injuries, not accidents. In so doing, he developed a generic approach to the analysis, management, and control of such injuries, which he treated as fundamentally a result of the rapid and uncontrolled transfer of energy (Haddon, 1970). Experience has shown that his approach can be applied to all types of occupational and environmental hazards, ranging from automobile accidents to oil spills to major accidents in nuclear power plants. In fact, the methods recommended by the International Commission on Radiological Protection (Chapter 17) for responding to nuclear power plant accidents are based on Haddon's approach (ICRP, 1984). The versatility of his approach is such that it can also be applied to the control of acts of violence.

To facilitate an analytic approach, Haddon divided accidents into three *phases:* the pre-event phase (the factors that determine whether an accident occurs), the event itself, and the post-event phase (everything that determines the consequences of the injuries received). The *factors* operating in all three phases are (1) the humans involved, (2) the equipment they are using or with which they come into contact, and (3) the environment in which the equipment is operated. Combining these three phases and the three factors yields a nine-cell matrix (Figure 11.3) that public health workers can use to determine where best to apply strategies to prevent or control injuries. Addressing these problems from a different perspective, Ralph Nader, a lawyer, published a paper in which he pointed out that although the U.S. auto industry placed a premium on style, chrome, and speed, they did not emphasize safety. This was followed by his book *Unsafe at Any Speed*, in which he noted that although ancient Roman chariots had padded dash panels, U.S. automobiles lacked not only padded dashes but also seat belts, collapsible steering columns, and side protection (Nader, 1965). Because vehicular accidents account for more than one-third of the deaths resulting from unintentional disabling injuries in the United States, they are used as examples in the discussion that follows.

Phases	Factors		
	Human	Equipment	Physical and socioeconomic environment
Pre-event	(1)	(2)	(3)
Event	(4)	(5)	(6)
Post-event	(7)	(8)	(9)

Figure 11.3   Matrix for the analysis of accidents

PRE-EVENT PHASE

The goal in this phase is to reduce the likelihood of a vehicular collision. Factors that should be considered include:

Humans involved. Driver impairment by alcohol or other drugs; the thoroughness of testing procedures for licensure; the degree of enforcement of traffic rules and regulations, including mandatory use of seat belts; and the availability of mass transportation as an alternative to the use of private vehicles.

Equipment. The condition of headlights, tire treads, traction control, and brakes (and whether they include antilock features); the size and visibility of brake lights; the speed the vehicle can attain; and the results of vehicular crash tests.

Environment. The presence of barriers and traffic lights to protect pedestrians; the design, placement, and maintenance of road signs for ready comprehension; and the design of bridge abutments to prevent or reduce impact damage.

EVENT PHASE

The goal in this phase is to reduce the severity of the "second collision," for example, when the driver is thrown against the steering column or windshield. Factors that can reduce the extent of injuries include the following:

Humans involved. Proper use of seat belts and child-resistant systems (average U.S. driver seat-belt use was only 83 percent in 2008; NSC, 2009), and driver abstention from alcohol (which affects cell membrane permeability, so that even in low-impact collisions people who have consumed alcohol are more likely to sustain severe or even fatal neurological damage). Unfortunately, laws requiring the mandatory use of helmets by motorcycle riders have been repealed by some states (Homer and French, 2008).

Equipment. Whether the vehicle was equipped with an airbag, collapsible steering column, high-penetration-resistant windshield, interior padding (i.e., on the dashboard), recessed door handles and control knobs, and structural beams in doors; low bumpers with square fronts to reduce the likelihood of pelvic and leg fractures to pedestrians who are hit; and a bar under the rear end of large trucks to prevent cars from "submarining" beneath them

Environment. Breakaway sign posts, open space along the sides of the road, wide multiple lanes, guard rails to steer vehicles back onto the road, and road surfaces that permit rapid stopping

POST-EVENT PHASE
The goal in this phase is to reduce the disabilities due to the injuries. Factors that can reduce or limit the effects of injuries include the following:

Humans involved. Rapid and appropriate emergency medical care, followed by adequate rehabilitation; properly trained rescue personnel; and injury severity scores to help medical personnel evaluate multiple traumas and predict outcomes

Equipment. Fireproof gasoline tanks to prevent fires after an accident

Environment. The "jaws of life" to extract victims from damaged vehicles; ambulances and/or helicopters for rapid transport of the victims to medical-care facilities; trauma centers equipped to handle injured victims; ramps and other environmental changes to reduce the real "cost" to the victims of being disabled; and rehabilitation of victims.

Vehicular Accidents

MAGNITUDE AND PERSPECTIVE

Motor vehicle crashes kill more children and young adults in the United States between the ages of 1 and 24 years than any other single source. In fact, such events are the leading cause of death from unintentional injuries for persons of all ages. Their significance is further demonstrated by the fact that they are the source of almost 36 percent of a total of 120,000 deaths involving unintentional injuries in this country each year (NSC, 2009). Nonetheless, progress is being made. For example, the number of deaths in vehicular accidents has been reduced on essentially a continuing basis from about 26.5 per 100,000 population in 1970 to 13.5 in 2007, a reduction of almost 50 percent. Concurrently, the number of deaths per 100 million vehicle miles has been reduced from 5.0 to 1.5, a reduction of 70 percent. Had these improvements not been achieved, the number of people dying each year in vehicular accidents in the United States today would have approached 120,000. Assuming that the current rates of associated injuries apply, this means that the number of disabling injuries from this source would have exceeded 6.4 million (Moeller, 2003). Even so, such accidents continue to account for more than 35 percent of all disabling injuries in this country, and the associated annual economic costs exceed $260 billion (NSC, 2009). Nonetheless, during the first half of 2009, the total number of deaths due to vehicle accidents was 16,626. This was primarily the result of the increased price of gasoline (Associated Press, 2010). If this trend continues, the total number of deaths will be 23 percent less than it was in 2007.

In the meantime, pedestrian deaths showed little change from 2003 through 2007, remaining at about 6,000 per year, In contrast, those due to pedal cycling (particularly bicycling) have increased from slightly more than 800 deaths in 2003 to 1,000 in 2007. The risk of serious head injury in such events can be reduced by as much as 40 percent if the cyclist wears a protective helmet. In fact, if all cyclists wore helmets, perhaps 500 lives could be saved and 135,000 head injuries prevented in this country each year (NSC, 2009).

IMPLEMENTING CONTROLS

The development of a program for preventing or reducing injuries suffered in vehicular accidents has political, social, behavioral, and economic aspects. It therefore requires a multifaceted approach involving new technical approaches as well as policies and strategies. One approach that has proved successful is to base the changes on increasing the safety consciousness on the part of the public. Several of these advances are discussed below.

Air bags. Studies show that a combination of lap/shoulder belts (Dimeo-Ediger, 2008) and air bags offers the best available protection for motor vehicle occupants. This combination, however, is not a cure-all. In fact, the added fatality-reducing effectiveness of the air bags is estimated to be about 11 percent over and above the benefits of using safety belts alone. Nonetheless, they are saving an estimated 1,300 U.S. lives each year. To increase the protection, a number of manufacturers now install air bags in the doors and outer walls of cars to protect against side impacts. Unfortunately, air bags have been shown to have potentially negative effects, especially for children and persons of short stature who may be sitting in the front seat and are too close to either the steering wheel or the dashboard at the time of the deployment of an air bag. To resolve this problem, the U.S. Department of Transportation issued a rule that permits vehicle owners who meet certain qualifying criteria to have air bag on/off switches installed in their vehicles.

Improved head restraints. These devices can be extremely effective in preventing whiplash injuries in rear-end collisions. In order to do so, however, the restraint must be directly behind and close to the heads of the occupants. Unfortunately, studies by the Insurance Institute for Highway Safety (IIHS, 1999) showed that a third of those installed in new cars did not meet these basic requirements. Until these problems are corrected, the full benefits of these devices will not be realized.

Antilock brakes. Early evaluations on the test track indicated that such a system would provide many benefits in emergency braking situations, especially on road surfaces that are wet and slippery. While this has been confirmed in the case of large trucks, especially tractor-trailer units, such systems have not reduced the frequency of automobile accidents. Although there may be several explanations, one is that emergencies on the road often involve complicated scenarios that differ significantly from test situations. In addition, it may be that individuals driving cars with antilock brakes place too much confidence in the system and take more risks.

Daytime running lights. Daytime running lights (reduced intensity headlamps) have proved to be particularly beneficial in reducing automobile accidents in urban areas, where traffic congestion is heavy and demands on driver attention are numerous. Although they have been approved for use in the United States, not all manufacturers have chosen to install them. This is in contrast to Sweden, Norway, Finland, and Canada, where they are mandatory. Such lights reduce collision damage by several percent. Due to the invention of light-emitting-diode (LED) head and tail lights, the accompanying energy consumption has also been significantly reduced.

Truck trailer visibility. During 2007, 4,808 fatalities occurred from crashes involving a large truck, and 75 percent of the deaths were occupants of vehicles other than the truck. An estimated 101,000 people were injured. Sixty-two percent of such events occurred in rural areas, 78 percent on weekdays, 82 percent involved multiple vehicles, and 62 percent occurred at night (NSC, 2009). In this regard, studies have shown that nighttime collisions of other vehicles with the trailers of large trucks can be reduced by increasing their visibility. One low-cost approach (about $100 per unit) is to add reflective material to their sides and rear. Recognizing this fact, the National Highway Traffic Safety Administration (NHTSA) requires that all new truck trailers in the United States be equipped in this manner. Estimates are that this reduces accidents involving such units by about 15 percent.

Speeding. Exceeding the posted speed limit or driving at an unsafe speed is the most common error contributing to fatal accidents. Results of surveys on urban interstates in 2007 revealed that the average speed of passenger vehicles exceeded the limits in every case. Right-of-way violations predominated in the injury accidents, as well as in all other accident categories. It is estimated that, during 2007, speeding was a factor in 32 percent of all traffic fatalities, causing 11,659 fatal crashes that resulted in an average of 36 deaths per day (NSC, 2009). Many drivers, including operators of commercial trucks, use radar detectors to alert them to the presence of police speed-monitoring units. Because the devices have only one purpose—to alert speeding drivers to slow down—they have been banned in many states. In fact, they have been banned nationwide by the Federal Highway Administration on all commercial vehicles, primarily trucks used in interstate commerce.

POLICY AND ETHICAL ISSUES

Multiple policies and strategies have been developed and applied in recent years to improve vehicular safety. Several examples are discussed below.

Alcohol. Drunk driving continues to be a major contributor to vehicular accidents and deaths. In spite of major efforts by groups such as Mothers Against Drunk Driving (MADD), it continues annually to cause almost 13,000 U.S. deaths, about 32 percent of all vehicular deaths. In addition, the annual cost of alcohol-related crashes is estimated to exceed $114 billion, and the cost per alcohol-related fatality is $3.5 million (MADD, 2009). Random stops of drivers at locations nationwide, particularly during the early hours on Saturday mornings, show that this is a time when a relatively large percentage of drivers are drunk (Editorial staff, 2009). At the same time, alcoholic beverage companies continue to promote the consumption of beer as representing "the good life," particularly during television sports programs that feature healthy, athletic young men and are watched by hosts of teenagers.

To combat this situation, MADD is supporting legislation that would make it mandatory that ignition interlocks be required on the cars of convicted drunk drivers in all 50 states (MADD, 2008). Prior to starting a vehicle, the driver would be required to breathe into the interlock, which is about the size of a cell phone. If his/her alcohol level is in excess of a predetermined level, the car will not respond. It is also possible to incorporate what are called "running retests" that require the driver to breathe into the device periodically. These devices are also designed to prevent a driver from asking a sober friend to start the car. If a driver does so and then fails a running retest, the vehicle's horn blows and the lights turn on, alerting law enforcement officials of the situation. Studies have shown that interlocks are, on average, 64 percent effective in reducing repeat drunk-driving offenses.

Legislative restrictions. Multiple approaches are being used to reduce vehicular accidents. Several of these are discussed below.

Drunk driving—Data show that (1) drivers who are convicted of drunk driving have generally done so at least 87 times prior to the first conviction, (2) two-thirds of drunk drivers continue to drive even when their license is suspended, and (3) one-third of all drivers arrested for drunk driving are repeat offenders. Based on this evidence, legislators in several states have proposed a stiffer set of drunk-driving penalties: (1) following a first offense, the license of a driver would be automatically cancelled for a year; (2) following a second offense, his/her license would be cancelled for 10 years; (3) the license of chronic drunk drivers would be suspended for life; and (4) the license of a drunk driver who causes the death of a person in another vehicle would be charged with murder (CBS, 2009). Led by New Hampshire, Massachusetts, and New York, this last step has now been adopted in several other states. Another recommendation that is gaining increasing support is to require that the installation of ignition interlocks be made mandatory in the cars of all drivers who have been convicted of drunk driving (MADD, 2008).

Traffic violators—One of the most common sources of accidents on urban and suburban streets is the motorist who "runs" a red light. To increase the efficiency of detecting such violations, many city authorities have authorized the installation of cameras that photograph the license plate of the offending motorist, with the ticket being mailed to him/her. Concurrently, paints have become available that will prevent a license being photographed. In essence, efforts to improve the efficiency in reducing one type of offense have created a new one.

Speed limits—Although the common assumption is that higher limits will pose an increased danger to vehicle occupants, the latest studies of the association between speed limits and crash fatalities among passenger vehicles have yielded mixed results. The same is true for studies of fatal crashes in the case of trucks. In fact, the imposition of lower speed limits on truck drivers has the potential of two streams of traffic flowing at different rates. This can, unto itself, be a source of accidents. This has been confirmed by data documenting that a higher difference in speed is associated with a significant increase in fatalities. Even so, models indicate that if all states had adopted a speed limit of 55 mph for all vehicles, around 560 fatalities would have been averted in 2005. The same analyses also implied that if the speed limit had been increased to 75 mph, an estimated 360 additional deaths would have occurred. Nonetheless, the savings in time represent an important factor to many people. Another factor of interest was that a higher permissible truck length would yield significantly higher fatality rates, while increased weight limits were not significant. To the contrary, the outcome implied that higher weight limits might be beneficial. At the same time, there are multiple other factors that need to be considered. One example is that truck drivers are subject to on-site safety reviews, required rest periods, and limits on the time traveled per day. Compliance is confirmed by checkpoints on interstate highways. Nonetheless, although large trucks account for less than 5 percent of the registered vehicles in the United States and only 8 percent of the total miles driven, they are disproportionately involved in passenger vehicle–occupant deaths compared to other types of vehicles (Neeley and Richardson, 2009).

Cell phones, tweeters, and other distractions—Studies show that 6 percent of drivers now use a handheld cell phone, a 5 percent increase compared to a similar survey in 2006. Compounding the problem is that such use was higher among drivers aged 16 to 24 years, and increasing numbers of them were using other types of handheld devices (NSC, 2009). In response, some state regulators are considering limiting the use of cell phones to passengers. A driver who needs to make or accept a call would be required to stop at the first opportunity prior to responding. Cell phone usage accounts for some of the most flagrant bad driving on our roads today (Davis, 2009). In fact, one study showed that cell phone usage, combined with adjusting the radio and other distractions caused a 28 percent increase in fatalities between 2005 and 2010. Other distractions include conversations with passengers, eating, smoking, manipulating controls, and in-vehicle route guidance systems. The extent of disruption increases in proportion to the degree of mental demand required by the task. During such times, critical hazards, such as a pedestrian on the side of the road, can go unnoticed. In fact, those who are distracted might not even be aware they have just missed a near crash. On average, it is estimated that drivers spend about 30 percent of their time in distracting tasks (NSC, 2009). Even after surviving a crash involving a fatality, drivers often fail to adjust their behavior (Lee, 2009).

Daylight saving time—Because it adds an hour of sunlight to the afternoon commuting time and increases the visibility of both vehicles and pedestrians, the adoption of daylight saving time is a proven method of reducing vehicular accidents. Although this step also eliminates an hour of sunlight in the morning, since more pedestrians and vehicles are on the road in the afternoon, the increase in accidents in the morning is not sufficient to outweigh the lives saved in the afternoon.

Vehicle size and body style—In response to the mandatory requirements for increases in corporate average fuel economy (Chapter 16), many manufacturers are reducing the size of their vehicles as one of the primary approaches to comply (Walsh, 2009). Unfortunately, accidents involving smaller vehicles can be a source of increased injuries and deaths. In fact, among the 11 U.S. vehicles with driver death rates at least twice as high as the average, 10 are small and 1 is midsize (IIHS, 1994). When essentially everyone is driving a smaller car, this problem should be significantly reduced.

OTHER CONSIDERATIONS

Teenage drivers. Studies by the Insurance Institute for Highway Safety show that motor vehicle accidents were responsible for more than half of all deaths (i.e., motor vehicles, homicide, suicide, and cancer) among teenagers in 2005. The maximum numbers of deaths were as a result of motor vehicle accidents that involved those in the 17, 18, and 19-year-old group, and that it steadily increased with age. This is due to several factors, the most significant of which are failure to use seat belts, underage alcohol consumption, and use of cell phones and other handheld devices. Since reaching a maximum of 5,540 in 2004, the total number of deaths in this age group has been reduced at a continuing rate to 4,460 in 2007. Nonetheless, this still represented 14 percent of all such deaths (Wilson, Stimpson, and Hilsenrath, 2009).

Motorcyclists. Although fatalities due to automobile accidents are being reduced, those due to motorcycle accidents are rapidly increasing and have become the leading cause of deaths among young adults in the United States. Between 2000 and 2007, the number of registered motorcycles increased from 3.8 million to 6.7 million, a gain of 75 percent. Strongly influencing this increase was the increase in gasoline prices—motorcycles average more than 56 miles per gallon. Concurrently, the number of associated fatalities increased from 2,116 to 5,154, a factor of more than 2.4. There are multiple other reasons for these increases. Analyses show that 40 percent of motorcycle riders involved in crashes have had no formal training. They are either self-taught or rely on family or friends for guidance (Wilson, Stimpson, and Hilsenrath, 2009). In addition, 27 percent of those involved in fatal crashes were intoxicated. This compares to 23 percent for motor vehicles. Although helmets have been documented as an effective method for reducing deaths, their use declined from an average of 71 percent in 2000 to 63 percent in 2008 (NSC, 2009). Also of note is that so-called street motorcycles have engines with sufficient power to provide acceleration equivalent to or in excess of that of the more expensive sports cars. It is no surprise, therefore, that two-thirds of motorcycle crashes are caused by errors by the rider. Another factor is climate. Motorcycles are far more difficult to manage than four-wheeled vehicles during rain and snow (Wilson, Stimpson, and Hilsenrath, 2009). As a result, the death rate per passenger vehicle mile for motorcycles in 2006 was 35 times higher than that for passenger cars and light trucks (NSC, 2009).

Pedestrians. During 2007, there were an estimated 5,000 pedestrian deaths and 85,000 pedestrian injuries in motor-vehicle–related accidents in the United States. About 50 percent of these occurred when pedestrians crossed or entered streets, and about 8 percent occurred when people were walking along the roadway. In 38 percent of the accidents that involved the death of a pedestrian, the driver, the pedestrian, or both were intoxicated (NSC, 2009). As might be anticipated, pedestrian deaths are a special problem among the elderly (Figure 11.4). As may be noted, the death rate for females 65 years of age or older is almost three times higher than that for those 35 to 64 years of age; the comparable difference for males is about 1.5 times higher. Even so, it is equally important to recognize that during 1995 about 15 percent of all motor-vehicle pedestrian-related deaths involved children. The highest fatality rates occurred on October 31, Halloween, between 4

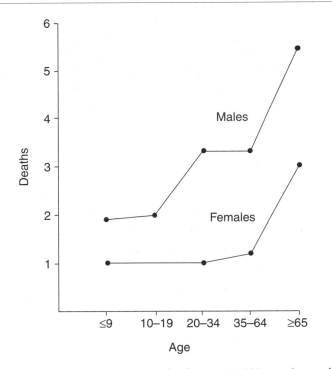

Figure 11.4 Male and female pedestrian deaths per 100,000 people as a function of age

and 10 P.M. and involved children aged 5 to 14 years (CDC,1997). Another group of "pedestrians" who require special consideration are highway and street construction workers, about 1,000 of whom were killed in the United States during the 1990s. Most of these deaths involved vehicles and moving equipment, and more than a third involved workers on foot who were struck by a vehicle. One of the most important measures for reducing such deaths is to require that such highway and street workers wear high-visibility apparel. Another step is to use fluorescent and retro-reflective material on head gear and on the gloves of people who wave flags to give directions (NIOSH, 2001, 2003). Steps that can be taken to protect pedestrians, in general, include providing separate pathways for walkers, placing sidewalks well back from the road, restricting on-street parking, and requiring that the exterior of motor vehicles have no sharp edges or protrusions (CDC, 1997, 2001).

Collisions with animals. In addition to increasing the spread of various insect and animal related diseases (Chapter 10), the encroachment of society upon wildlife habitats (or vice versa) has led to a dramatic increase in the number of accidents caused by collisions between motor vehicles and wild animals, particularly deer. Nationwide, such collisions annually cause the deaths of 100 to 150 people. States with high deer populations, such as Michigan, North Carolina, Pennsylvania, and Wisconsin, each annually experience from 40,000 to 50,000 deer–vehicle collisions. Nationwide, the annual cost of the associated damages exceeds hundreds of millions of dollars. One countermeasure that is being applied is a "deer whistle" that is mounted on a vehicle and is activated by the onrushing air. Although the resulting ultrasonic sound is supposed to repel deer, the effectiveness of such devices has not been confirmed. Another approach is to install specially designed roadside reflectors to try to prevent animals from crossing in front of vehicles. Presumably, the reflector, illuminated by the headlights of the oncoming vehicle, will frighten the deer and cause them to stop.

Trends in motor-vehicular deaths, as related to miles traveled and deaths per 100 million VMT (vehicle miles traveled), are summarized in Figure 11.5.

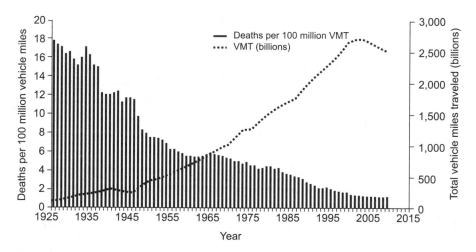

Figure 11.5 Trends in total vehicular miles traveled (VMT) and death rates per hundred million miles per VMT, United States, 1925–2009

New Developments and Technologies

In the early 1990s, Congress appropriated about $650 million to be spent in stimulating the development of what would become known as intelligent vehicle highway systems (IVHS). Often referred to as "smart cars" or "smart highways," these encompassed a range of high-technology approaches specifically designed to reduce vehicular injuries and deaths. Products developed through this program include the previously discussed antilock braking systems, electronic message boards that alert drivers to upcoming road conditions, sensors in the rear bumpers of cars that sound an alarm when the driver is backing up and about to collide with an obstacle, related alternate or backup small TV-like screens that show the driver what is behind the car, and on-board navigation systems that enable drivers to determine exactly where they are and to select the best route to reach their destination.

Another product is a pulse-transmitting system that can determine the location, size, and distance of objects in front of a car and transmit that information back to sensors that will alert the driver. In the event of an impending collision, the system will apply the brakes. Studies are also underway to determine if computers can teach us how to drive (Wilson, 2009).

Also being explored is the design of improved safety belts. These include four-point X-shape devices based on the full harnesses worn by race-car drivers. Since such harnesses will move the mounting points for the belt outward, their use should eliminate the chafing inherent in the designs being used today. Another device, now being installed in U.S. cars, is a sensor that informs drivers when one or more of the tires are under-inflated. Another system that has been developed in Germany will activate certain safety features just prior to a crash. Through detectors designed to sense the probability of a crash, the system will not only activate the air bags, move the steering wheel forward, and pretension the safety belt, but it will also extend the front bumper to help dissipate the energy of the crash (Phillips, 2001). Another important factor to consider is the increasing price of gasoline (Wilson, Stimpson, and Hilsenrath, 2009). As this leads to the sale of more small vehicles, the fatality rates could significantly increase.

Sports and Recreational Injuries

Of the 33,256 medically consulted injuries in the United States during 2006, nearly 40 percent were related to leisure or sports activities (Table 11.2). In

Table 11.2   Primary sources of sports-related injuries treated in emergency
hospital departments in the United States, 2007 [a]

Sport	Number of injuries	Percent of participants
Bicycle riding	505,413	13.5
Basketball	481,011	2.0
Football	455,193[b]	2.6
Soccer	198,679	1.4
Exercise	194,431	—[c]
Swimming	171,704	0.3[d]
Baseball	167,661	1.2
Skateboarding	143,682	1.4
Softball	110,086	1.1
Horseback riding	78,527	—[c]
Weight lifting	72,369	0.2
Fishing	71,615	0.2
Roller skating	67,734	—[c]
Snowboarding	57,792	1.1
Volleyball	57,039	0.5
Cheerleading	26,786	0.7

a. Includes data ranging from 2004–2007.
b. Includes both touch and tackle football.
c. Data not available.
d. Includes injuries associated with swimming, swimming pools, diving and diving boards, and other swimming pool equipment.

fact, these accounted for 70 percent of all injuries among children under the age of 12, and for about 75 percent of similar injuries among teenagers between 13 and 17 year old. During 2007, bicycle riding accounted for more than 500,000 emergency department visits, primarily among children in the 5–14 age range; basketball for more than 480,000, primarily in the 15–24 age range; and football for more than 450,000, primarily in the 5–14 age range (NSC, 2009). Preventive measures include ensuring that the equipment is appropriate for the age group using it and that it is maintained in a safe condition (CDC, 1999a).

Concurrently, the nature of the sources of injuries to children is constantly changing. A good example is the small foot-propelled scooter that was introduced into the U.S. market in the late 1990s. These devices, which had small, low-friction wheels similar to those on in-line skates, proved to be extremely popular and led to a dramatic increase in the number of scooter-related injuries. Exacerbating the problem was the introduction in the spring of 2000 of a new aluminum version that weighed less than 10 pounds and could be folded for easy portability and storage. Shortly there-after, the number of children being treated for scooter-related injuries had increased to more than 40,000. About 85 percent involved children less than 15 years old, and almost 25 percent were less than 8 years old (CDC, 2000). Many of the injuries could have been prevented or reduced in severity if protective equipment (helmets, elbow pads, and knee pads) had been worn by the users (NSC, 2001).

When all age groups are considered, one of the major recreational sources of death in the United States is drowning, which accounted for some 4,700 fatalities during 2007. Although 750 of these occurred among children less than 4 years old, by far the largest number were among people older than 75 (NSC, 2009). Most of these occurred at home and involved either swim-ming pools or bathtubs. Second were those that occurred during recre-ational boating, with alcohol consumption reported to be a contributing factor in about 25 percent of the cases. About 70 percent of these could have been avoided if the victims had been wearing life jackets.

Intentional Injuries (Violence)

Although the public health community was slow to acknowledge the im-portance of unintentional injuries, it has been even slower in recognizing acts of violence (suicides, homicides, and assaults) as sources of intentional injuries. The modern age of international terrorism (Chapter 17) has tremen-dously expanded both the nature of such acts and the number of people who can be affected by a single event. Nonetheless, the discussion here will concentrate on suicides, homicides, and assaults. In this regard, certain facts are apparent. One is that the personal environment plays a prominent role in determining the extent and nature of violence in a community. Neigh-borhoods that inspire people to befriend one another, that are protective of local children, and that share resources appear to provide the kind of sup-port that fosters healthy development. These characteristics may explain why some poor urban neighborhoods escape the violence that takes an enor-mous toll only a few blocks away. In contrast, neighborhoods that are so-cially and politically disorganized create conditions that contribute to anti-social behavior. When such behavior begins to dominate, there is an exodus of the small businesses that typically provide the glue that holds a neighbor-hood together. The way is then paved for illegal economies, such as drug dealing and gambling. This is one of the reasons that the brownfields pro-grams for restoring contaminated sites in abandoned urban areas (Chapter 9) are receiving such widespread support. Another relevant factor is the situation within the home, particularly the relationship between husband and wife or other heads of households. Studies repeatedly demonstrate that more violence is caused by family and former friends than by strangers. In fact, 2 percent of the women murdered in the United States in 2002 were killed by a husband, ex-husband, lover, or suitor.

Two examples of violence-related events are discussed below. The first emphasizes that acts of violence are not restricted to the community and the home; the second emphasizes the increasing recognition that too little effort is being directed to identifying the causes and developing programs for the prevention of such events.

WORKPLACE HOMICIDE

During 2005, almost 14,000 people died as a result of homicide in the United States. A sizeable portion of these were killed while at work (NSC, 2009). Workplaces with the highest number of deaths are grocery stores, eating and drinking places, taxicab services, and justice or public-order establishments. Occupations with the highest rates are taxicab drivers/chauffeurs, law enforcement officers, gas station or garage workers, and security guards. Factors that increase the risk for homicide among workers include the exchange of money with the public, working alone or in small numbers, and working late at night or in the early morning hours (NSC, 2001). Control measures include the installation of physical barriers, such as bullet-resistant enclosures with pass-through windows on critical services counters, alarm systems and panic buttons, video surveillance with closed-circuit television, bright and effective lighting, and training of employees in the identification of hazardous situations and appropriate responses (Mandelblit, 2001).

FIREARMS (GUNS)

The unintentional discharge of firearms led to 3,159 U.S. deaths in 2005 (NSC, 2009). Firearms also serve as a major source of violence, accounting for the deaths of almost 65 percent of the people who are murdered and almost 60 percent of those who commit suicide. If all types of firearm-related deaths in this country are considered, the total number of people killed each year equals about 70 percent of those killed in motor-vehicle accidents. Although this may seem to be surprising, in reality it is not. The estimated number of guns owned by civilians is almost one per person, and guns are present in about 25 percent of U.S. households (Miller, 2002).

The role of guns in violence-related deaths is illustrated in other ways, one being a nationwide study of suicide deaths among children in a group of five states with the highest rates of household gun ownership (so-called high-gun states), versus those in five states with the lowest rates of such ownership (so-called low-gun states). Although the number of non-gun suicides in the two groups of states were similar, seven times as many children killed themselves with guns in the five "high-gun" states than in the five "low-gun" states. Likewise, the comparable number of children murdered with guns was three times higher. It is little wonder, then, that gun-related deaths of children aged 5 to 14 in the U.S. now ranks as the third-leading cause of mortality, being exceeded only by motor-vehicle accidents and cancer. The overwhelming contribution of the prevalence of firearms as a factor in gun-related suicides, homicides, and other types of violence among this age group is vividly illustrated by the data presented in Figure 11.6 (Miller, Azrael, and Hemenway, 2002).

Studies show that deaths from suicides account for about 30,000, and those from homicides for about 18,000 annually in the United States. The combined number of deaths exceed those due to motor-vehicle accidents. Although

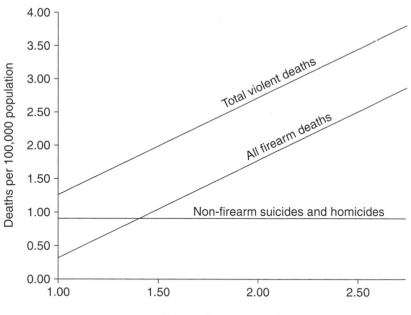

Figure 11.6   Rates of firearms prevalence and violent deaths among 5–14 year olds, United States, 1988–1997

federal data systems exist to inform policymaking about motor-vehicle–related deaths, only the most rudimentary data are available on deaths due to violence. Public debates on policies to reduce violence and suicide rely largely on anecdote and rhetoric in the absence of objective data (Harvard School of Public Health, 2009).

OTHER INJURIES INVOLVING CHILDREN

Another source of unintentional deaths and injuries among children is the family farm. In all, there are more than 2 million children who reside on U.S. farms. These farms range from those operated by a single family and part-time retirement farms to large production facilities. Nonetheless, the family farm remains the dominant entity, especially in the southeastern part of the country. Included are nearly 1 million children ages 14 and under who are exposed to a range of potential sources of unintentional injury-related death. These include exposures to tractors, chemicals, machinery, pesticides, and large animals. Exacerbating the situation is that it may be difficult for parents, who are performing farming duties, to provide adequate supervision. In addition, children may perform work-related tasks inappropriate for their ages (NSKC, 2004).

Data show that about 70 children under 15 died due to injuries that occurred on farms in 2001. Almost 40 percent of these deaths were due to machinery, and 23 percent were due to drowning. More than 40 percent of all farm machinery–related deaths and 45 percent of all farm-related drownings are among children under 5. In 2001, 22,600 children were injured while visiting, working on, or living on farms. Although the total number of injuries nationwide has been reduced significantly in recent years, the rate of childhood farm-related injuries has been reduced only slightly (CDC, 1996). The overall injury rate was 10.9 per 1,000 youths. The comparable rate for non-work-related injuries was 10.3 per 1,000. Leading causes were machinery, motor vehicles, electrocution, environmental hazards, and falling objects (NSC, 2002). Also of note (Chapter 4), agricultural firms with fewer than 11 workers are normally exempt from inspections. Only if unsafe conditions are found and reported (usually after an accident with fatalities or major injuries to workers) will they be inspected and, if conditions warrant, be fined.

FALLS

As noted earlier, falls are the third leading cause of unintentional deaths in the United States. Not including those that are work or transportation related, falls caused 20,600 U.S. deaths in 2007. These represented 2 percent of all unintentional injury deaths that occurred in the home and community, and they occurred primarily among persons 73 years of age or older (NSC, 2008). More than half occurred in the home or on the premises, and 30 percent involved leisure activities. In the case of children less than 12 years of age, more than two-thirds of these types of injuries occurred at home. As with many other sources of injuries, increased effort needs to be made to identify and evaluate the origins and causes of these events, particularly among the elderly. Wheelchair users represent a specific group in need of attention. A recent study showed that almost 40 percent of such users had suffered at least one fall during the previous year, with about half being injured. In many cases, a contributing factor was that the home in which the person resided had not been modified to accommodate the use of a wheelchair (Berg, Hines, and Allen, 2002).

FIRES

Fires in U.S. residences accounted for an estimated 383,000 (almost three-quarters) of all structure-related fires during the 1990s. Almost three-quarters occurred in one- and two-family dwellings. Death rates were highest from December through February, reflecting the seasonal use of heating devices (e.g., portable space heaters and wood-burning stoves). The most vulnerable age groups were children under 5 and adults over 65, their death rates being two to six times the average for all ages (CDC, 1998). In all, these events accounted for an estimated 3,600 deaths and about 18,600 injuries in the United States during 1995. About 150 of these were caused by children playing with cigarette lighters. Arson is also an important factor, serving as the leading cause of fire-related injury (10 percent) and economic loss (18 percent; NSC, 1990).

As with most environmental and public health problems, the control of deaths and injuries from fires requires a systems approach. Increased fire-fighting capabilities, stricter enforcement of building and housing codes, and intensified pursuit of arsonists are all helpful, but these approaches alone will not control the problem. These efforts must be supplemented by the installation and continued maintenance of smoke detectors and sprinkler systems in buildings, and increased attention to the design, installation, operation, and maintenance of heating systems. Also important is that sleeping garments, especially those worn by children, be fire resistant and that bedding and upholstered furniture be not only fire resistant but also incapable of releasing toxic gases when exposed to heat and flame. As a result of these and other efforts, annual deaths from fires were reduced from more than 4,000 to 2,700 (more that 30 percent) between 1990 and 2007 (NSC, 2009), even though the U.S. population has increased 20 percent (Chapter 1).

The General Outlook

As noted earlier, there has been a dramatic reduction in the deaths per mile due to motor-vehicle accidents. How this was accomplished is an outstanding example of applying the systems approach. Motor vehicles were modified to include headrests, energy-absorbing steering wheels, shatter-resistant windshields, safety belts, and air bags. Roads were improved by better delineation of curves, adding center line stripes and reflectors, converting to breakaway sign and utility poles, illuminating many key sections of roadways at night, installing barriers to separate oncoming traffic lanes, and designing guardrails to guide vehicles back onto the road should the driver lose control. Other measures included improved driver licensing and testing and vehicle inspections, better enactment and enforcement of traffic safety laws, and reinforced public education. Also contributing were the previously discussed better enforcement of safety belt and child safety seat requirements, motorcycle helmet laws, and the establishment of the graduated system for licensing young drivers.

If comparable approaches are designed and applied to other sources of injuries, similar progress can be made. At the same time, however, this is not to imply that the problem of motor-vehicle injuries and deaths has been solved. Many challenges remain. While annual motor-vehicle crash-related fatalities involving alcohol were reduced from about 52 percent in 1990 to about 40 percent in 2000, it remains a major contributor to the associated injuries and deaths. Another challenge is the continuing relatively high rate of deaths among young drivers and passengers. Since these deaths, many of which could readily be avoided, occur at such a young age, the years of life lost are not only a tragedy for the families concerned but also in terms of the contributions these people could have made to society had they been able to lead full and productive lives.

These and many other injury-related problems emphasize once again how important it is that members of society recognize their obligations in helping to ensure their own safety as well as that of their families. Parents, for example, need to ensure that doors to cabinets for the storage of household cleansers and other toxic agents are child resistant, that stairs are equipped with handrails and padding, that play yards are fenced, and that, as noted previously, the ground beneath swings, slides, and other playground

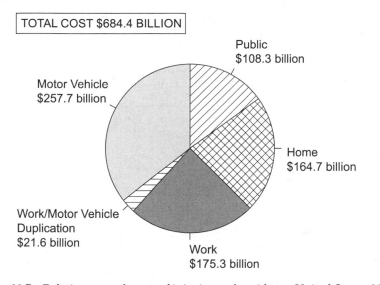

TOTAL COST $684.4 BILLION

Public
$108.3 billion

Motor Vehicle
$257.7 billion

Home
$164.7 billion

Work/Motor Vehicle
Duplication
$21.6 billion

Work
$175.3 billion

Figure 11.7   Relative annual costs of injuries and accidents, United States, 2007

equipment is covered with soft dirt or other similar material. Obviously also of importance is close parental supervision of small children at all times.

At the same time, legislated codes and standards can also contribute to the reduction in the numbers of childhood injuries and deaths. Examples include the requirement that all toxic materials be sold in containers with childproof caps, that hot-water heaters have temperature limits to prevent scalds and burns, that barriers be installed on upstairs windows of residential buildings, that fences be erected around swimming pools, that electrical outlets near the floor be covered and that control knobs on stoves be located out of reach of children, and that paint used on indoor walls, furniture, and equipment for children be lead-free.

Finally, the annual economic costs of injuries and deaths should not be overlooked. The latest estimates show that it is in excess of $684 billion (Figure 11.7; NSC, 2009).

12

ELECTROMAGNETIC RADIATION

R ADIATION MOVES through space at the speed of light in the form of what are called photons (Roentgen, 1895). These, which interact with matter at the atomic level, are classified into two types: *ionizing* and *nonionizing*. Only those in the higher energy range are capable of causing ionization. The energy ranges for the various types of radiation have not been precisely defined in every case; overlaps are common.

- *Nonionizing radiation* does not possess sufficient energy to ionize atoms. This includes radiation in the lower-energy ranges, such as the lower-frequency ultraviolet portion of the spectrum, as well as infrared waves, microwaves, and radio waves (Figure 12.1).

- *Ionizing radiation* is exemplified by those in the higher-energy ranges, such as cosmic, X-rays, and gamma rays, that have sufficient energy to disrupt atoms. They do so by interacting with the orbital electrons and stripping them away. Once an electron is removed, it exhibits a unit negative charge, and the residual atom shows a net unit positive charge. The two products are known as an ion pair (Figure 12.2). The accompanying transfer of energy can result in chemical and biological changes that are harmful to health.

All humans, as well as all matter, are exposed to radiation from several sources: that from natural sources on a continuous basis, and that from sources of human origin on an infrequent basis, or both. It is propagated through space in the form of packets of energy called photons, which travel at the speed of light (3×10^{10} centimeters per second). Each photon has an

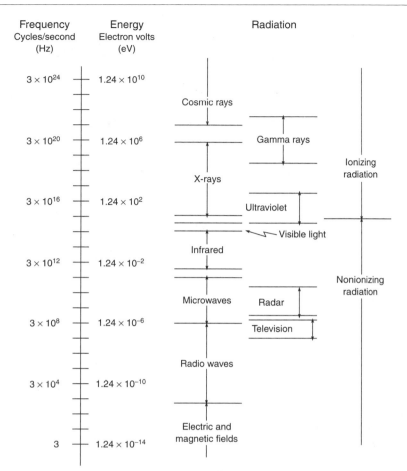

Figure 12.1   The electromagnetic spectrum

associated frequency and wavelength. Its energy is directly proportional to its frequency and is expressed in terms of electron volts (eV—the energy that an electron would acquire in being accelerated across an electrical potential of 1 volt. Higher-energy photons, such as cosmic rays, have frequencies of 10^{21} hertz (Hz, cycles per second) or more, and energies of 10^7 eV or more; lower-energy photons, such as those associated with electric and magnetic fields, have frequencies of 1–10^3 Hz and energies only a tiny fraction of an eV. Photons in the intermediate energy range (10^{-2}–10 eV), such as those associated with infrared and visible light, have frequencies of 10^{12}–10^{15} Hz. High-energy photons are extremely penetrating and can have

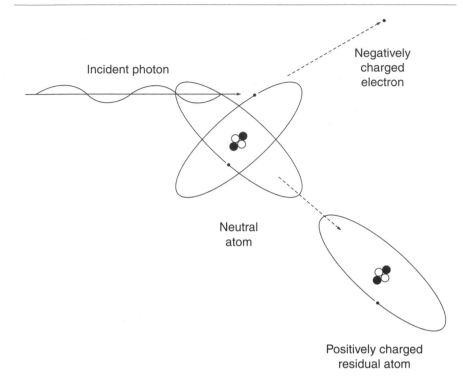

Figure 12.2 Interaction of an X-ray or gamma-ray photon with a neutron atom to produce an ion pair

effects far from their source; the effects of lower-energy photons are concentrated near the source. Only intermediate-range electromagnetic radiation, such as infrared radiation like that of the sun, can be detected by the human senses.

Ionizing Radiation

The first source of human-generated ionizing radiation (X-rays) was discovered by Wilhelm Roentgen (1895) when he noted changes in a photographic plate that was accidentally exposed to radiation from a high-voltage discharge tube. Radioactive materials, the disintegration of unstable nuclei of atoms, were discovered a few months later in 1896 by Henri Becquerel while trying to induce X-ray fluorescence in a uranium phosphor with sunlight. Follow-up studies by Pierre Curie and Maria Sklodowska-Curie led to the discovery of polonium and radium in 1898. Three types of radiation (alpha, beta, and gamma) were subsequently identified. Studies by Ernest Rutherford, in turn, enabled him to describe the atom as composed of a central core—a very small, massive, positively charged nucleus—surrounded by relatively distant, negatively charged orbiting electrons (Lapp and Andrews, 1948). The interaction of alpha particles with beryllium resulted in a nuclear reaction that emitted nucleons with zero charge. This resulted from experiments by James Chadwick in 1932, and the new atomic particle was called a neutron. Seven years later, medium-weight atoms were produced in experiments designed to create new chemical elements by irradiating uranium. Since this was similar to the process by which biological organisms multiply, it was defined as *fission*.

Normally, fission yields two fission products. Since the sum of the weights of the two products is less than that of the original uranium atom, enormous amounts of energy are yielded. The fission products, in turn, are radioactive and continue to release energy (heat) as they decay. This led to the invention of the nuclear reactor by Enrico Fermi in 1942, the construction of massive reactors for the production of plutonium, and the development of the nuclear weapons that brought an end to World War II.

BIOLOGICAL EFFECTS

Biological effects in living organisms exposed to ionizing radiation involve a series of events. The first is the previously discussed ionization. The residual molecule, left with a positive charge, is highly unstable and will rapidly undergo chemical changes. One such change is the production of "free radicals," which are extremely reactive chemically. The ensuing reactions may, in turn, lead to permanent damage of the affected molecule—or the energy may be transferred to another molecule and the free radicals may recombine. The time required for this chain of physical and chemical events to take place is on the order of a microsecond or less. The subsequent development of biochemical and physiological changes, however, may require hours; in the case of latent cancers, it may require years (Little, 1993).

All cells are susceptible to damage by ionizing radiation, and only a very small amount of energy needs to be deposited to produce significant biological changes. For example, if all the deposited energy were converted to heat, a dose of radiation sufficient to be lethal to human beings would raise the temperature of the body by only 0.001°C. Fortunately, the exposure from ionizing radiation can be accurately assessed at levels multiple orders of magnitude below those required to produce measurable biological effects, such as leukemia and cancers in various body organs.

As a result of the long-term epidemiological studies of the survivors of the nuclear weapons detonations in Japan, it has been shown that, at doses of the order of 1 Sv, radiation may lead to heart disease, stroke, digestive disorders, and respiratory disease. Additional evidence comes from cancer patients who have received radiotherapy. Whether there is a threshold for such effects is not certain. If there is, it is probably in the range of 0.5 Sv. The data are not sufficient to estimate the potential detriment at doses less than 100 mSv (ICRP, 2007b). The units of dose due to exposures to ionizing radiation are discussed below.

UNITS OF DOSE

The most common unit is that for the *equivalent dose,* the *sievert* (Sv). Since this amount is far in excess of the doses usually encountered, subunits have been developed, such as the millisievert (mSv), which, as the name indicates, is one-thousandth of the Sv. As commonly applied, the Sv and mSv express the dose to an individual. To express the full range of the impacts of radiation exposure, the health effects are classified as either *deterministic* or *stochastic. Deterministic* effects are those for which the severity of the effect varies with the dose and for which a threshold may therefore exist. *Stochastic* effects are those that will not appear until some years after being exposed at lower levels of radiation. For certain purposes, such as comparing the relative societal impacts of several sources of ionizing radiation, the International Commission on Radiological Protection (ICRP, 1991) has developed what is called the *collective dose.* As noted (Table 12.1), care must be exercised in the application of this concept, particularly in terms of assessing the risk to large population groups who have received very small doses. The range of health effects, as a function of dose, is summarized in Table 12.2.

For decades, there has been an ongoing controversy on whether radiation has health effects, regardless of how low the dose. This has largely been the result of continuing challenges by nuclear critics (Figure 3.3, page 58). Such critics fail to recognize several things (NCRP, 2001; NRC, 1990):

- Radiation is a very weak carcinogen, and there is a delay of 10 to 15 years before any cancers that may develop will appear

- The nature of the deposition of the energy absorbed from many internally deposited radionuclides—the sources of human-made

Table 12.1   Units of dose for ionizing radiation

Unit	Description
Roentgen	The roentgen, the *unit of exposure doses*, now obsolete, was first introduced at the Radiological Congress held in Stockholm in 1928. It was based on the quantity of electrical charge produced in air by ionizing radiation (such as X-rays or gamma rays). One roentgen (R) of exposure will produce about 2 billion ion pairs per cubic centimeter of air. Later it was noted that the exposure to soft tissue or similar material to 1 R resulted in the absorption of about 100 ergs of energy per gram. By multiplying the amount of energy absorbed by what is called a radiation-weighting factor (to account for the nature of the radiation, plus other factors), it is possible to estimate the accompanying biological effects. For X-rays, gamma rays, and beta radiation, the radiation-weighting factor is 1. For alpha radiation, it is 20. Due to its explanation of the fundamentals of dose units, as well as historical interest, it is included here.
Gray	The gray is the unit of *absorbed dose* and is equivalent to the absorption of 10^4 ergs of energy per gram. If soft tissue or similar material is exposed to 100 roentgens, the amount of energy absorbed is equivalent to about 1 gray.
Sievert	The sievert is the unit of *equivalent dose* (often simply called the *dose*). One sievert (Sv) is equal to 1,000 millisievert (mSv). It was designed to provide a means for expressing the biological effects of all types of ionizing radiation on an equivalent basis, and it is commonly used for expressing the dose to all or a portion of the body of an individual.
Person-Sv	After the sievert was developed, a need for a unit for expressing the societal risk associated with doses to large population groups was recognized. This led to the development of the unit of *collective dose*, the product of the number of people exposed and their average dose. Imbedded in the application of this unit is the assumption that the relationship between the dose and the accompanying health effects is linear. Following this approach, an average dose of 1 mSv to 1,000,000 people (yielding a collective dose of 1,000 person-Sv) would, by definition, have the same societal impact as a dose of 1 Sv to 1,000 people. Due to misapplications such as this, it is recommended that this unit not be used for expressing the risks of very small doses to large population groups.

Table 12.2　Summary of radiation induced health effects

Dose	Effects on individuals	Consequences for an exposed population
Very low dose: about 10 mSv (effective dose) or less	No acute effects; extremely small additional cancer risk	No observable increase in the incidence of cancer, even in a large exposed population
Low dose: toward 100 mSv (effective dose) or less	No acute effects; subsequent additional cancer risk of less than 1%	Possible observable increase in the incidence of cancer, if the exposed group is large (perhaps more than 100,000 people
Moderate dose: toward 1,000 mSv (acute whole body dose)	Nausea, vomiting possible, mild bone marrow depression; subsequent additional cancer risk of about 10%	Probable observable increase in the incidence of cancer, if the exposed group exceeds a few hundred people
High dose: above 1,000 mSv (acute whole body dose)	Certain nausea, likely bone marrow syndrome; high risk of death from about 4,000 mSv of acute whole body dose without medical treatment. Significant additional cancer risk	Observable increase in the incidence of cancer

exposures that most members of the public will receive—are often far less harmful than estimated. To cause harm, they must be uniformly deposited within the tissue. In reality, most of them will be deposited in the nature of uneven, less harmful, hot spots (Bair, 1997).

- Difficulties in interpreting data on the health effects of ionizing radiation are enormous. This is vividly illustrated by the results of the epidemiological studies of the survivors of the atomic bombings in Japan at the close of World War II. Even as late as 2000, 55 years after they were exposed, 44 percent of them were still alive. Of those who had died due to cancer and leukemia (~10,000), the number whose deaths were attributed to radiation was about 5 percent (Clarke, 2008). This is discussed in more detail in Chapter 15. Had they been exposed at a lower dose rate the number of deaths would have been considerably less. Even with careful statistical analysis, quantification of the health effects of exposures to ionizing radiation is difficult, if not impossible.

Even so, U.S. organizations such as the NCRP (2001) and National Research Council (NRC, 1972, 1980, 1990, 2006) have spent hundreds of thousands of dollars on such assessments, only to conclude that "the available data do not suffice either to exclude or confirm a LNT (linear nonthreshold) dose-incidence relationship for mutagenic and carcinogenic effects of radiation in the low-dose domain" (NCRP, 2001). This has been eloquently expressed by Roger Clarke (2008), Chairman Emeritus, ICRP, in his statement, *"In this zone [domain] the relationship is irrelevant"* (Figure 3.3, page 58).

RELATIONS BETWEEN DOSE AND HEALTH EFFECTS

The NCRP (2003) has provided details on the possible relationships over the full range of doses (Figure 12.3). In general, mutations induced by radiation in cultured human cells follow a linear model (graph a). This also appears to be true for the induction of solid tumors. In contrast, cell death

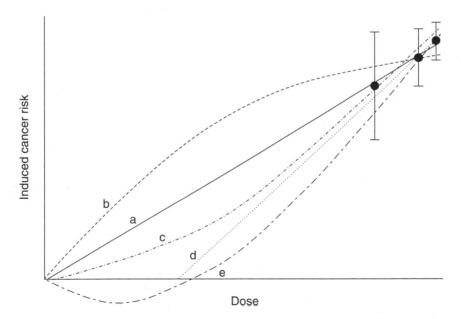

Figure 12.3 Dose-response models for quantifying the effects of ionizing radiation

is related exponentially to dose (graph b). Since the response appears to vary with the nature of the radiation, it has been suggested that for X-ray and gamma-ray photons, the relationship appears to follow linear and quadratic models (graph c). For certain internally deposited alpha-emitting radionuclides, (i.e., ^{226}Ra and ^{239}Pu), long-term epidemiological studies have documented that they have thresholds for intake that are higher than any worker, much less a member of the public, could possibly exceed (Chapter 13). The relationship for these is illustrated in graph d. Finally, the assumption that, in some cases, low doses of radiation might be beneficial (referred to as *hormesis*) is illustrated in graph e. This is also known as the "J-shaped" curve (Figure 2.3, page 36). The existence of such effects, however, has not been confirmed.

Nonionizing Radiation

Although the biological effects of *ionizing* radiation have been recognized and reasonably well understood for some time, questions remain concerning the nature and effects of certain types of *nonionizing* radiation, in particular the photons associated with lower-energy electric and magnetic fields. Consequently, techniques for assessing such radiation, especially in terms of the specific parameters that need to be measured to evaluate their potential impacts, continue to evolve.

ULTRAVIOLET RADIATION

The principal natural source of *ultraviolet radiation* (UVR) is the sun. Since the atmosphere serves as an absorbent, the amount of UVR reaching the Earth at a particular location increases with altitude. Since the stratospheric ozone layer also serves as a protective barrier, its depletion is a matter of considerable interest (Chapter 18). As a result of technological developments, there are now many artificial sources of UVR. These include the electric arcs used in lights, welding torches, plasma jets, germicidal lamps, and tanning lamps.

High short-term (acute) exposures to UVR can produce marked systemic effects, including fever, nausea, and malaise. Cumulative (chronic) effects include aging of the skin and pre-malignant or malignant changes. The effect of most concern is the development of malignant melanoma, the most serious form of skin cancer. Based on reviews by the American Cancer Society (ACS, 2009), it was the source of more than 65,000 deaths in the United States during 2008, and the rate is increasing. Although ordinary window glass will remove most of the higher frequencies, the protection afforded by clothing depends on its composition and color. Whereas cream-colored cloth woven of 100 percent cotton provides a protection factor of less than 10, the same cloth in bright pink or turquoise provides a protection factor of more than 30 (Agnew, Grainger, and Driscoll, 1998).

Because studies show that exposure to tanning lamps, especially by young people, can lead to a doubling of the risk of several types of skin cancer, some health officials have recommended that tanning salons be closed to minors.

VISIBLE LIGHT, LASERS

The health effects of visible light may be direct or indirect. An example of the former is a retinal burn caused by looking at the sun during an eclipse without adequate filtration. Examples of the latter are injuries from an accident that is caused by insufficient lighting (i.e., a fall), or excessive lighting (i.e., the bright headlights of an oncoming car). Other health problems can result from the use of devices in which beams of visible light can be focused both temporally and spatially. These devices, called lasers (for *l*ight *a*mplification by *s*timulated *e*mission of *r*adiation), have found wide application (e.g., in video disc players, supermarket scanners, and facsimile and printing equipment). In most cases, laser units are totally or partially enclosed to prevent exposure to direct or scattered radiation. A common exception, however, is the use of laser pointers that lecturers use to focus the attention of the audience to relevant portions of projected slides. Unless care is exercised, such use can readily lead to harm, the reason being that even minute quantities of laser light can burn a small hole in the retina and permanently impair the vision of any person whose eyes may be subjected to the direct beam (O'Hagan and Hill, 1998). A summary of the mechanisms of interaction and examples of the adverse effects of different parts of the optical spectrum, including visible light and lasers, is presented in Table 12.3.

INFRARED RADIATION

All objects emit infrared radiation to other objects that have a lower surface temperature. Examples are the heat that reaches the Earth from the sun and that produced by a stove or by the radiant heating units used in many dwellings. Fortunately, the sensation of heat quickly provides adequate warning of extreme conditions. Infrared radiation does not penetrate deeply into tissues, but, if not controlled, it can burn the skin surface and produce cataracts in the lens or retina of the eye (which have poor heat-dissipating

Table 12.3   Mechanisms of interaction and examples of adverse effects of exposures to radiation within different portions of the optical spectrum

Part of spectrum	Mechanisms of interaction	Adverse effects
Ultraviolet radiation (180–400 nm)	Photochemical alterations of biologically active molecules, such as DNA, lipids, and proteins	Acute erythema, keratitis, conjunctivitis, cataracts, photoretinitis, accelerated skin aging, skin cancers
Visible radiation (380–600 nm)	Photochemical alterations of biologically active molecules in the retina	Photoretinitis ("bluelight hazard")
Visible and near-infrared radiation (400–1,400 nm)	Thermal activation or inactivation; photocoagulation	Thermal injury, skin burns and retinal burns, thermal denaturation of proteins, tissue coagulation/necrosis
Middle and far-infrared radiation (3 μm–1 mm)	Thermal activation or inactivation; coagulation	Thermal injury, skin and corneal burns, cataracts, thermal denaturation of proteins, tissue coagulation/necrosis
Laser radiation (180 nm–1 mm)	Photochemical, photothermal, photoacoustic, exposure duration <1 μs; photoablative, exposure duration <1 ns; bubble or plasma formation (change of phase); nonlinear optical effects	Tissue damage, skin burns, ocular burns, tissue vaporization

mechanisms). Cataracts can readily be prevented by wearing protective glasses.

MICROWAVE RADIATION

Sources of this type of radiation include radar, radio and television transmitters, satellite telecommunication systems, and microwave ovens. Microwaves are used in industry to dry and cure plywood, paint, inks, and synthetic rubber, as well as to control insects in stored grain. They are also used in medicine to provide deep-heat therapy for the relief of aching joints and sore muscles.

As noted earlier, the human body is largely transparent to the lower frequencies of microwaves. Those in this energy range, including those emitted by television sets, produce no biological effects. As the frequency increases, however, the energy is increasingly absorbed, reaching a maximum at about 3×10^8 Hz, the ultra-high-frequency (UHF) television range. At still higher frequencies ($>10^9$ Hz), less of the energy is absorbed, and above 10^{10} Hz the skin acts as a reflector. Potentially the most hazardous microwaves are those in the range of 10^8–10^9 Hz since, at these frequencies, there is little or no heating of the skin and the thermal receptors are not stimulated. As a result, the person is not able to recognize that he or she is being exposed.

One of the most common sources of microwaves is the microwave oven. In fact, such appliances are being used in essentially all U.S. homes, as well as in restaurants and many hotel and motel guest rooms. To perform their function, microwaves agitate water molecules, causing them to vibrate millions of times each second and rub against one another. The accompanying friction manifests itself as heat. Although the window in the microwave oven door is relatively clear, it contains a thin metal wire mesh with holes large enough for visibility but too small for the microwaves to escape. Due to earlier fears that microwave ovens could interfere with pacemakers and other implanted medical devices, such devices now contain shields to protect users from most ambient radiation (Ropeik and Gray, 2002).

During the 1990s, increasing concern was expressed about two widely used devices that are possible sources of microwave radiation—handheld cell phones and traffic radar units. Based on the consensus view of the scientific and medical communities, the World Health Organization (WHO, 2010) has stated that cancer is unlikely to be caused by cellular phones or their base stations. Nonetheless, due to the fact that an estimated more than 4 billon cell phones are in use worldwide, some national radiation advisory authorities have recommended measures to minimize exposures. Similarly, the director general of the WHO has recommended that parents limit the use of cellular phones by children (Kirschner, 2002; WHO, 2010). Interestingly, the increasing use of texting phones could significantly reduce such risks, if they exist.

Although epidemiological data have not confirmed the association, radar units have been cited as a possible source of testicular cancer among police officers.

ELECTRIC AND MAGNETIC FIELDS (EMFS)

As noted earlier, all atoms contain positively and negatively charged particles. Since most objects contain a balance of particles with such charges, they are electrically neutral. When this balance is upset, an *electric field* is produced that leads to effects that can be readily observed, a common example being the attraction between a comb and a person's hair. When electric charges (i.e., electricity) flow through a wire, a *magnetic field* is generated. In both cases, however, the field is confined to the vicinity of the source. As such, any home appliance that has an electric motor can be a source of a *magnetic field*. Typical examples are refrigerators, clothes washers, vacuum cleaners, electric mixers, as well as personal items such as electric shavers, hair dryers, electric blankets, and magnets used to hold notes and pictures on refrigerators (Ropeik and Gray, 2002). Interestingly, the magnetic fields produced by many of these sources are comparable to, or far in excess of, those present under electric-power transmission lines. Even the Earth itself produces a magnetic field (Valberg, 2001). The mechanisms of the interaction and examples of the biological effects of exposures to electric and magnetic fields are summarized in Table 12.4.

While several investigators have claimed to have observed a link between childhood cancer and EMFs, particularly those associated with transmission lines, laboratory evidence in support of these findings has not been confirmed, and an accepted mechanism by which such fields can cause disease has not been identified. The same is true in terms of assessments of the possible health impacts of cell phones. Nonetheless, the National Institute for Environmental Health (NIEHS, 2009) has recommended continued education on practical ways for reducing such exposures. Similarly, some public health personnel are recommending that cell-phone users practice certain precautions, for example, limiting exposures of younger age groups and pregnant women since children and developing fetal organs are more sensitive to EMFs, and, for adults, limiting the duration of their calls, using a speakerphone or opting for text messaging, and selecting a phone with the lowest specific absorption rate, or SAR (Herberman, 2008).

STANDARDS FOR CONTROL

Globally, the primary source of guidelines for limiting exposures from sources of nonionizing radiation (NIR) is the International Commission on Non-Ionizing Radiation Protection (ICNIRP). In this regard, it plays a role similar to that of the ICRP in terms of ionizing radiation. Specific sources addressed to date include optical radiation (i.e., ultraviolet, visible, and infrared, including lasers), and the nonionizing portions of the electromagnetic spectrum (i.e., microwaves, as well as other radio-frequency fields including those down to the range encompassing static electric and magnetic fields; ICNIRP, 2009a). These also include guidelines on the application of medical magnetic resonance procedures in medicine (ICNIRP, 2009a, 2009b).

Table 12.4 Mechanisms of interaction and examples of adverse effects of exposures to electric and magnetic fields

Part of the spectrum	Mechanisms of interaction	Adverse effects
Static electric fields	Surface electric charges	Annoyance of surface effects, shock
Static magnetic fields	Induction of electric fields in moving fluids and tissues	Effects on cardiovascular and central nervous systems
Time-varying electric fields (<10 MHz)	Surface electric charges	Annoyance of surface effects, electric shock, burns
	Induction of electric fields and currents	Stimulation of nerve and muscle cells, effects on nervous system functions
Time-varying magnetic fields (<10 MHz)	Induction of electric fields and currents	Stimulation of nerve and muscle cells, effects on nervous system functions
Electromagnetic fields (100 kHz–300 GHz)	Induction of electric fields and currents, absorption of energy within the body	Excessive heating, electric shock, burns
	>10 GHz: surface absorption of energy	Excessive surface heating
	Pulses <30 μs, 300 MHz–6 GHz: thermo-acoustic wave propagation	Annoyance from micro-heating effect

Comparable guidance for the protection of workers has been provided by the American Conference of Governmental Industrial Hygienists (ACGIH, 2008).

Assessments of External and Internal Exposures to Ionizing Radiation

Regardless of the assumptions regarding the dose-response model, in the case of ionizing radiation sources *external* to the body, the resulting biological effects will depend on (1) the total dose, (2) the dose rate, and (3) the percentage and region of the body exposed. In general, the potential for harmful effects increases in response to increases in each of these three factors. In the assessment of internal exposures, the major factors that determine the potential for harm are the types and quantities of material taken in and the length of time they remain, that is, the combination of their biological and radiological (i.e., physical) half lives. Although those that emit alpha particles deposit more energy per unit pathway of travel than those that emit beta particles, their potential effects now appear to be ameliorated due to the confirmed threshold for their effects, described earlier.

The quantity of a radioactive material is expressed in terms of the Becquerel (Bq), which is an amount that contains sufficient radioactive material such that one atom is decaying per second. The ICRP, in turn, has developed tables listing the dose (Bq) per unit of intake, separately for ingestion and inhalation, for more than 700 individual radionuclides (i.e., the individual radioisotopes of all the radionuclides). Since these must be applied to members of the public as well as radiation workers, doses per unit intake are provided for adults as well as children 1, 5, 10, and 15 years old (ICRP, 1996). Most of this information has, in turn, been converted into so-called *secondary* limits that are expressed as corresponding limits in terms of their limiting concentrations in food and water. Since both workers and the public generally consume the food and water from the same sources, these are based on dose-rate limits for members of the public. In cases of occupational exposures, where the primary source of intake is frequently airborne, secondary limits are expressed as annual limits on intake (ALIs) and, in turn, converted into equivalent derived air concentrations (ICRP, 1994).

Techniques for ensuring compliance with the regulations for occupational exposures include analyses of intakes via inhalation and possible ingestion (Chapter 14), as well as bioassays of excreta, supported by whole-body counting (Chapter 4). Although many sources of external radiation expose essentially the whole body, individual radionuclides (exclusive of ^{3}H and ^{137}Cs) that are taken into the body, primarily irradiate individual body organs. Details on the conversion of partial body doses to an equivalent whole-body (or effective) dose are discussed in Chapter 13.

Natural Background Radiation

Humans have always been exposed to significant levels of ionizing radiation from two sources: naturally occurring radioactive materials in the Earth and cosmic radiation from outer space. While both of these expose the human body externally, radioactive materials can also cause internal exposures if they are ingested or inhaled. People in high-flying aircraft, and those who participate in space missions, are also externally exposed to rapidly moving charged particles.

COSMIC RADIATION

The annual dose from *cosmic radiation* at sea level is about 0.3 mSv. Since the Van Allen belts provide protection along the equator, the cosmic dose rate increases as one moves closer to the north and south poles. In addition, as was the case with UVR, because the atmosphere between the Earth and outer space serves as a shield, the accompanying dose rate increases with altitude. At sea level, the annual dose rate is about 0.32 mSv; at an altitude of ~1600 meters (1 mile), it is about 0.5 mSv; at ~3660 meters (12,000 feet), about 1 mSv; and at ~9,150–12,200 meters (30,000–40,000 feet)—where commercial jet aircraft operate—the range is 45–70 mSv per year. In the case of supersonic passenger aircraft, which no longer operate, the annual dose rates ranged from 80–90 mSv. Nonetheless, because supersonic aircraft were able to complete a given trip in less time than a subsonic jet flying at a lower altitude, the dose for a given trip was about the same (Table 12.5) (Alverez, 2010; Eisenbud, 1973).

TERRESTRIAL RADIATION

As in the case for cosmic radiation, the external dose rates that individual members of the population of the world receive from naturally occurring radionuclides in the soil vary widely. In some areas of the United States, for example, they may be as low as a few tenths mSv per year; in other areas, they may be as high as 1 mSv per year (Figure 12.4). As may be noted, the regions with the highest terrestrial dose rates are those associated with

Table 12.5 Cosmic radiation dose to airline passengers on subsonic flights between various cities

Trip	Duration (hours)	Altitude (Meters)	(Feet)	Dose (μSv)
Los Angeles to Honolulu	5	10,700	35,000	14
London to New York	7	11,300	37,000	37
Athens to New York	>9	12,500	41,000	62
Tokyo to New York	12	12,500	41,000	70
Hong Kong to New York Transpolar route	17–20	12,500	41,000	100

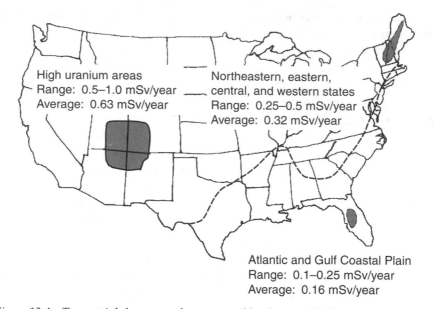

Figure 12.4   Terrestrial dose rates from natural background radiation in the coterminous United States

uranium deposits in the Colorado plateau, granitic deposits in New England, and phosphate deposits in Florida; those with the lowest rates are the sandy soils of the Atlantic and Gulf coastal plain. In contrast, the coastal monazite-bearing areas in the states of Rio de Janeiro and Espirito Santo, Brazil, yield external dose rates up to 30 mSv per year. The same is true for coastal areas in Kerala and Tamil Nadu, India, and certain areas in the People's Republic of China (ICRP, 1999). The corresponding estimated average dose rate to members of the U.S. population is slightly less than 0.2 mSv per year (NCRP, 2009).

INGESTED RADIONUCLIDES

As noted earlier, humans are exposed to ionizing radiation due to naturally occurring radioactive materials within the body. The primary radionuclide of consequence is ^{40}K, an extremely long-lived radioisotope of potassium. Since the dose rate is influenced by the amount of muscular tissue in the body, it tends to decrease with age. Although one might be inclined to avoid foods containing potassium, maintenance of a proper balance of this element is essential to health. Significant doses, although obviously harmless, also occur as a result of the ingestion of other naturally occurring radionuclides, such as radium (primarily ^{226}Ra) through the consumption of Brazil nuts, and those of hydrogen (^{3}H, commonly known as tritium) and carbon (^{14}C). While ^{226}Ra is continually produced through the decay of naturally occurring radionuclides, beginning with uranium, ^{3}H and ^{14}C are continually replenished through the interaction of cosmic rays with various atoms in the atmosphere (NCRP, 1987a). Other consumer products are also sources of exposure (NCRP, 1987b).

INHALED RADIONUCLIDES

Another very important avenue of intake is inhalation, the major contributor in this case being radon (^{222}Rn). Although radon itself is a short-lived gas, it decays into a solid radioactive radionuclide that, in turn, decays in sequence into other solid radioactive materials. Since radon is continuingly released into the air from the ground (Chapter 5), it and its decay products are present in many homes and buildings. Although the radon that is released outdoors is diluted with copious quantities of air, that which is released into the home is not. The associated risk to members of the U.S. public is estimated to be equivalent to that resulting from a whole-body effective annual dose of about 2.3 mSv (NCRP, 2009). This is the highest contributor to the dose from natural background sources.

TOTAL DOSE FROM NATURAL SOURCES

Globally, the average dose rate from natural background radiation sources is estimated to be 2.4 mSv per year. For members of the U.S. public, the dose rate is estimated to be slightly more than 3.2 mSv per year. This includes external exposures to cosmic and terrestrial sources and internal exposures due to natural radionuclides ingested and inhaled into the body (i.e., radon decay products). Although natural background dose rates in various parts of the world cover a wide range, the vast majority incur rates in the range of 2.2 to 2.4 mSv per year (ICRP, 1999). For people living in the 48 conterminous U.S. states, the dose rates from all natural-background radiation sources are estimated to range from <1 mSv for people living in radon-free houses at sea level on the Atlantic and Gulf coastal plain to 5 to 10 times this value or more for those living at high elevations on the uranium-bearing lands of the Colorado plateau.

ASTRONAUTS—DOSES IN OUTER SPACE

With the continued interest in space exploration, as exemplified by the establishment of the International Space Station and initial planning for human missions to planets such as Mars, more attention is being directed to the radiation doses that astronauts participating in such activities will receive (Chapter 15). In fact, concern about radiation exposures is becoming one of the dominating factors in developing plans for such missions. While the dose to crew members spending 90 days on the former Soviet *Mir* station was about 70 mSv, estimates are that the dose to those participating in a 1,000-day round trip mission to Mars will be about 1,000 mSv (1 Sv). One reason is that not only will the dose rates encountered in missions into deep space be a factor of about three higher, but limits on the size and weight of the spacecraft will also restrict the amount of shielding that can be provided. Furthermore, the nature and effects of the radiation fields will be unique and not well known. This is exemplified by the fact that astronauts and cosmonauts on the International Space Station typically received from 0.5 to 1.2 mSv d^{-1}, with ~75 percent coming from the galactic cosmic ray background (GCR) and ~25 percent from protons encountered in passages through the South Atlantic Anomaly region of the Van Allen belts. Once that hurdle has ended, those going to Mars will be exposed to galactic cosmic ray background (GCR) and sporadic solar particle events (SPEs), which vary significantly during the 11-year solar cycle. The accompanying health effects could include early effects on the brain and peripheral nervous system, cardiovascular disease, and effects on immunological functions that could contribute to life shortening and diminished quality of life. Other potential harmful effects include biophysical factors such as microgravity and exposure to UV light or to microwaves (NCRP, 2000).

Technologically Enhanced Natural Exposures

Dose rates from certain radiation sources of natural origin can be increased by human activities. When this occurs, the doses are said to have been *technologically enhanced.* One example, described earlier, is the increase in concentrations of radon inside buildings that have been constructed to be more energy efficient and therefore have low indoor-to-outdoor air exchange rates. Another is the use of tobacco. This will be discussed later in the section "Consumer Products and Related Sources." The classic example of technologically enhanced exposures was the generation of uranium mill tailings in the course of separating uranium from the original ore during World War II. When in the ground, the ore does not constitute a significant source of exposure. Once it is mined and the uranium removed, the so-called tailings (which contain the radium and other long-lived naturally occurring radioactive decay products that were in the original ore) remain. If left exposed on the surface of the ground, they represent a readily available source not only for the release of radon into the atmosphere but also for the leaching and transport of radionuclides into nearby surface waters. Since these materials were similar to clean sand, many people unknowingly used it to fill under and around nearby houses and in public projects such as road construction. Once the problem was recognized, remediation programs were initiated to remove the tailings from the homes (Eisenbud, 1973).

Sources of Human Origin—Radiation Machines

The principal artificial sources of ionizing radiation are radiation machines, primarily X-ray generators, and radioactive materials produced in nuclear reactors and particle accelerators.

MEDICAL APPLICATIONS

One of the major sources of human exposures from X-rays is from their applications in medical diagnosis (i.e. chest X-ray examinations). In fact, about 2 billion diagnostic medical X-ray examinations, including more than 500 million dental X-ray examinations, are performed worldwide each year. In the case of the United States, which has only about 5 percent of the world's population, half a billion diagnostic X-ray examinations, plus 19 million nuclear medicine procedures, are performed each year. Typical doses resulting from such procedures are summarized in Table 12.6. The most important steps for limiting the doses from these types of examinations is to ensure that the X-ray beam is properly collimated, filtered, shielded, and the time of exposure is limited. The accompanying increase in the doses to the U.S. population has been dramatic: from an average per capita dose of 0.54 mSv and collective dose of 124,000 person-Sv in 1982 to 2.2 mSv and about 450,000 person-Sv in 2006, an increase in the collective dose by a factor of more than 3.6 (Mettler et al., 2008). To place this in perspective, the global annual collective dose due to occupational exposures is less than 30,000 person-Sv, and that due to exposures to the workers involved in combating radionuclide releases from the Chernobyl nuclear power plant following the accident in the former USSR in 1986 was about 20,000 person-Sv (NRC, 2006).

The major increases in dose were due to computed tomography (CT) scanning, which escalated from an annual 18.3 million procedures in 1993

Table 12.6   Effective doses for diagnostic examinations of various body organs
using multiple detector CT systems, 2007

Examination	Effective dose (mSv)
Abdomen and pelvis	14.4
Liver/kidney	11.5
Coronary CTA	10.5
Aorta, abdominal	10.3
Lumbar spine	8.1
Pelvis	7.2
Aorta, thoracic	6.7
Chest	5.7
Pulmonary vessels	5.4
Cervical spine	2.9
Brain	2.8
Face and neck	2.0
Face and sinuses	0.8

to 62 million in 2006. This accounted for 15 percent of the total number of procedures and more than half of the collective dose. From 1995 to 2005, the annual number of nuclear medicine visits increased from 10 to 17 million. This accounted for about 4 percent of the procedures and 26 percent of the collective dose. The corresponding number of computed tomography visits increased from 20 million to about 60 million (Figure 12.5). As a result, for the first time the annual dose rate from diagnostic medical radiation procedures and nuclear medicine procedures was equivalent to 48 percent of that that contributed from natural background (Moeller, 2010). Long active in this field, the ICRP (2004, 2007a, 2007d, 2008) continues to provide guidance on managing doses to patients during diagnostic radiology and in nuclear medicine (Table 12.7). Properly applied, the guidance in these reports will reduce the doses both to the patients and the medical staff.

The doses from Computed Assisted Tomography (so-called CAT scans) are of special concern. Included in this category are cardiac catheterizations, whose use has increased significantly because it enables a diagnosis to be made without invading the body and avoids traditional surgical exploration and other invasive testing. In addition, it enables surgeons to conduct their preoperative planning in a more leisurely and detailed manner. Nonetheless, there have been several episodes in which patients have received acute burns to their skin (ICRP, 2000; Goans and Christensen, 2009).

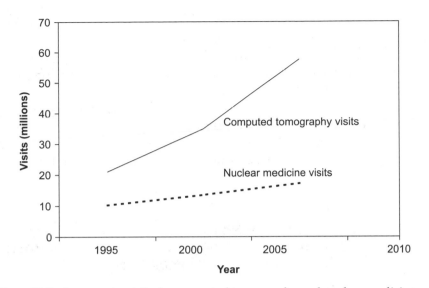

Figure 12.5 Increase in visits for computed tomography and nuclear medicine, United States, 1995–2005

Table 12.7 Examples of ICRP and NCRP documents on radiation protection in medicine

ICRP	Topic	NCRP	Topic
Publication: Date		Report: Date	
86: 2000	Accidental exposures in radiation therapy	140: 2002	Medical ultrasound
93: 2004	Managing dose in computed tomography	145: 2003	Radiation protection dentistry
97: 2005	Accidental exposures in high-dose brachytherapy	147: 2004	Structural shielding for medical imaging
102: 2007	Multi-detector computed tomography	148: 2004	Radiation protection in veterinary medicine
104: 2007	Radiological protection in medicine	149: 2004	Mammography and breast imaging
106: 2008	Pharmaceuticals	151: 2005	Structural shielding for therapy facilities

INDUSTRIAL APPLICATIONS

Industrial X-ray devices include radiographic and fluoroscopic units used to detect defects in castings, fabricated structures, and welds, and fluoroscopic units used to detect foreign material in items such as food products. Today there are about 16,000 active industrial radiographic installations in the United States, and 40,000–50,000 people are occupationally exposed in their operation. The primary concern is the control of exposures to the X-ray machine operators. The same techniques of filtration, coning, shielding, and limits on the time of exposure apply here as in the applications of medical X-rays.

COMMERCIAL APPLICATIONS

As noted earlier (Chapter 6), various types of X-ray devices are being used to irradiate food products to extend their shelf life and/or destroy disease-causing organisms. Although such applications in the United States have been hampered by lack of public education and support, analyses show that this practice could dramatically reduce the number of foodborne diseases. In contrast, such devices, as well as sealed sources of radioactive materials (discussed below), are routinely used to sterilize medical supplies and equipment.

SECURITY APPLICATIONS

Since the 1970s, X-ray machines have been increasingly used to inspect luggage at U.S. airports for purposes of security. In fact, tens of thousands of such units are currently in operation. Although travelers in the boarding area often pass close to these units, their advanced design restricts the dose per inspection (in the range of a few thousandths of a mSv). The metal detectors used for checking passengers are not a source of radiation exposure. At the same time, however, there is an increasing trend on the part of U.S. government agencies and other institutions to expand the use of various types of X-ray devices to screen members of the public. The accompanying dose per screening, however, is very low (from about 0.1 to as much as 4 μSv), the magnitude depending on the nature of the equipment used and the purpose of the screening (NCRP, 2003).

Newer computed tomography fluoroscopy units are also being applied to conduct full-body scans of passengers. These, however, apply what is called back-scatter radiation, which significantly limits the dose. To ensure privacy, the images are seen only by security personnel and are immediately deleted. This, however, emphasizes once again the need for care in balancing the risks and benefits of such applications of radiation (Alverez, 2010).

RESEARCH APPLICATIONS

High-voltage X-ray machines and particle accelerators are common equipment in the laboratories of universities and research organizations. More than 1,000 cyclotrons, synchrotrons, Van de Graaff generators, and betatrons are in operation in the United States, plus about 3,000 electron microscopes and some 20,000 or more X-ray diffraction units. Modern electron microscopes are shielded to protect the operators, but diffraction units still account for a significant number of radiation injuries (primarily burns on the hands).

Applications of Radioactive Materials in Medicine, Research, and Industry

More than 20,000 hospitals, academic, industrial, and research organizations in the United States have been licensed to use radioactive materials. Such materials, for example, are used in almost 20 million medical diagnostic procedures each year. Worldwide, the number is about 32 million. The increasing trend in the applications in nuclear medicine in the United States was depicted in Figure 12.5. Radioactive materials are also used in the United States to perform some 200,000 medical treatments and about 100 million laboratory procedures each year. Although the latter provide diagnostic information, they do not require the administration of radioactive materials to the patient. Prominent examples are radioimmunoassay tests on blood and bodily fluids from patients. Although the doses to patients in some cases (e.g., heart scans using ^{99m}Tc) can be higher, on average the dose *per procedure* is about 4.4 mSv. The accompanying dose rate to nuclear medicine technical personnel who prepare and administer these materials is about 4 mSv *per year*, less than 10 percent of the limit (Mettler et al., 2008).

Artificially produced radioactive materials are also widely used in universities and other institutions for teaching and research. One of the best-known research applications is the use of ^{14}C and other radionuclides for dating artifacts. Radionuclides are also routinely used as tracers in chemistry experiments and in chemical and polymer synthesis. In addition, both portable and fixed devices, such as thickness, level, and moisture-density gauges and static eliminators, are used in a wide range of industrial operations. Sealed capsules containing radionuclides that emit gamma radiation and combinations of radionuclides that produce neutron radiation are used to log wells during explorations for oil and gas. As discussed earlier, sealed radionuclide sources that emit gamma radiation are used for sterilizing medical supplies and equipment, disinfesting food products, and extending the shelf life of poultry and other perishable products (Chapter 6).

Nuclear Power Operations

As of 2010, there were 104 nuclear power plants operating in the United States (Figure 12.6). These units had a generating capacity in excess of 100,000 megawatts and were producing more than 20 percent of the nation's electricity. This amounts to about 31 percent of the world's net nuclear-generated electricity. Although other countries, such as France, produce a higher percentage of their electricity from nuclear power, the output in the United States in 2008 (more than 800 billion kWh) was almost twice that of France (USNRC, 2009). In addition, more than 150 additional power reactors are being used by the military services as propulsion units in submarines, cruisers, and aircraft carriers. Worldwide, about 450 nuclear power plants, with a total generating capacity of over 350,000 megawatts, are producing about

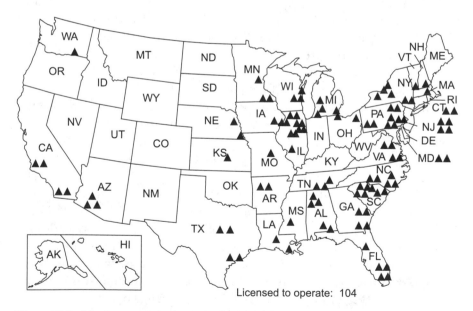

Figure 12.6   Nuclear power plants in the United States

17 percent of the global electricity supply (Chapter 16). During 2006, the annual average dose to workers at U.S. nuclear plants was about 1 mSv (i.e., about 2 percent of the annual limit). The collective dose per plant was 1 person-Sv. These results reflect the regulatory efforts of the USNRC and the supporting activities of the Institute of Nuclear Power Operations, a self-evaluation organization supported by the nuclear industry. A major reason for achieving such low doses is the fact that the operators of all such facilities maintain the doses they permit their personnel to receive and the quantities of radionuclides they release into the environment to "as low as reasonably achieveable" (ALARA; Chapter 14).

Consumer Products and Related Sources

Consumer products and related sources cause a variety of radiation exposures to the general population.

- Brazil nuts, which contain relatively high quantities of naturally occurring ^{226}Ra, are a major source of internal exposure. The accompanying collective dose, due to their ingestion, is more than that to the workers at all the nuclear facilities of the U.S. Department of Energy, those at all the U.S. commercial nuclear power plants, and the crews on all the nuclear-powered surface vessels and submarines of the U.S. Navy (Moeller and Sun, 2009). Because ^{226}Ra has an extremely high threshold for producing any health effects (Evans, 1974), the ingestion of Brazil nuts is not of concern.

- Cigarettes are a major source of radiation exposure due to the presence of naturally occurring ^{210}Po and ^{210}Pb in tobacco. When a smoker lights up, these radionuclides are volatilized and deposit preferentially in the bronchial epithelium of his/her lungs. The annual dose to critical lung tissues is as high as 160 mSv (Little, 1993; Little et al., 1965). Exacerbating the biological effects is the fact that there is a synergistic effect between the effects of the radiation and the chemical carcinogens in cigarettes. As noted earlier (Chapter 3), the combination of these two carcinogens leads to an effect that is up to 25 times the sum of the two alone (Darby et al., 2005; EPA, 2003).

- Building materials, such as masonry, bricks, and concrete, include naturally occurring radionuclides. The major contributors are coal ash and granite, both of which contain relatively large quantities of uranium, thorium, and radium. These yield external dose rates, to people living in such houses, of 0.07 mSv per year (NCRP, 1987b). Since an estimated 150 million members of the U.S. population live in masonry houses, the accompanying collective dose was about 10,500 person-Sv.

- Exposures from cosmic and solar sources during air travel can also be a source of external dose. Exposures depend on the flight routes, altitude, and total hours in the air. During 2006, about 660 million members of the U.S. population flew on domestic flights, and almost 85 million on international flights. The total collective dose for passengers on domestic flights was more than 6,000 person-Sv, and that for international flights was more than 4,000 person-Sv (NCRP, 2009).

- Other sources include external exposures from uranium and its decay products in dental prostheses, ophthalmic glass, luminous watches and clocks, highway and road construction materials, and glass and ceramics. Also contributing are internal exposures from airborne ^{222}Rn due to the domestic combustion of natural gas and coal, as well as agricultural products produced using fertilizers containing phosphates and sulfates containing ^{40}K and decay products of uranium and thorium. The estimated combined annual collective dose from these sources was almost 4,700 person-Sv (NCRP, 2009).

- Smoke detectors contain ^{241}Am. The dose, however, is miniscule, and the benefits are enormous. Since they became available several decades ago, they have saved several thousand U.S. lives each year. This radionuclide is also a component of static eliminators that are widely applied in photocopying and newspaper printing, and in brushes for processing photographic film to reduce the associated buildup of electrical charges. Without this benefit, these processes would not be possible. Another example is the radioactive thickness gauge. This enables manufacturers to ensure that the thickness of sheet steel, aluminum, and other metal products is maintained within standards. This, in turn, saves money and improves the quality of these products.

Summarized in Figure 12.7 are the contributions of various ionizing radiation sources to the annual dose to the average member of the U.S. public. As may be noted, the total annual effective dose from all sources of human origin is slightly less than that from those of natural origin.

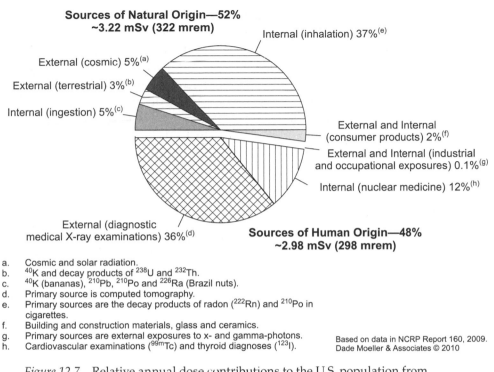

Sources of Natural Origin—52%
~3.22 mSv (322 mrem)

Internal (inhalation) 37%[e]

External (cosmic) 5%[a]

External (terrestrial) 3%[b]

Internal (ingestion) 5%[c]

External and Internal
(consumer products) 2%[f]

External and Internal (industrial
and occupational exposures) 0.1%[g]

Internal (nuclear medicine) 12%[h]

External (diagnostic
medical X-ray examinations) 36%[d]

Sources of Human Origin—48%
~2.98 mSv (298 mrem)

a. Cosmic and solar radiation.
b. ^{40}K and decay products of ^{238}U and ^{232}Th.
c. ^{40}K (bananas), ^{210}Pb, ^{210}Po and ^{226}Ra (Brazil nuts).
d. Primary source is computed tomography.
e. Primary sources are the decay products of radon (^{222}Rn) and ^{210}Po in cigarettes.
f. Building and construction materials, glass and ceramics.
g. Primary sources are external exposures to x- and gamma-photons.
h. Cardiovascular examinations (^{99m}Tc) and thyroid diagnoses (^{123}I).

Based on data in NCRP Report 160, 2009.
Dade Moeller & Associates © 2010

Figure 12.7 Relative annual dose contributions to the U.S. population from radiation sources of natural and human origin, 2006

The General Outlook

Perhaps one of the most significant messages in this chapter is that in spite of the many environmental and occupational sources that can potentially cause exposures to radiation, the primary concerns arise through the personal environment. This is true for both ionizing and nonionizing sources. Those people who desire to keep their dose rates to a minimum should avoid overexposure to the sun, ensure that the doctors examining them use X-ray equipment that is up-to-date and well maintained, and, above all, avoid smoking cigarettes. Another message is the multitude of ways in which applications of radiation are helping to solve various types of problems in our daily lives. As noted, smoke detectors, which contain radioactive material, have reduced fire-related deaths significantly. Similar benefits have resulted from applications of nonionizing sources; witness, for example, the widespread benefits of microwave ovens and the improvements in airline safety made possible by radar.

Applications of radionuclides in agricultural research have provided similar benefits. Through the use of radionuclides to identify the best methods for applying nitrogen fertilizers in rice production, the quantities of such fertilizers required have been reduced by up to half. Nuclear techniques such as gamma or neutron irradiation have induced mutations in seeds and thus created new varieties of food crops. As previously described (Chapter 10), radiation is also being used to control insects such as the screwworm fly, the Mediterranean fruit fly, and the tse-tse fly. In short, although radiation can be harmful, with care it need not be feared (ICRP, 2007c).

13

OCCUPATIONAL, POPULATION, AND ENVIRONMENTAL STANDARDS

MANY PROFESSIONAL organizations and federal and state agencies have developed recommendations, guidelines, standards, and regulations for limiting exposures to a variety of occupational and environmental contaminants and physical agents. Prominent among these are the American Conference of Governmental Industrial Hygienists (ACGIH), which has issued recommendations for limiting exposures to chemical and physical stresses in the workplace (Chapter 4), and the International Commission on Radiological Protection (ICRP) and the National Council on Radiation Protection and Measurements (NCRP), which have developed similar guidelines for protection against ionizing radiation (Chapter 12). Based on reviews of these recommendations, as well as evaluations of a range of other factors, federal agencies, such as the Environmental Protection Agency (EPA), the Food and Drug Administration (FDA), the Occupational Safety and Health Administration (OSHA), and the U.S. Nuclear Regulatory Commission (USNRC), have promulgated standards and regulations for the control of various contaminants and physical stresses within the workplace, the home, and the ambient environment.

Types of Standards

The basic goal in establishing standards is to protect human health. These are generally referred to as *primary* standards. Another, equally important goal is the protection of the environment. These are generally referred to as *secondary* standards. In the case of air pollution, *secondary* standards are designed to protect agricultural crops and property, such as buildings and statues. In the case of water pollution, they are designed to ensure that rivers and streams are acceptable for fishing and swimming. In the case of drinking water, they are designed to ensure acceptability of the aesthetic qualities of the product, such as temperature, color, taste, and odor. Interestingly, the *secondary* standards for many contaminants are more stringent than the *primary* standards. Another category is what are called *derived* or *tertiary* standards or guides that specify the limiting concentrations of various toxic agents in air, water, food, and so on, which, if not exceeded, should ensure that the *primary* standards will be met.

In this regard, the EPA has for many years stipulated limits on releases of various contaminants into rivers, lakes, and streams to ensure that sufficient concentrations of dissolved oxygen are maintained for fish and other aquatic life. In a similar manner, the ICRP and NCRP have initiated action to develop a framework for evaluating the impacts of ionizing radiation on nonhuman species and plants (ICRP, 2003; NCRP, 1991). As a component of these endeavors, the ICRP (2008) has proposed the concept and use of reference animals and plants, similar to representative individuals for humans. This is discussed in more detail later.

History of the Development of Radiation Protection Standards

The history of the development of radiation protection standards is summarized in the discussion that follows. A unique feature of these efforts is the global framework that has been established by the ICRP for the coordination of the establishment and ongoing revisions of radiation protection standards throughout the world. These efforts include the application of a risk-based approach, which considers effects other than those to health (ICRP, 2007a).

SCIENTIFIC BASES
Shortly after the discovery of X-rays (Roentgen, 1895), radiation injuries were reported. Recognizing the need for protection, physicists recommended limits on the allowable doses from X-ray generators. Their initial concern was to avoid direct physical symptoms. As early as 1902, however, scientists recognized that radiation exposures might also have latent effects, such as the development of cancer. This hypothesis was confirmed for external sources during the next two decades, and for internally deposited radionuclides during the late 1920s, when bone carcinomas were reported among

workers who applied relatively large quantities of ^{226}Ra luminous paints to the dials of clocks and watches (Evans, 1974).

Initial recommendations for the control of exposures were developed on an informal basis. This changed in 1928 with the establishment of the International X-Ray and Radium Protection Committee (known today as the ICRP) and the U.S. Advisory Committee on X-Ray and Radium Protection (known today as the NCRP). One of the major benefits of the international committee and its successor was that it provided a forum for radiation protection experts to discuss the latest information on the biological effects of ionizing radiation and to propose appropriate radiation protection standards. Since there are close ties between the ICRP and the NCRP, the recommendations of the two groups are essentially identical. Similar relationships exist among many other countries. The credibility of the ICRP and NCRP recommendations is documented by the fact that the USNRC has indicated that it will never knowingly promulgate regulations that violate any of the ICRP or NCRP recommendations.

A year prior to the formation of the ICRP and the NCRP, Herman J. Muller reported on experiments with *Drosophila* flies that aroused concern about possible hereditary effects of radiation exposures in humans (Muller, 1927). This consideration shaped radiation protection guidelines and standards from about 1930 to 1960. Within a decade after the end of World War II, epidemiologic studies revealed an increase in leukemia among the survivors of the nuclear explosions in Japan. These studies did not, however, confirm the anticipated hereditary effects. As a result, somatic effects, primarily leukemia, became the basis for radiation protection standards. In 1972 the Committee on the Biological Effects of Ionizing Radiation (BEIR I Committee) reported that solid tumors (cancers of the lung, breast, bone, and thyroid), not leukemia, were the dominant biological effects of human exposures to ionizing radiation (NRC, 1972). Since that time, solid tumors have remained the primary basis for the development of radiation standards. Concurrently, new information on hereditary effects provided by the ongoing Japanese studies led to significant reductions in the concern for those effects (ICRP, 2007a).

As might be expected, the initial focus of radiation protection guidelines was on the establishment of dose-rate limits for radiation workers. As public concern increased due to worldwide exposures to fallout from atmospheric nuclear weapons tests during the early 1950s and continuing for a decade thereafter, attention shifted to the need for dose-rate limits for the public. This led to the establishment of the Federal Radiation Council (FRC) in 1959 with the specific assignment to develop U.S. policy on such expo-sures and to establish standards for their control (FRC, 1960).

Establishment of Standards

One of the initial requirements in the development of primary standards is to identify or define the exposed person who is to be protected. Is it a worker or a member of the public and, if the latter, is it an adult, a child, an infant, or a fetus? Aware of the differences in their characteristics, the ICRP has published guidance for six different age groups, with specific intake limits for individual radionuclides for each (ICRP, 1989, 1993, 1995). Since the processes for absorbing materials into the body vary, intake limits for materials that are inhaled are different than those for ingestion. In the case of inhaled particles, for example, there are also specific limits, de-pending on the age/size of the exposed individual (Chapter 5). In addition, where the health effects of an exposure depend on the short-term con-centration (as is the case for ozone) or intensity (as is the case for noise), the ACGIH has recommended maximum levels for workers as a function of its duration. These, and other refinements, illustrate the range of factors that are to be considered in the establishment and application of the stan-dards (Table 13.1).

Another consideration, in terms of the public, is to designate, in a generic sense, the person who is to be protected. Early on, the EPA regulations re-quired protection of the "maximally exposed individual." Those promul-gated by the USNRC stipulated that no "individual member" of the public receive dose rates in excess of the limits. Soon it was recognized that the identification of such an individual would place an arduous burden on or-ganizations seeking to document compliance with the applicable regula-tions. Recognizing this fact, the EPA modified its regulations to require pro-tection of the "reasonably maximally exposed individual," or RMEI (EPA, 2001), and the USNRC adopted the policy that compliance could be based on the dose rate to an average member of the "critical group." This was de-fined as the group of people who, because of their location or living habits, would be most highly exposed. Subsequently, the EPA stated that their use of the RMEI was intended to imply the same approach. This has the advan-tage of ensuring not only that average members of the public do not receive unacceptable doses but also that decisions on the acceptability of a given practice are not prejudiced by a small number of individuals with unusual living habits. To avoid accusations of chauvinism, the ICRP has changed its

Table 13.1   Factors to consider in the establishment and application of standards

Understand the conditions under which the standards are to be applied, including whether they are to serve a *primary* or *secondary* role

Identify what is to be protected: humans (their age ranges), other animals (including birds), plants, trees, aquatic and marine life, etc.

Review all possible pathways of exposure: external, internal (inhalation, ingestion, adsorption) through the solid, liquid, and/or airborne pathways

Review the applicable human epidemiologic data to understand the effects, short- and long-term, chronic and acute, neurologic, carcinogenic, noncarcinogenic, etc., of the toxic chemical and physical agents under consideration

Review the applicable data on the negative impacts of various toxic chemical and physical agents on nonhuman animals, plants, and the ecologic environment

Establish, if possible, a schedule for the completion of any studies required to gain the necessary information to fill the gaps in the available information and ensure that the full range of human and non-human species to which the standards are to be applicable are considered

Evaluate existing standards and their source, concentrating on those provided by authoritative organizations such as the Agency for Toxic Substances and Disease Registry, American Conference of Governmental Industrial Hygienists, Centers for Disease Control and Prevention, Environmental Protection Agency, the International Commission on Radiological Protection and the National Council on Radiation Protection and Measurements, National Institute for Occupational Safety and Health, the Occupational Safety Health Administration, and the U.S. Nuclear Regulatory Commission

Develop guidance for those who apply the standards, including guidance on the qualifications, limitations, uncertainties, and precautions to be observed

terminology from "reference man" (ICRP, 1975) to "representative individual" (ICRP, 2006). In addition, to ensure protection of the environment, the ICRP has sought to identify "representative plants and animals" that would be comparable to the "representative individual" (ICRP, 2003). In essence, the methodologies for assessing the risks of exposures are undergoing rapid change with the application of new approaches and concepts to what has, for years, remained a rather static field of endeavor. As will be recognized, these efforts require not only input from essentially every facet of the field of science and technology but also the need to continue such evaluations as more detailed information is obtained.

LIMITATIONS

The underlying assumptions and scientific bases for occupational and environmental standards are subject to a host of limitations.

1. The most common endpoints are different types of cancer. Since cancers produced by one agent are virtually indistinguishable from those induced by another, including a host of "natural" causes, the induction can be inferred only on statistical grounds, that is, from an analysis of the dose-dependent increase in their frequency within the exposed population.

2. Compounding the problem is that in only a very few instances have dose-response data been developed for the purpose of establishing standards for protection. Most such data are by-products of epidemiologic studies (Chapter 3) and descriptive and analytic studies designed to test specific scientific hypotheses. Even where the necessary data are available, they contain a range of uncertainties—not only those commonly encountered in the study of any biological system but also those involved in extrapolating data from laboratory studies of animals to humans.

3. With few exceptions, the dose to the affected tissues is not known well enough to define the dose-incidence relationship except in a general way. Analyses of dose-incidence relationships for chemical carcinogens are also complicated by the fact that the dose of a chemical at its biological site of action depends on a number of metabolic and pharmacokinetic factors that can vary with route of exposure, age, sex, genetic constitution, physiological state, action of other chemicals, and other variables.

4. Few standards were developed to account for differences in the weight, size, diet, and lifestyle of various population groups (for example, Japanese and Americans). One solution might be to express the total intake limit for a given contaminant in terms of body weight. As noted earlier, the ICRP has sought to accomplish this by establishing age-dependent guidance for estimating the dose rates to various individuals due to the intake of radionuclides. Similar information needs to be developed for a range of toxic chemicals, as well as for physical and biological stresses.

5. Despite a consensus that standards for the general public should be more stringent than those for workers, the agencies responsible for

occupational (OSHA) versus environmental (EPA) standards seldom coordinate their standards-setting efforts. Moreover, except for radon, standards for limiting concentrations of airborne contaminants inside dwellings and office buildings are essentially nonexistent (Chapter 5).

6. Some standards do not apply to all sources of a given contaminant or physical factor. For example, guidelines for acceptable radiation doses to workers and the public do not include those they receive from natural background and medical applications (ICRP, 2007a; NCRP, 1993).

7. Standards for workers and the public are commonly set for individual contaminants in specific environmental media—air, water, food, and soil—even though it is the total intake of the contaminant that is critical. In addition, few standards are set with consideration of the effects of exposures to a combination of occupational and environmental contaminants and stresses (i.e., noise, heat, etc.), let alone potential synergistic effects, such as those due to combinations of chemical and radiological carcinogens, for example, cigarettes (Chapter 12).

8. In some instances, such as exposures to electric and magnetic fields, definitive data or sound epidemiologic evidence on which to base the standards is lacking. Yet public pressure, and the fact that exposures are occurring, make standards necessary even when their bases are suspect.

9. The risks associated with the exposures permitted by many standards have not been quantified. Unless they are, it is not possible to compare the relative stringency of, or protection afforded by, standards developed for different environmental contaminants or stresses.

10. With the exception of air and water, few limits have been derived for protection of the natural environment, property, or aesthetic features.

Another major limitation (Chapter 2) is that the biological effects of high-level acute doses of most any agent have a threshold (i.e., only if the dose is sufficiently high will immediate effects be observed). The primary question is whether there is a similar threshold for the latent effects that sometimes appear following low-level doses received over a long period of time.

Complicating the situation is that while there may be no observable effect in the case of certain chemicals taken into the body at low dose rates, a range of nonobservable precursors to these effects may be occurring. In fact, the published literature contains data from an increasing number of studies that show that the relationship between effects and exposure/dose at low levels for a host of toxic agents, including some that are carcinogenic, follows a "J-shaped" curve, implying that such doses may in some cases be beneficial (see Figure 2.3, page 36). Some scientists claim that this type of benefit (Chapter 12) is confirmed by (1) the lack of convincing evidence that indoor exposures to radon cause lung cancer (Cohen, 1995; Parsons, 2002), (2) the fact that application of an earlier preparatory external dose to humans will reduce the health effects following subsequent doses, and (3) the basic process through which vaccines prevent disease—by causing an earlier mild case and creating immunity. Nonetheless, it is the conclusion of the ICRP (2007a) that although the health effects of low levels of radiation exposure are still not known, there is no weight of evidence to support the concept that there is either a supra-linear response in this region or that one for a low-dose threshold should be preferred for purposes of radiological protection. In addition, the ICRP has never supported the concept of possible benefits at low doses, the so-called radiation hormesis effect.

Even so, some long-held concepts relative to the health effects of internally deposited radionuclides are being dramatically revised. The most prominent is whether long-lived alpha-emitting radionuclides have a threshold for the production of bone carcinomas. These changes are based on the long-standing but neglected epidemiologic studies of the effects of ^{226}Ra ingested by workers who applied radium-based luminous paints to the dials of clocks and watches (Evans, 1974) and similar studies of workers at the Los Alamos Scientific Laboratory (1995) who processed ^{239}Pu for nuclear weapons during World War II and the "cold war" that followed. In the case of ^{226}Ra, the data showed that although a hundred or more dial painters developed bone carcinomas, no one who received a cumulative dose of less than 1,000 rad (200 Sv) was affected (Figure 13.1). In the case of the ^{239}Pu workers, the change was based on the fact that the limit for intake by the ^{239}Pu workers was based on the ^{226}Ra studies and that not a single cancer due to ^{239}Pu intake has been observed among the thousands of workers involved in these activities. For these and other reasons, multiple ICRP members believe that there is strong evidence for the existence of a practical threshold for cancer induction by both ^{226}Ra and ^{239}Pu, as well as ^{237}Np and ^{241}Am. Should the ICRP accept the results of these studies, this would rep-

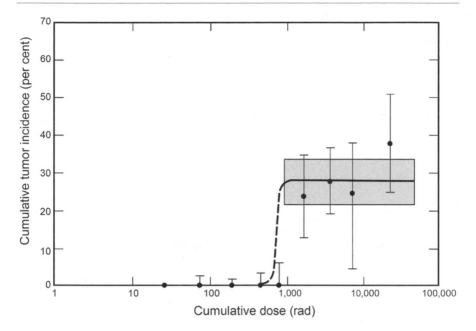

Figure 13.1   The observed radiogenic tumor cumulative incidence or occurrence in "epidemiologically suitable cases" due to the ingestion of ^{226}Ra

resent a significant change in one of their most basic fundamentals related to the dose-effects relationships of radiation exposures and the associated standards (Osborne, 2009).

Trends in Standards

Based on the gradual increase in knowledge, standards for protection against exposures to radiation have been gradually reduced during the past 80 years. This is exemplified by those for occupational exposures recommended by the ICRP (Figure 13.2), as well as those for members of the public and the environment. Included in the figure are the primary observations on which the limits during each designated time period were based. This is a result of several factors: (1) the shift from observations of acute to chronic effects, (2) the results of ongoing epidemiologic studies, (3) advances in technological methods for control, and (4) recognition of the need to consider potential health effects on vulnerable workers as well as members of the public. Much the same trends have occurred in the limits for

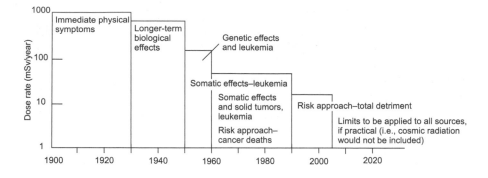

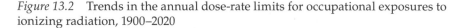

a. The limits since 1925 are based on recommendations of the International Commission on Radiological Protection.
b. The statements on the figure are the key factors on which the limits during each era were based.

Figure 13.2 Trends in the annual dose-rate limits for occupational exposures to ionizing radiation, 1900–2020

exposures to chemical and physical agents. This has been due to multiple examples of unanticipated health effects due to exposures to toxic chemicals, such as polychlorinated biphenys, lead, and mercury, and to small particulates in the air. In fact, a review of the publications of the ACGIH (2008) reveals continuing reductions in the threshold limit values for occupational exposures to multiple chemical substances and physical agents as new information continues to be developed.

At the same time, there is a need to adopt different strategies in the development of occupational standards and those for the public and the environment. This emphasizes the need to consider the fact that the public includes pregnant women, infants and children, the sick, and the elderly, each of whom may represent a group at increased risk. Also to be considered is that members of the public may be exposed 24 hours a day, 7 days a week, throughout their lives, whereas workers are generally exposed only during the time they are on the job. Members of the public also have no choice about their exposures to most environmental sources, and they may receive no direct benefit. In addition, they are not subject to the selection, supervision, and monitoring afforded workers. For this and other considerations, the recommended average annual dose limits to members of the public are well below those for workers.

Application and Enforcement of Standards

There have traditionally been two approaches for enforcing standards: (1) the "top-down" strategy adopted by some federal regulatory agencies and (2) the "bottom-up" strategy adopted by others. The former involves setting an upper bound or limit, then reducing the limit (on the basis of site-specific considerations) to a reasonably achievable lower level. The USNRC and U.S. Department of Energy have consistently favored this approach. In contrast, the "bottom-up" strategy has been used to control a variety of other environmental exposures. This involves initially setting a lower, relatively stringent dose or risk goal, the understanding being that the goal is to be considered a desirable target, not a limit. If the goal is not achievable on the basis of technical feasibility, cost, or other factors, the regulatory agency may decide to accept a less stringent level. This strategy is reflected in certain of the EPA regulations, such as those pertaining to toxic air pollutants and the cleanup of contaminated soil and groundwater. With two such opposite strategies, standards that ultimately result in comparable risk control may, on the surface, appear to be quite different.

Another challenge is the increasing frequency with which federal agencies incorporate considerations of ethical and social factors in the establishment of dose and risk limits. This may require consideration not only of the economic costs of serious health effects but also of less quantifiable factors, such as the equitable sharing of costs and benefits, perceived public aversion to the given contaminant at any exposure level, and the costs and benefits that could accrue to those outside the at-risk population. Adding to these complications is the lack of interagency consensus on the amount of risk that is acceptable to the public and/or the environment. These types of unresolved issues raise serious questions about the precision, credibility, and overall effectiveness of federal standards and guidelines. Clearly, interagency guidance is needed to derive a structured approach for incorporating cost and benefit considerations into various protective strategies and to base their standards on a uniform limit on lifetime risk. Examples of existing disparities are illustrated in Figure 13.3.

Assessing Compliance with the Standards

Once the basic standards have been developed, derived or tertiary guides must be established for determining, through monitoring programs, whether the basic standards are being met. Since the standards include limits on

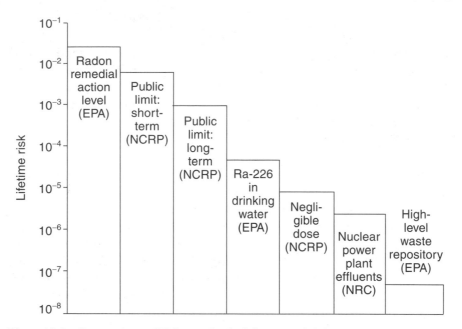

Figure 13.3 Comparison of U.S. standards (lifetime risks) for exposures to various sources of ionizing radiation

intakes of individual contaminants by the exposed population, their concentrations of individual contaminants within various environmental media (air, water, food), and similar limits to protect plants and nonhuman animals, confirmation of their acceptability will require detailed analyses of a wide range of samples from both the exposed populations and the environment, including the air, water from both aquatic and marine sources, soil, and agricultural products (Chapter 14).

Standards Based on an Acceptable Level of Risk

In 1977, the ICRP broke new ground by proposing a mathematical system to enable the commission to establish radiation protection standards based on what was considered an acceptable level of risk, namely, the probability of death due to the development of latent cancer (ICRP, 1977a). Just a little more than a decade later, this approach was broadened to include impacts other than the cancer fatality rate. Examples were the nature of the cancers, the relative importance of the affected organ, the probability of surviving, the associated health-care costs, and the years of productive life lost (ICRP, 1991). This led to the concept of *total detriment.*

The Concept of Total Detriment

To implement this concept, it was necessary to develop a method for esti-mating the total risk from all sources due to a specific stress. In the case of ionizing radiation, this included accounting for the doses from whole-body external exposures and those due to intakes of radionuclides that prefer-entially exposed single organs. This, in turn, required an understanding of the nature of exposures from internal, versus external, sources. To accom-plish this task, the ICRP developed the concept of *effective dose*—that is, the dose to the whole body that carries with it the same overall risk as a higher dose to a single portion of the body. One difference that had to be resolved was that external doses cease as soon as the radiation source is removed or the person leaves the area. In contrast, doses due to internally deposited radionuclides continue until they either decay and/or are biologically elim-inated. To account for this fact, the ICRP next developed the concept of *com-mitted dose.* In the case of workers (i.e., adults), this includes the estimated dose that would be received (accumulated) over the next 50 years following intake; for children, it is computed for the period up to 70 years of age (ICRP, 2007a). This accounts for the fact that members of the public may be exposed from the beginning of life, whereas most workers are not exposed until they are in the range of 20 years of age. An additional complication is that for internally deposited radionuclides with long biological and radio-active half-lives, the percentage of the calculated dose that a person will actually receive will depend on the age at which he/she is exposed. In fact, the NCRP (1993) estimates that, on average, for radionuclides with a long effective half-life (i.e., one for which both its radioactive and biological half-lives are long), in comparison with the remaining years of life for the ex-posed individual, the committed dose will overestimate the actual dose by a factor of approximately two, or more.

Implementing the Concept of Total Detriment

As would be anticipated, the probability coefficient for total detriment will be higher than that for fatal cancer alone (i.e., the increase will depend on the nature and characteristics of the cancer and many other factors). For this reason, it will be less for a cancer that has a low lethality rate, as contrasted

Table 13.2   Tissue-weighting factors and organ doses comparable to 20 mSv, the annual occupational dose limit recommended by the ICRP

Tissue or organ	Tissue-weighting factor (w$_T$)	Comparable organ dose rate limit (mSv)[a]
Bone marrow	0.12	150
Breast	0.12	150
Colon	0.12[b]	150[b]
Lung	0.12	150
Stomach	0.12	150
Remainder	0.12[c]	150
Gonads	0.08[c]	250[d]
Bladder	0.04	500
Esophagus	0.04	500
Liver	0.04	500
Thyroid	0.04	500
Bone surface	0.01	200
Brain	0.01	200
Salivary glands	0.01	200
Skin	0.01	200
Total	1.00	—[e]

a. Based on average annual dose rate during a specified 5-year period; rounded off to avoid implying more accuracy than warranted.

b. The dose to the colon is the mass-weighted mean of the upper and lower large intestine.

c. The values for the tissue-weighting factors for the remainder and the breast were increased by 0.07 each, plus 0.01 for the salivary glands and brain, for a total increase of 0.16. To compensate, those for the bladder, liver, esophagus, thyroid, bone surface, and skin were each reduced by 0.01, and that for the gonads by 0.12, for a total reduction of 0.16.

d. The dose to the gonads is the mean to the testes and ovaries.

e. Not applicable.

to one with a high rate, and for one that causes fewer, as contrasted to more, years of productive life lost (ICRP, 1977b). Since the task of implementing this concept was formidable, the ICRP turned to Sir Edward Pochin, an ICRP member who was globally acknowledged for his leadership in the field of risk assessment, to lead this effort. Based on an exhaustive review of the literature, he prepared what can best be described as a masterpiece. The outcome of his efforts (ICRP, 1985) provided a procedure that, with the input of factors to account for the rate of mortality, the years of productive life lost, and other considerations, provided a basis on which to accomplish this task.

To implement the concept, the ICRP developed what are defined as *tissue-weighting factors,* one for each of the full range of tissues and organs in the body. Each of these reflect the importance of the tissue relative to that of the total body and enable radiobiologists to establish a dose-rate limit for that tissue that would yield a dose whose risk would be equivalent to the recommended occupational limit (effective dose) for the whole body. To avoid implying more accuracy in the calculations than was justified, the factors were assigned to one of four groups, and the distribution of organs assigned to each was such that the sum of all the factors was 1.00. A summary of the latest tissue-weighting factors is presented in Table 13.2 (ICRP, 2007a).

Current Radiation Protection Standards

The latest ICRP recommended dose-rate limits in planned exposure situations for workers and members of the public are summarized in Table 13.3. As may be noted, the latest suggested occupational dose rate limit is 20 mSv per year, averaged over a defined period of 5 years. That for members

Table 13.3 Recommended dose limits in planned exposure situations[a]

Type of limit	Occupational	Public
Effective dose	20 mSv per year, averaged over defined periods of 5 years[e]	1 mSv in a year[f]
Annual equivalent dose in:		
Lens of the eye[b][c]	150 mSv	15 mSv
Skin[c][d]	500 mSv	50 mSv
Hands and feet	500 mSv	—

a. Limits on effective dose are for the sum of the relevant effective doses from external exposure in the specified time period and the committed effective dose from intakes of radionuclides in the same time period. For adults, the committed effective dose is computed for a 50-year period after intake, whereas for children it is computed for the period up to age 70 years.

b. This limit is being reviewed by an ICRP task force.

c. The limitation on effective dose provides sufficient protection for the skin against stochastic effects.

d. Averaged over 1 cm² area of the skin regardless of the area exposed.

e. With the further provision that the effective dose should not exceed 50 mSv in any single year. Additional restrictions apply to the occupational exposure of women.

f. In special circumstances, a higher value of effective dose could be allowed in a single year, provided that the average over 5 years does not exceed 1 mSv per year.

of the public is 1 mSv per year. Also specified are annual equivalent dose rates for the lens of the eye, the skin, and the hands and feet for workers, and the eye and skin for members of the public. Comparable recommended occupational dose rate limits for other organs of the body were provided in Table 13.2. Even though the annual dose rate limit stipulated in the regulations of the U.S. Nuclear Regulatory Commission remains at 50 mSv, their regulations also required that all exposures be maintained *as low as reasonably achievable*. The result is that the average annual dose to workers at the 104 U.S. commercial nuclear power plants in 2005 was less than 1 mSv (Chapter 12).

Other Guidance and Recommendations of the ICRP

During the latter part of the twentieth century, the ICRP recognized that the existing radiation protection recommendations were not only more complicated than necessary, they were also not "user friendly." To correct the situation, they adopted a new approach in the development of such standards that is outlined in Publication 104 (ICRP, 2007b). In an opening editorial, Abel J. Gonzelez, chair of the task group that prepared the report, made the following statement that reflects a portion of the proposed changes:

> Why does society require control of some endeavors and disregard others? Radiological protection has not been immune to this dilemma: Why have relatively modest radiation exposures been subject to rigorous control, while relatively high (and controllable) exposures not been uncontrolled (sometimes on the dubious grounds that the exposures were "natural" in origin)?

In preparing Publication 103, the ICRP made a special effort to conduct the process in the open. In so doing, they adopted as a basic philosophy that protection of the individual member of society should be the primary goal, the reason being that, if this goal were accomplished, society, as a whole, would be protected. Additionally, they decided to express all limits not only in units of dose rates but also in terms of how these compare to those from natural background radiation. The importance of this last recommendation is illustrated by the fact that more than half of the dose being received by average members of the U.S. public is from natural sources (Figure 12.7, page 293).

In response, many national regulatory agencies have initiated action to revise their regulations and standards. This includes the USNRC. In this case, the primary effort is being directed to an update of the regulatory guides for implementing its "Standards for Protection Against Radiation" (USNRC, 2005). As might be expected, this effort could require several years or more to complete. Although there may be minor differences in the regulations developed by various countries, the international community is united in its approach to this effort. In fact, representatives from multiple nations are directly involved in ICRP activities, and drafts of Publication 103 received detailed reviews by and comments from several international organizations, including the International Atomic Energy Agency (Osborne, 2009).

Progress in Cancer Prevention and Cure

One favorable development is the rapid progress being made in the development of methods for the prevention and medical therapies for curing the more common cancers that afflict humankind (Chapter 15). Between 1993 and 2001, for example, the overall annual death rate from cancer among members of the U.S. population was reduced by an average of 1.1%. Between 2002 and 2004, the rate of reduction had been increased to 2.1%, and this rate of reduction continued through 2006. Since cancer is the primary health impact of radiation exposures and chemical carcinogens, this will lead to corresponding reductions in the risks per unit dose for all types of such agents. Although this does not mean that radiation exposures and the release and control of carcinogenic agents can be ignored, it is certainly encouraging news. This is discussed in more detail in Chapter 15.

Continuing Challenges

EXPLORATION OF OUTER SPACE

One of the latest challenges is to develop the methodology for establishing standards for astronauts in outer space. This is important not only if the Earth-orbiting shuttle program continues to operate or is replaced by more advanced such systems, but also if plans materialize for establishing a permanent colony on the moon or for traveling to distant planets such as Mars. All of these need to be addressed because radiation exposures in the environments that will be encountered consist of a wide range of incident particles, some of very high energies. As might be expected, these produce a wide range of deposition patterns in tissue. Also to be recognized is the potential impacts of solar flares (NCRP, 2006). Dose limits are currently being expressed in terms of effective dose and Gray equivalent, applying

Table 13.4 Ten-year career dose limits for astronauts in low Earth orbit

Age at exposure (years)	Effective dose limit (Sv)[a]	
	Female	Male
25	0.4	0.7
35	0.6	1.0
45	0.9	1.5
55	1.7	3.0

a. These limits are based on an excess lifetime risk of cancer mortality of 3 percent.

appropriate weighting factors for the types of radiation and health effects of concern. Data are being gathered that should lead to significant improvements in the accuracy of the associated risk estimates. A summary of recommendations for the 10-year career dose limits for astronauts in low Earth orbit is presented in Table 13.4 (NCRP, 2002). As may be noted, the limits for females are less than those for males. This is due to the increased susceptibility of women to breast cancer.

The General Outlook

Clearly emerging from this review, especially in the case of the United States, is the need to harmonize the methodologies for developing and the procedures for applying occupational and environmental standards. This is true in terms of methods for calculating doses to the public and estimating the associated risks, as well as for determining what levels of risk are acceptable. As noted earlier (Figure 13.3), there are differences in the lifetime risks associated with the U.S. standards that have been established for various environmental contaminants. These diverge by orders of magnitude. This is due to several causes: the work of many federal agencies is conducted independently, and, in some cases, the standards were developed without adequate input and review by the scientific community. One of the immediate benefits of a harmonized system would be an enhancement in the cost-effectiveness of environmental controls. Indeed, if all standards had a common risk basis, the current practice of trade-offs (tradable emission permits) among various contaminants might be expanded to include trade-offs among releases of the same contaminant to different components of the environment. Ultimately, it might even become possible to effect trade-offs among different types of contaminants and/or stresses—for example, reducing releases of chemical or radiological contaminants to compensate for those of biological and/or physical agents. Due to the wide range of associated implications, however, trade-offs of occupational and environmental stresses in this manner would require considerable review, evaluation, and deliberation.

These efforts, however, should not end there. They need to be expanded so that ultimately all countries of the world use the same approaches and methodologies. Only with such uniformity can one be sure that evaluations of the risks and applications of control, especially in terms of the environmental contaminants, are conducted on a comparable basis worldwide. One immediate advantage internationally would be that the exportation of hazardous industries from one country to another to avoid regulatory controls would no longer be a viable choice.

Also needing to be addressed is the continuing public misconception that compliance with environmental standards will ensure risk-free conditions. Although many standards are expressed as single numerical limits, organizations such as the ACGIH stress that, in the case of occupational exposures to airborne toxic chemicals, these limits do not represent "fine lines between safe and dangerous concentrations" nor do they represent "a relative index of toxicity" (Chapter 4). Similar qualifying statements accompany the limits recommended by the ICRP (2007a) and the NCRP (1993).

Finally, there is a need to recognize that the establishment of limits for various environmental physical, biological, chemical, and radiological agents will require continuing updates as methods are developed for the prevention and cure for many of the maladies they can cause. Also to be considered, as noted earlier, is the increasing acknowledgment (Osborne, 2009) on the part of the ICRP and other organizations that there is a relatively high threshold for the health effects of ^{226}Ra and ^{239}Pu, as well as ^{237}Np and ^{241}Am. Obviously, the establishment of standards for occupational, population, and environmental limits will continue to be a challenge for years to come.

14

EFFLUENT MONITORING AND ENVIRONMENTAL SURVEILLANCE

ENVIRONMENTAL monitoring programs were initially conducted on a local basis and had two basic objectives: (1) to estimate exposures to people resulting from certain physical stresses (such as noise and radiation) and from toxic materials that are being, or have been, released and are subsequently being ingested or inhaled and (2) to determine whether the resulting exposures complied with the limits prescribed by the regulations. Such programs were either *source* related or *person* related. Source-related programs are designed to determine the exposures or dose rates to a specific population group resulting from a defined single source or practice. Person-related programs are designed to determine the total exposure from all sources to a specific population group. The various pathways of exposure, as a result of the release of toxic materials into the environment, are illustrated in Figure 14.1. The key factors that should be considered in the design of an environmental monitoring and surveillance program and the performance of the associated dose assessments are outlined in Figure 14.2.

Guidance on Environmental Monitoring

Although multiple federal agencies are active in this field, the U.S. leaders are the Environmental Protection Agency (EPA), the National Laboratories of the U.S. Department of Energy (USDOE), and the U.S. Nuclear Regulatory Commission (USNRC). Also contributing are independent organizations, such as the International Commission on Radiological Protection (ICRP), the National Council on Radiation Protection and Measurements

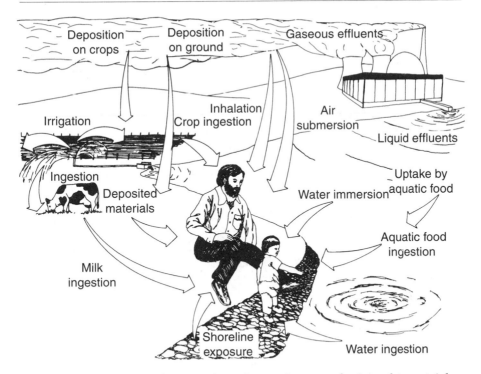

Figure 14.1 Exposures to humans through aquatic, atmospheric, and terrestrial pathways

(NCRP), and the American National Standards Institute (ANSI). The EPA standards are based on multiple laws that have been passed by the U.S. Congress. These include those pertaining to air and water pollution, in general, and for the control of releases of toxic chemicals and radionuclides into the atmosphere, onto land, and into surface and groundwater. In recent years, specific attention has been directed to the protection of wetlands and endangered species (Wikipedia, 2009). In the case of the USNRC, the regulations for radionuclides are expressed in terms of limits on the total annual intake of specific radionuclides. In cases where a specific radionuclide is being consumed from several sources (i.e., air and water), *secondary* limits are developed to express the limits for each source in terms of concentrations (USNRC, 2005).

Industrial organizations, licensed to operate nuclear facilities by the USNRC, have also been proactive through the conduct of their RETS-REMP *(Radioactive Effluent Technical Specifications—Radiological Environmental*

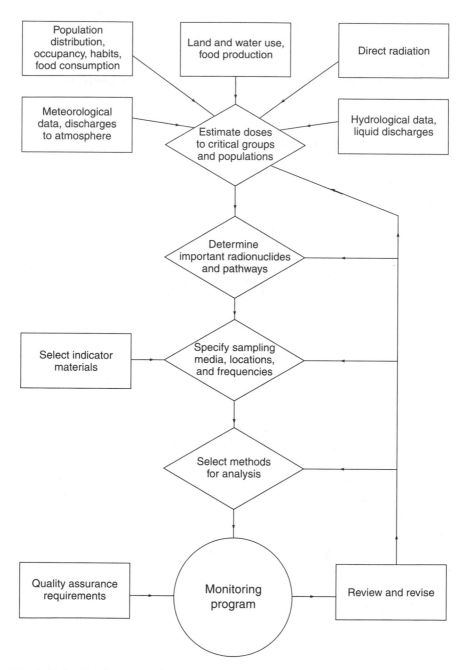

Figure 14.2 Key factors in the design of an environmental monitoring and surveillance program and the performance of the associated dose assessments

Monitoring Programs) forums. The most comprehensive guidance, however, is that provided by the U.S. Department of Energy (USDOE, 1991). As part of their activities, each of its facilities issues an annual Site Environmental Report (PNNL, 2008). The ANSI and its Accredited Standards Developer organizations also prepare useful information for environmental monitoring. For example, ANSI/HPS N13.1-1999 (HPS 1999) contains guidance on sampling and monitoring airborne releases. This consensus standard is recognized worldwide and served as the model for a similar standard prepared by the International Organization for Standardization (EPA et al., 2001; ISO, 2009).

Related guidance is provided on the adequacy of the cleanup of Superfund sites (i.e., those being converted into brownfields) (Chapter 9) and commercial nuclear power plants being decommissioned by nuclear utilities, as well as facilities being remediated by the USDOE. To meet the latter two needs, the *Multi-agency Radiation Survey and Site Investigation Manual* (MARSSIM) has been developed collaboratively by four federal agencies (the EPA, USDOE, USNRC, and the U.S. Department of Defense) having authority and control of radioactive materials. MARSSIM provides comprehensive guidance on the basic methods for (1) translating regulatory criteria, expressed in dose rate or risk limits, into derived guides, expressed, for example, in terms of contaminant concentrations in the soil; (2) acquiring scientifically sound and defensible data, using proper field and laboratory equipment and techniques, on the contaminants present at the site; and (3) documenting that the resulting data confirm that the sites meet the regulatory criteria. It also provides specific guidance for radiation surveys. Organizations that apply the prescribed techniques can do so with the knowledge that they are following practices that will be acceptable to regulatory agencies.

Scope of Environmental Monitoring Programs

It is increasingly being recognized that the assessment of risks solely to human health or focusing on problems only on a local scale may be inadequate. The purposes and goals have expanded far beyond these earlier objectives. The long-standing assumption had been that if humans are protected, biota and animals will also be protected. Scientists are now assessing impacts to biota (USDOE, 2002; ICRP, 2008; NCRP, 1991). It is now also recognized that environmental impacts should be examined on a regional, local, and global basis (EPA, 1993). This includes assessments of the impacts on ecosystems and evaluations of factors that may have wide-scale,

Table 14.1 Types and purposes of environmental monitoring and surveillance programs

Type of program	Purpose
Based on nature of the stress	
Physical stress	To assess the impact of environmental stresses such as noise and external radiation, where the evaluation is based primarily on exposure measurements made in the field, not on samples collected and returned to the laboratory for analysis
Chemical stress	To assess exposures resulting from the ingestion and inhalation of chemical and radioactive contaminants
Based on geographic (spatial) coverage	
Local	To evaluate the impact of a single facility on the neighboring area
Regional	To evaluate the combined impact of emissions from several facilities on a large area
Global	To determine worldwide impacts and trends, such as acidic deposition, depletion of the ozone layer, and potential for global warming
Based on temporal considerations	
Preoperational	To determine potential contamination levels in the environment prior to operation of a new industrial facility; to train staff; to confirm operation of laboratory and field equipment
Operational	To provide data on releases; to confirm adequacy of pollution controls
Postoperational	To ensure proper site cleanup and restoration
Based on monitoring objectives	
Source related	To determine population exposures from a single source
Person related	To determine total exposure to people from all sources
Environment related	To determine impacts of several sources on features of the environment such as plants, trees, buildings, statues, soil, water, and ecosystems
Research related	To determine transfer of specific pollutants from one environmental medium to another and to assess their chemical and biological transformation as they move within the environment; to determine ecological indicators of pollution; to confirm that the critical population group has been correctly identified and that models being applied are accurate representations of the environment being monitored
Based on administrative and legal requirements	
Compliance related	To determine compliance with applicable regulations
Public information	To provide data and information for purposes of public relations

long-range effects. The types and purposes of environmental monitoring and surveillance programs are summarized in Table 14.1.

Local environmental monitoring programs for industrial facilities are generally conducted by plant personnel or environmental service contractors, whereas regional programs are the responsibility of state and local environmental health and regulatory authorities. The planning and coordination of national programs is managed by federal agencies. Close coordination between the facility operator and the local agencies is necessary if all objectives of the monitoring program are to be met. A well-planned program will usually involve some overlap in the activities of the several monitoring groups, including exchanges of samples and cross-checking of data.

Monitoring Physical Stresses and Toxic Materials

Because of differences in the nature of the exposures, the monitoring of physical stresses and toxic materials requires different approaches. In the case of the physical stresses, monitoring may simply involve identifying their sources and measuring their magnitude. Expanding this to include measurements of the distribution of their energies can provide data for estimating the accompanying dose as a function of tissue depth and specific body organ. For some stresses, such as electric and magnetic fields, where there is a lack of information on which attributes of the sources may cause harm, there will be questions as to what types of measurements should be made. Measurements to determine exposures from stresses such as noise and external sources of ionizing and nonionizing radiation must commonly be made on a real-time basis. Generally, instruments are placed near the people being exposed or in concentric rings at various distances from the source. It is important to recognize, however, that the presence of people and monitoring equipment may alter the environment in such a way as to make accurate measurements difficult. In addition, the position and location of the people being exposed (for example, whether they are standing on the ground or near a tree, or sitting inside an automobile) can nullify the usefulness of the resulting data.

Assessments of toxic materials in airborne or waterborne releases are commonly made at the points of release. Additional steps include assessing the movement or transport of specific contaminants within given environmental media (air, water, soil), their transfer from one medium to another, and their chemical and biological transformation as they move within the environment. This is necessary to estimate their deposition within the body and uptake by various organs. Because many contaminants can cause exposures by several avenues, most environmental monitoring specialists try to identify and trace the movement and behavior of several representative contaminants through several environmental pathways so as to identify those contaminants and pathways that are most important.

From the standpoint of health effects, the physical and chemical properties of airborne particles and gases are noteworthy since they are the primary factors that determine both their behavior within the environment and their health impacts. As noted earlier (Chapter 5), this is especially important in the case of the size distribution and chemical composition of airborne particulates. In addition, it is important to determine what other chemicals are associated with a given contaminant, since certain combinations are synergistic. In a similar manner, certain factors are important from the standpoint of gases. For example, sulfur dioxide, a ubiquitous acidic gas that is highly soluble and is ordinarily taken up entirely in the throat and upper airways (where its effects on health are minimal), acutely impairs the functioning of the lungs when carried to the alveoli as an acid condensed on the surfaces of small airborne particles.

Measuring Waterborne and Airborne Exposures

In most cases, air and water serve as the principal pathways for direct exposures (through inhalation and the consumption of drinking water) and as a vehicle for the transport of contaminants from the point of release to other environmental media (such as milk and food). Measurements of the airborne and waterborne contaminants leaving a plant can also provide advance information on pending problems in other environmental media. Since critical contaminants can be missed if only the obvious and easily measured effluents are monitored, or if monitoring is discontinued during shutdowns for repairs and maintenance, sample collection and analysis should be conducted during all phases of plant operations.

ASSESSING WATERBORNE RELEASES
A range of samples can be collected to assess the impact of waterborne releases. These include:

> *Grab samples.* Since these are collected on a one-time basis, they
> represent at best a snapshot of the characteristics of the effluent

stream. Unless its composition is relatively uniform with time, such samples will not provide useful information.

Composite samples. These are composed on a blending of a series of smaller samples. Although these represent a combination of the characteristics of the effluent stream at the times of collection, they are better than single-grab samples.

Timed-cycle samples. These are collected in equal volumes at regular intervals. Nonetheless, they represent the characteristics of the effluent stream only if the flow rate is relatively constant and its characteristics are relatively uniform.

Continuous flow-proportional samples. These are collected on a continuous basis in proportion to the volume of flow. As a result, they should provide a sample that is representative of the effluent stream.

Indicator samples. These include living organisms and plants. Since they concentrate various contaminants, they can provide useful data in identifying contaminants whose concentrations in lakes and streams are below the limits of analytical sensitivity. Information on the history of releases can often be obtained through similar analyses of bottom deposits and their impacts, for example, the performance of a global inventory of coral reef stressors based on satellite-observed nighttime lights (Aubrecht et al., 2008). Quantifying the amounts of the identified contaminants, however, can be difficult (Griffith and Hunsaker, 1994), and the results can have significant special uncertainties (Hunsaker et al., 2001).

The quantity of sample collected depends on the number and nature of the parameters being tested. It should be sufficient to permit all the desired analyses, allowing for possible errors, spillage, and sample splitting for purposes of quality control. If, after collection, the sample is placed in a bottle for transport to the laboratory for analysis, care must be taken to ensure that ionic species or small particles suspended in the waste do not agglomerate or attach themselves to the walls of the container. This problem can be avoided by an appropriate choice of bottle, by adjusting the pH, or by adding stabilizing chemicals to the sample prior to placing it in the container. Similar steps, including refrigeration, will ensure that the samples are properly protected against deterioration due to either chemical or biological processes. Having addressed all these issues, it is important, as mentioned earlier, that they be analyzed using approved methodologies.

ASSESSING AIRBORNE RELEASES

As with liquid wastes, the initial step is to sample the various release points at the facility being monitored. Once released, airborne contaminants can be rapidly dispersed, so it is important that the characteristics of the effluent be determined prior to their release. This is accomplished through what is called stack sampling. Selection of the sites where environmental samples are collected should be based on the best available meteorological information coupled with data on local land use. Sites for monitoring the impacts on various ecosystems will require a similar approach, the exposed entity in this case being the environment, not people.

Samplers are generally of two types: those that employ (1) a filter or electrostatic precipitator to collect airborne particles and (2) an appropriate set of adsorbers to collect gaseous and volatile contaminants. To ensure accuracy in the resulting data, all sampling pumps should be calibrated so that the flow rate is well known. The size distributions of airborne particles can be determined by cascade impactors or other mechanical separation devices. The choice of sampler depends on the desired sample volume, sampling rate, power requirements, servicing, and calibration. The minimum amount of air to be sampled is dictated by the sensitivity of the analytical procedure. The amount is often a balance between sensitivity and economy of time. As in any monitoring program, care must be exercised to ensure that the samples are representative.

At the same time, rapid advances in technology have led to a variety of new techniques for monitoring contaminants in real time on a localized basis. These include portable gas chromatographs for monitoring airborne contaminants in the field. One of the advantages of this technology is that it permits the analyst to identify and quantify the concentrations of organic compounds, such as benzene, toluene, and xylene, on an individual basis. Since the data are immediately available, more samples can be processed and the boundaries of contamination better defined. In some cases, as many as nine instruments can be linked by radio to a central command post at a site. Should the concentration of volatile organic compounds exceed a preset action level, this information will be automatically reported so that appropriate action can be taken (Ebersold and Barker, 2003).

Measurements of air pollution can also be performed through the use of optical remote-sensing technology, one example being the use of reflected infrared waves to measure the rate at which an atmospheric plume absorbs energy (EPA, 2001b). By applying techniques similar to those used in medical CT scans (Chapter 12), this enables scientists to measure the concentrations of hazardous air emissions in multiple single-plane (sliced) images within a plume. These data are then analyzed using computer models to determine the concentrations of contaminants both vertically and horizontally. This technique has proved especially useful in monitoring area source emissions (for example, those from motor vehicles) that are either too numerous to measure individually or are released from a single source over a large area (for example, large municipal landfills). Other applications include refineries, where this technique permits expensive and troublesome regulation of fugitive releases to be replaced by a "cap" for the entire facility that, in turn, can be monitored continuously (EPA, 2001b). Summarized in Table 14.2 are the advantages and disadvantages of various sampling methods for the principal types of environmental contaminants and receptor media. Current guidance on environmental surveillance and environmental protection has been prepared by the New Jersey Department of Environmental Protection (Fucillo, 2009).

MONITORING THE INDOOR ENVIRONMENT

Although the earlier discussion has been directed to monitoring the ambient environment, there is an equivalent need to monitor the indoor environment, particularly in terms of various airborne contaminants. Two such sources are the airborne contaminants generated by molds and outdoor pollens. As previously noted (Chapter 5), molds can grow on any substance, including wood, paper, carpet, and food, so long as moisture is present. Molds can often be detected by their characteristic odor or a discoloration of growth on ceilings, walls, floors, or within the heating, ventilating, and air conditioning (HVAC) system. Since they can grow behind a wall or in sections of the HVAC system, their presence may not be recognized. For these reasons, confirmation of their presence generally requires the collection of air samples or wipes of room surfaces, followed by analysis at a qualified environmental microbiological laboratory.

Whether monitoring the indoor or outdoor environment, it is often difficult to interpret the observations in terms of doses to those exposed. Few people, for example, spend significant amounts of time at specific locations either indoors or outdoors. For this reason, increasing efforts have been devoted to the development of methods to evaluate the concentrations of contaminants in the air actually being breathed by people. This has been made possible by the development of personal samplers that can be worn

Table 14.2 Advantages and disadvantages of various environmental sampling methods

Type of sample	Advantages	Disadvantages
Atmospheric environment		
Direct measurement		
Real-time field measurements of physical stresses such as noise and radiation	Monitors can be put in place to assess time-integrated exposures	Monitor often disturbs field being monitored; some monitoring equipment (e.g., for assessing electric and magnetic fields) is expensive and complex
Airborne particulates		
Respirable fraction via air sampling	Direct-dose vector; provides data on potential effects on lungs	Omits larger particles that may be significant when deposited in nose, mouth, and throat
Total particulates via air sampling	Provides data for assessing doses to lungs as well as possible effects on skin and intake through ingestion	Not all measured contaminants are respirable
Collection of settled particulates	Represents an integrated sample over known time and geographical area	Weathering may alter results; only large particles are collected by sedimentation
Gases		
Integrated (concentrated) sample	Concentration of samples permits detection of lower concentrations in air	Samples must usually be analyzed in laboratory; chemical reactions may change nature of collected compounds
Direct measurement	Provides data on real-time basis	Lower limit of detection may not be adequate
Terrestrial environment		
Milk	Direct-dose vector, especially for children; data easily interpreted	Milk samples are not always available

Foodstuffs	Direct-dose vector; data easily interpreted	Samples are not always available from areas of interest; weathering and processing may affect samples
Wildlife	Direct-dose vector	High mobility; not always available; data difficult to interpret
Vegetation	Samples readily available; multiple modes available for accumulating contaminants (by direct deposition and leaf and root uptake)	Data are difficult to interpret; weathering can cause loss of contaminants; not available in all seasons
Soil sampling	Good integrator of deposition over time	High analytical cost; data difficult to interpret in terms of population exposure and dose
Aquatic environment		
Surface water (nondrinking)	Readily available; indicates possibility of contamination by aquatic plants and animals	Not directly dose related; difficult to interpret data
Groundwater (nondrinking)	Indicator of unsatisfactory waste-management practices	Not always available; data difficult to interpret because of possibility of multiple remote sources
Drinking water	Direct-dose vector; consumed by all population groups	Contaminant concentrations are frequently very low
Aquatic plants	Sensitivity	Data are difficult to interpret; not available in all seasons
Sediment	Sensitivity; good integrator of past contamination	Data are difficult to interpret because of possibility of multiple remote sources
Fish and shellfish	Direct-dose vector; sensitive indicator of contamination	Frequently unavailable; high mobility
Waterfowl	Direct-dose vector	Frequently unavailable; high mobility; data are difficult to interpret

by individual members of the public and are designed to evaluate the quantity of various contaminants being inhaled (Chapter 5). Such samplers are similar to the personal monitoring devices available for assessing doses from external radiation sources. Important characteristics of such samplers are that they have minimal power requirements and are relatively quiet and lightweight.

Designing an Environmental Monitoring Program

One of the first steps in designing an environmental monitoring program is to define its objectives and how the data are to be handled and evaluated (Table 14.3). Not only must the program be planned so that the relevant questions are asked at the right time but also so that the data necessary to answer these questions are obtained. As a result of the increasing sophistication of our understanding of the environment, and accompanying technological developments, it is mandatory that these questions be addressed on a continuing basis. In fact, the answers to them will never become final. Other attributes of a successful environmental monitoring program are that (1) its cost is low enough to survive unexpected reductions in supporting funds, (2) it is simple and verifiable so that it is not significantly affected by changes in personnel, and (3) it includes measurements that enable it to be modified to respond to changes within the environment.

Tracing the movement of all contaminants through all potential pathways would be physically and economically impossible. Fortunately, in most cases, the primary contributors to population dose consist of no more

Table 14.3   Questions to be answered when implementing an environmental monitoring and surveillance program

Program stage or component	Question
Purpose	What is the goal or objective of the program?
Method	How can the goal or objective be achieved?
Analysis	How are the data to be handled and evaluated?
Interpretation	What might the data mean?
Fulfillment	When and how will attainment of the goal or objective be determined?

than half a dozen toxic agents moving through no more than three or four pathways. Once these are identified, along with the habits of the people living or working in the vicinity, it should be possible to identify a representative individual whose exposures would provide information on the average dose to the exposed population (ICRP, 2006). Another approach is to identify the reasonably maximally exposed individual (RMEI), whose location and habits and accompanying doses would be representative of the maximum received. This, for example, served as the basis for setting the standards for the control of radionuclide releases from the High-Level Radioactive Waste Repository in Yucca Mountain, NV (EPA, 2001a).

Most environmental monitoring programs have at least five stages: (1) gathering background data, (2) identifying and evaluating the various pathways of exposure, (3) collecting and analyzing samples, (4) establishing temporal relationships, and (5) confirming the validity of the results. Although this would imply that the various steps are applied sequentially, in reality decisions relative to each step must be made on an iterative basis. Also important is that the analytical procedures be supported by a formal program for quality control. This will be discussed later.

BACKGROUND DATA

Before monitoring begins, background information is needed on other facilities in the area, the distribution and activities of the potentially exposed population, patterns of local land and water use, and the local meteorology and hydrology. These data permit identification of potentially vulnerable groups, important contaminants, and likely environmental pathways whose media can be sampled. This requires an evaluation of natural features (climate, topography, geology, hydrology), artificial features (reservoirs, harbors, dams, lakes), land use (residential, industrial, recreational, dairying, farming of leaf or root crops), and sources of local water supplies (surface or groundwater). Results from a monitoring program conducted before a facility begins operation can be used to confirm these analyses and establish baseline information for subsequent interpretation.

EVALUATION OF PATHWAYS OF EXPOSURE

Contaminants released from an industrial facility may deposit in many segments of the environment, and their quantity and composition will vary both with time and the nature and extent of facility operations. For example, a secondary lead smelter has the potential to release elemental lead and associated compounds into the atmosphere, whereupon they may become a source of exposure from inhalation. These releases may subsequently contaminate the soil, groundwater, and surface water and be taken up by fish and agricultural products (Figure 14.3). This can lead to higher-than-normal concentration of lead in milk from cows and beef from cattle grazing on adjacent pastures. Children playing on the ground have likewise shown elevated lead concentrations in their blood. Arsenic emitted by copper smelters follows identical pathways of contamination and human exposure. Contaminants discharged to the liquid pathway can pose similar problems. It is also important to recognize that surface water can become groundwater through percolation. If the water is used for irrigation of agricultural crops, they will also be contaminated.

Even so, unsuspected pathways can prove to be important. For years, operators of a major nuclear facility in the United Kingdom disposed of low-level liquid radioactive wastes into the Irish Sea, theoretical evaluations

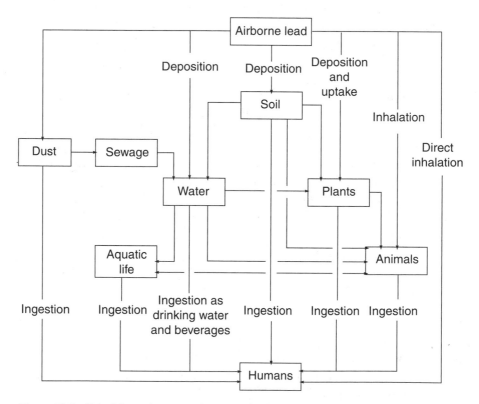

Figure 14.3   Possible pathways to humans for lead releases into the atmosphere

having shown that this would be acceptable. Later, they discovered to their dismay that certain population groups were consuming larger quantities of radioactive material than had been anticipated, the source being seaweed that they converted into flour to make bread. Related studies revealed that pigeons that fed in areas close to a nearby nuclear fuel-reprocessing site were serving as a pathway for the offsite transport of radionuclides. This was due to the fact that they were attracted to local gardens and their droppings proved to be a source of contamination. Resolving the problem included the removal and replacement of the topsoil in the gardens and the establishment of measures to prevent access by the pigeons to contaminated areas, combined with reductions in the pigeon flock through culling (Wilkins, 1999).

SAMPLE COLLECTION AND ANALYSIS

In most environmental monitoring programs, trade-offs must be made to obtain adequate coverage of critical contaminant-pathway combinations at satisfactory analytical sensitivities and costs. Samples collected directly from effluent streams discharging into the air and water generally contain the largest number of contaminants at the highest concentrations. Analyses of these samples can provide information on the specific contaminants being released as well as the amounts anticipated to be present in the neighboring environment. Once sampling and analyses shift to the environment, it is better under essentially all conditions to collect and carefully analyze a small number of well-chosen samples taken at key locations to provide a reliable index of environmental conditions than to process larger numbers of poorly selected, nonrepresentative samples.

At the same time, there may be a need to measure certain other contaminants because of the history of operations at the site or because of specific concerns of the local population. For example, the collection and analyses of possible contaminants in oranges in a major citrus-producing area may be necessary regardless of the concentrations anticipated; the same is true with respect to contaminant levels in cranberries if they are a major source of income to local farmers. Other examples are analysis for radioactive iodine in milk-producing areas near commercial nuclear power plants and analysis for plutonium in the agricultural land near nuclear facilities operated by the USDOE. With the increased interest in ensuring that the environment is being adequately protected, it may also be wise to collect and analyze samples that will indicate the range of exposures to plants and animals within various ecosystems.

Because they are faster and less expensive than analyses for specific contaminants, gross measurements of the concentrations of "total suspended particulates" in the atmosphere are sometimes used as a surrogate for, or indicator of, trends in the concentrations of particles in the PM_{10} or $PM_{2.5}$ size range. This is acceptable only so long as the relationships between the surrogate and the specific particle size group remain reasonably constant. To ensure that this is the case, the relationship should be verified on an intermittent basis, and especially when the nature of industrial operations or traffic patterns in an area change. Another example of this type of practice is the use of *Escherichia coli* (coliform organisms) as indicators of the possible presence of feces and accompanying disease organisms in drinking water (Chapter 7). Here, again, it is important to recognize the limitations associated with such measurements.

Another factor to be considered is the wide range of impacts resulting from continuing reductions in the permissible limits for various contaminants in essentially every type of environmental media. One of these is the need for more sensitive and more accurate analytical capabilities. Although one way to solve this problem is to collect larger samples, this is not always necessary, since technological developments have in many cases enabled the more sensitive measurements to be made on existing size samples. At the same time, however, the ability to detect almost any amount of a contaminant, regardless how small, will raise fears in certain segments of the public. Until agreement is reached on what levels of exposure to people are sufficiently small to be considered negligible, this situation will continue.

APPLICATION OF STANDARD PROCEDURES FOR SAMPLE ANALYSIS

For many years, several professional environmentally related societies have been active in the development of standard procedures for analyzing environmental samples. Three are the American Public Health Association, the American Water Works Association, and the Water Environment Federation. The American Public Health Association has for more than 60 years published a combination of these recommendations in *Standard Methods for the Examination of Water and Wastewater* (Eaton at al., 2005). This book provides step-by-step procedures for more than 350 testing methods, including detailed procedures for sampling and analyzing a full range of biological and chemical contaminants. Another guide is an intersociety committee report, *Methods of Air Sampling and Analysis* (Lodge et al., 1988). These efforts complement the development of the MARLAP, or *Multi-Agency Radiological Laboratory Analytical Protocols Manual* (EPA et al., 2001). Enhancing the value of this report is that the techniques presented have been approved by the EPA for generating the types of data required in documenting compliance with federal regulations. Similar approval has been granted to the two professional society reports.

TEMPORAL RELATIONSHIPS

In the case of nuclear facilities, each type of release generates a characteristic pattern between the time of its occurrence and the time of the subsequent exposures. Similar relationships exist with respect to certain aspects of the exposures to toxic chemicals. The primary exception is that in the case of radioactive materials, releases to the atmosphere will subject people almost immediately to direct radiation from the cloud. Exposures resulting from the deposition of airborne releases onto the Earth will take longer, and uptake by agricultural crops and pasture grass longer still. Exposures to the bone and other organs such as the thyroid will be delayed, pending uptake and transfer of the material from the lungs. Likewise, exposures to specific organs (other than the stomach and gastrointestinal tract resulting from the ingestion of toxic agents in milk or food) will be delayed until the material is taken up by the blood and deposited in specific body organs. Acute effects from environmental exposures will appear within hours to weeks (depending on the dose); delayed effects (such as latent cancers) from lower-level exposures will not appear for some years. These relationships are depicted in Figure 14.4.

QUALITY-CONTROL REQUIREMENTS

To be effective, an environmental monitoring program must be supported by a sound quality-control program. This must include (1) acceptance testing or qualification of laboratory and field sampling and analytic devices, (2) routine calibration of all field-associated sampling equipment and flow-measuring instrumentation, (3) a laboratory cross-check program, (4) replicate sampling on a systematic basis, (5) procedural audits, and (6) documentation of laboratory and field procedures and quality-assurance records.

To facilitate meeting these requirements, the EPA, NIOSH, and the National Institute of Standards and Technology (NIST) make available a variety of standard and cross-check samples and provide guidance for the establishment and operation of such programs. The USNRC, for example, requires that all laboratories performing analyses of environmental samples from commercial nuclear power plants participate in the EPA program or its equivalent. Similar procedures are mandatory in support of data collected for demonstrating compliance with environmental regulations.

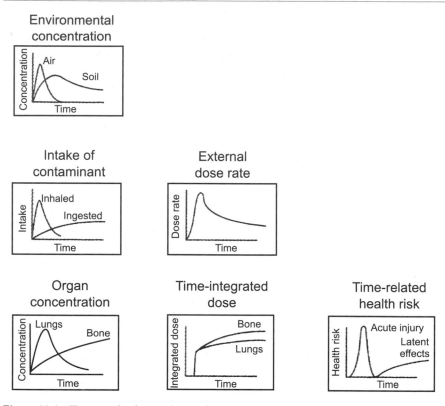

Figure 14.4 Temporal relationships of various types of contamination and accompanying exposures due to environmental releases

Meeting Special Needs

In certain cases, monitoring programs must be designed or modified to meet special needs. Several examples are described next.

CONFIRMING SOURCES OF CHEMICAL AND RADIOACTIVE CONTAMINANTS

Careful analysis of specific contaminants can provide data to confirm their source. Analyses for vanadium, a characteristic component of fuel oils, can be used, for example, to determine whether airborne contaminants in the environment resulted from the combustion of oil or coal. Another example is the use of the ratios of ^{134}Cs to ^{137}Cs in samples to determine whether a radioactive release resulted from discharges from a nuclear power plant or the detonation of a nuclear device. The energy source in both cases is nuclear fission, one of the products of which is radioactive ^{137}Cs. Also produced is ^{133}Cs, a stable isotope of cesium. If the ^{133}Cs is produced through detonation of a nuclear device, it remains in that form. If the ^{133}Cs is produced within a nuclear reactor, it is subjected to intense neutron bombardment, and a portion of it is converted into radioactive ^{134}Cs. Unless the analyses of an environmental sample reveals the presence of ^{134}Cs, it should not be attributed to releases from a nuclear power plant.

CONFIRMING SOURCES OF BACTERIOLOGICAL CONTAMINANTS

As with chemical compounds, it is often a challenge to determine the source or sources of bacteriological contaminants, for example, in lakes and streams. Although the use of E. coli, as noted earlier, is indicative of the possible presence of disease organisms, this leaves unanswered the question of whether the source is humans, farm animals, or wildlife. A solution to this problem is under development. It is based on identifying the specific bacterial contaminant and matching it with a similar organism in a previously created library of bacteria from known sources in the area. Another approach is to identify the specific contaminant based on the fact that strains of bacteria in people, farm animals, and wildlife respond differently to antibiotics (Malakoff, 2002).

IDENTIFYING SOURCES OF UNANTICIPATED RELEASES

When disruptions in facilities that process valuable metals, such as silver, gold, or plutonium, lead to the release of unusually large amounts of the product in the liquid waste stream, it is important to know exactly when and where the release occurred. In cases where the processing system is automated, workers may either not be aware that a system failure has occurred or reluctant to acknowledge that something went wrong. Another complicating factor may be that there may be more than one possible source (i.e., several individual processing systems from which the waste streams later enter a common discharge pipe). Under these circumstances, one method to confirm the source is to install equipment to collect a separate aliquot of the waste-stream flow from each possible source on an hourly basis. The next time such an incident occurs, it will be possible to determine where and when the release occurred. The ability to provide such information will also demonstrate to plant officials the benefits of a well-designed and adequately supported monitoring program (Moeller, 2005).

Computer and Screening Models

Computers offer an enormous capacity for collecting, organizing, and storing information that can assist in understanding the environment and the accompanying impacts of human activities. In fact, computer models have for some years been used to estimate the transport and accompanying concentrations of various pollutants in the atmosphere near emission sources. When information on the meteorology, terrain, and other factors for the given site have been incorporated in the models, the only additional input data required are the nature and quantities of the contaminants being released. So long as ambient air concentrations are periodically measured to validate the estimates being generated, this has become accepted practice (Weinhold, 2002).

As the field has matured, programmers have incorporated a wide array of features and capabilities into such models, including the ability to analyze dynamic transfers of contaminants within and between the main biotic and abiotic components of freshwater ecosystems (Beaugelin-Seiller et al., 2002). Unfortunately, this has created situations where the models are not applicable to many everyday situations. Recognizing this fact, the NCRP has developed a series of what are called *screening techniques*. Their primary purpose is to provide methods for evaluating, on a broad basis, whether releases from a given facility comply with the applicable regulations. Although originally designed to apply only to analyses of releases from point sources, these techniques can readily be modified for analyses of releases from multiple sources (NCRP, 1996) and for use as regulatory tools. For example, the EPA has incorporated similar models within its CAP-88 computer tool (EPA, 1988) for determining Clean Air Act compliance by Department of Energy facilities.

National and Global Monitoring Systems

The preceding discussion has been directed at programs to evaluate the impacts of environmental contaminants on a local or regional basis. Recognizing that many contaminants have widespread effects, major efforts are under way to develop systems capable of monitoring releases that have impacts on a national and global scale. Several of the more prominent such programs are discussed below.

NATIONAL MONITORING PROGRAMS

From the standpoint of the atmosphere, one of the primary U.S. monitoring networks is the National Atmospheric Deposition Program (NADP). Developed by the EPA, its primary goals are to characterize geographic patterns and temporal trends in atmospheric chemical deposition and to apply the resulting data in support of research on the productivity of managed and natural environmental systems. These include surface water and groundwater chemical interactions and pollutant source-receptor relationships. Ultimately, such data should provide a foundation for assessing other environmental impacts, such as visibility and materials degradation, and effects on the health of humans as well as domestic animals, wildlife, and fish. NADP has more than 300 monitoring stations located throughout the continental United States, Alaska, Hawaii, Puerto Rico, the Virgin Islands, and parts of Canada. Enhancing support for the program is widespread support by the scientific community for both the data being generated and the accuracy of the program in tracking air-quality trends in wet deposition, including acid rain (Lambert and Bowersox, 2002).

NAPD, in turn, has led to the development of the following three additional networks:

Clean Air Status and Trends Network. This is designed to measure gaseous and particulate matter as well as ground-level ozone concentrations and dry material that contributes to acid rain. In a cooperative venture with the U.S. Geological Survey, this network was used during 1990–1991 to examine the presence of commonly used herbicides, such as atrazine and alachlor, in precipitation in 26 states ranging from the upper Midwest to the East Coast. Interestingly, herbicides were detected at essentially every site in the study area, primarily during the late spring and summer (Lambert and Bowersox, 2002).

Regional Environmental Monitoring and Assessment Program (REMAP). This is designed to determine the extent (numbers, miles, acres) and geographic distribution of each ecosystem class of interest, assess the proportions of each such class that are in good or acceptable condition, evaluate what proportions are degrading or improving, in what regions, and at what rate, and appraise the likely causes and identify methods for improvement (EPA, 1993). As is the case with NADP, this program depends on ground-based remote-sensing and fast-response instruments to gather the required data. Some monitoring specialists have described EMAP's goal as the determination of the health of an ecosystem, in much the same way as a doctor determines the health of a patient (Table 14.4). Proponents believe that such a system, properly applied, would provide information that would enable environmental scientists to anticipate the point at which ecosystems might begin to break down. Other scientists oppose the analogy because ecosystems, unlike organisms, are not consistently structured, do not behave in a predictable manner, and do not have mechanisms such as the neural and hormonal systems of organisms to maintain homeostasis (Griffith and Hunsaker, 1994).

National Human Exposure Assessment Survey (NHEXAS). As contrasted to a typical monitoring program, this is a multiple-component effort that includes (1) the distribution of questionnaires to provide baseline information on the lifestyles, activities, and socio-demographics of population groups; (2) the collection of soil, house dust, indoor air, tap water, and diet samples; (3) the analysis of these samples for some 30 compounds, including airborne particulates in specific size ranges; and (4) the collection of samples of blood, urine, and hair as biological indicators of human uptake of individual contaminants (Newman, 1995). The last component is closely aligned with comparable efforts under way within the Agency for Toxic Substances and Disease Registry (Chapter 2). Although NHEXAS is labor intensive, the data it generates are essential for making longer-range and fuller-scale assessments of the impact of environmental pollutants on an individual basis.

As progress is achieved, other national needs and possible applications of such programs are being identified. One is a system for monitoring the use of water on a national basis. While it is widely recognized that there are major water shortages in many areas of this country, the data necessary to quantify the extent of these shortages are not available. Today, for example, government officials predominantly rely on stream gauges, river run-off monitors, and similar tools to assess the amount of water available locally. In most cases, no one measures or maintains accurate records on how much water is being withdrawn from wells, rivers, and aquifers. In fact, consumption is often estimated on the basis of how many hectares of agricultural crops are being irrigated, and national estimates are a compilation of what the individual states report, the quality of which varies consider-

Table 14.4   Comparison of ecological health research and human health diagnosis

Ecologic research issue	Analogous human health area
Early warning of ecosystem transformation (e.g., localized fish kill in a river)	Early warning of disease, as the PSA (prostate-specific antigen) test for prostate cancer
Exotic plant/animal/virus invasion or outbreak of native indigenous pathogens	Epidemiologic studies of disease outbreaks within a population group
Presence of "sensitive zones" in ecosystems	Study of certain body organs that are crucial to the functioning and well-being of the whole
Possible development of ecosystem immunity to particular classes or combinations of stress	Immune antibody responses to foreign antigens

ably from state to state. If the supply of water is to be effectively used, much more accurate data are needed (Brown, 2002).

GLOBAL SATELLITE MONITORING SYSTEMS

While satellite monitoring systems are finding increasing applications at the national level, they are almost mandatory for monitoring key environmental factors on a global basis. Such systems are now used to gather data on the rate at which the world's humid rain forest cover is disappearing. These data confirm that almost 6 million hectares of such forests were annually destroyed from 1990 through 1997. An additional 2 million hectares were annually being degraded (Achard et al., 2002). The significance of this, based on the evaluations of Guerney and Raymond (2008), is that this process must be stopped. Otherwise, the current trend will lead to the destruction of all forests. Satellite systems are also being applied by forestry personnel for forest-fire monitoring, the identification and control of invasive species, and the analysis of the beneficial effects of reforestation.

During the widespread floods due to the rains that accompanied Hurricane Floyd in September 1999, in eastern North Carolina, satellite systems were used to determine which of hundreds of confined animal feeding operations in eastern North Carolina were in flooded areas. This enabled researchers to assess the potential for dispersion of animal wastes and the possible pollution of ground- and surface-water supplies (Wing, Freedman, and Band, 2002). In addition, remote sensors that incorporate high-resolution hyper-spectral imaging systems, such as digital airborne imaging spectrometers, are being used to assess contaminated sites under consideration for transfer from one owner to another and reclamation (Howard, Pacific, and Pacific, 2002).

The General Outlook

One of the major changes in effluent monitoring and environmental surveillance systems during recent years has been to adopt both an ecosystem approach and a more global perspective. This has led to basic changes in the identity of the environmental factors or trends that need to be assessed and changes in the types of data that are required. Although the need for fundamental information on the concentrations of various contaminants in the air, water, and soil will remain, this still leaves open the question as to what contaminants should be measured, where, and how frequently. Even more difficult will be to determine the data required to assess the condition of equally important but less visible indicators, such as the ecological processes and conditions that yield food, fiber, building materials, and those that provide "services," such as water purification and recreation.

Specific factors that might be monitored to meet these needs include land cover and productivity, species diversity, and key ecological processes, such as those exemplified by the condition of wetlands, marshes, coastal zones, coral reefs, and marine estuaries. Also needed is a better definition of the indicators needed as input parameters for the various environmental models used to assess the condition of the environment, the meaning of the data that are generated, how best to gather the data, and the effects of emerging technologies on the measurements. The enormity of these challenges is illustrated by the monitoring requirements associated with efforts to assess the impacts of climate change. Until the required models are further developed and more of the key factors that influence temperature trends are identified, scientists will remain in a quandary as to exactly what types of data are required (NRC, 2000).

In the meantime, enormous amounts of data are being generated through existing environmental monitoring programs, and applications of these data continue to expand. One example is the previously discussed increasing use of information obtained by Earth-orbiting satellites. In accordance with a 1995 agreement, officials in Russia and the United States are exchanging data obtained through the operation of spy satellites during the Cold

War. Scientists believe that long-term records could provide valuable information on the effects of clouds on heating and cooling of the Earth, as well as insights into the effects of airborne contaminants on cloud formation and on the development of better methods for modeling these effects. Other information provided by satellites may lead to methods for providing warnings about the occurrence of volcanic eruptions, earthquakes, and other types of natural disasters.

With the increase in magnitude and comprehensiveness of the databases being established, far more efforts will need to be directed to their integration and the development of more systematic methods of their compilation and storage. Otherwise, some of the key potential benefits of these data, especially in terms of assessing the long-term effects of various environmental contaminants, may not be achieved. Computer models also need to be refined not only to predict human exposures but also to determine the interrelationships among environmental factors and, as noted above, to evaluate their potential impacts on ecosystems.

15

RISK ASSESSMENT
AND MANAGEMENT

IN A PERSONAL SENSE, risk can be defined as the probability that an individual will suffer injury, disease, or death under a specific set of circumstances. In the realm of environmental health, this definition must be expanded to include possible effects on other animals and plants, as well as on the environment itself. Knowing that a certain risk exists, however, is not enough. People want to be provided an estimate of how probable it is that they or their environment will suffer and, if they do, what the effects will be. Determination of the answers to these questions involves the science of risk *assessment*.

Risk assessment ranges from evaluation of the potential effects of toxic and/or radioactive chemical releases known to be occurring to evaluation of the potential effects of releases due to events whose probability of occurrence is uncertain. In the latter case, the risk is a combination of the likelihood that the event will occur and the likely consequences if it does. In essence, the process of risk assessment requires addressing three basic questions (Kaplan and Garrick, 1981):

- What can go wrong?
- How likely is it?
- If it does happen, what are the consequences?

As might be anticipated, there is a spectrum of endpoints that can be considered in evaluating whether a risk is acceptable (Table 15.1). Once the endpoint has been defined and the associated risk has been assessed, it can be expressed in qualitative terms (such as "high," "low," or "trivial"), or in

Table 15.1 Spectrum of adverse consequences and factors to consider in risk assessments[a]

1	Shortening of life (mortality)
	Cancer versus other causes
2	Illness or injury leading to disability
	Acute versus chronic
	Permanent versus temporary disability
	Serious versus minor disability
3	Illness or injury with temporary disability followed by recovery
	Chronic versus acute
	Serious versus minor disability
4	Physical discomfort without disability
5	Psychological disorder with behavior consequences
	Post-traumatic stress disorder
	Anxiety reaction
	Stress reaction
	Chronic frustration and anger
6	Emotional discomfort

a. Each category should be weighted by the number of people involved.

quantitative terms, ranging in value from zero (certainty that harm will not occur) to one (certainty that harm will occur). At the same time, it must be recognized that a given assessment provides only a snapshot in time of the estimated risk of a given toxic agent and is constrained by our current understanding of the relevant issues and problems. To be truly instructive and constructive, risk assessment should always be conducted on an iterative basis, being updated as new knowledge and information become available.

Once a risk has been quantified, the next step is to decide whether it is sufficiently high to represent a public health concern and, if so, determine the appropriate means for control. Such control, which falls under the rubric of what is called risk *management,* may involve measures to prevent the occurrence of an event as well as appropriate remedial actions to protect

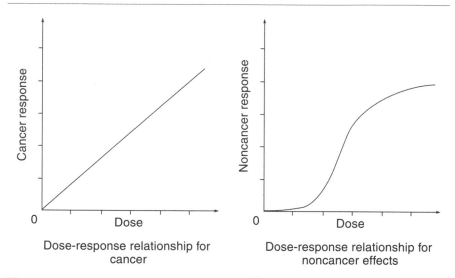

Figure 15.1 Dose-response relationships for cancer and noncancer agents

the public and/or the environment in case the event occurs. Because of the nature of these processes, each step in the risk-management process is accompanied by a multitude of uncertainties. The same is true for risk assessment.

As described in Chapter 2, two basic concepts are applied in assessing risks from toxic agents. These are exemplified by the two graphs shown in Figure 15.1. The graph on the left represents the linear nonthreshold dose-response curve that is generally applied to carcinogenic agents. In accordance with this model, any dose, regardless of how small, is assumed to have an associated risk. The graph on the right represents the threshold type of response—this is assumed to apply to many noncarcinogenic agents. Although a person is assumed to be "safe" as long as the dose is below the threshold, there is still a need for a quantitative estimate of the risk for any dose above the threshold. It is in this range that scientists apply risk-assessment procedures.

History, Development, and Early Applications

The concept of anticipating and preparing for risks was initiated more than 2,000 years ago. The Egyptians, for example, recognized that although the Nile River might provide a continuous supply of water for a period of perhaps 30 years, this could be followed by as much as 2 years of drought, during which their agricultural crops would fail. Based on this observation, they constructed storehouses to provide food during times of distress. Today, these concepts are being applied in a more systematic and sophisticated manner to anticipate, and prepare to avoid, mishaps involving increasingly complex facilities, operations, and events. Achieving success, however, has not been easy.

This is exemplified by the early history of the National Air and Space Administration (NASA). Unfortunately, a formal detailed risk-assessment program was not initiated until a fire occurred inside the *Apollo 1* command module during a mission simulation test on the *Saturn I* launchpad in 1967 and caused the deaths of three astronauts. Prior to that time, NASA had relied on its contractors to incorporate the design features necessary to provide quality assurance, quality control, and means for responding to unexpected events. Follow-up investigations revealed that the fire was fed by the 100 percent oxygen atmosphere inside the module, leading to a near explosive burn rate of the accompanying flammables. The belief that it was initiated by an electrical short led to the development of Kevlar and other fire-retardant materials that were incorporated into subsequent spacecraft. Another major defect was that the crew was unable to escape because the hatch opened inward.

Even so, the Apollo event was followed in 1986, less than 2 decades later, by the explosion of the *Challenger* shuttle shortly after launch. In this case, one of the reasons for the accident was that low temperatures on the morning of the launch caused a contraction of the rubber "O" rings that provided a seal between the segments of the solid booster rockets. This, in turn, permitted hot gasses to penetrate the wall of the expendable (hydrogen-oxygen) fuel tank (Moeller, J. A., 2005).

Another important contributor to this history was the use of applied risk analyses in the design and construction of commercial nuclear power plants. Initiated during the early 1970s by the U.S. Atomic Energy Commission (AEC), these efforts were based on a philosophy of risk management derived from the concept of a "maximum credible accident." Because credible accidents were already covered by the plant design, the goal was to estimate the residual risk by analyzing the consequences of what were called "incredible accidents." In so doing, the AEC also incorporated the concept of "defense in depth." Following this approach, the primary safety systems in all nuclear power plants are required to have a secondary backup system that would prevent an accident if the primary system failed. Unfortunately, the outcome of this effort was that no one was able to determine in a realistic manner the magnitude or probability of such an accident. This led to these efforts being redirected by reducing the probability of a catastrophic release. The focus in this case was the radionuclide inventory of the reactor core. This culminated in a decision by the U.S. Nuclear Regulatory Commission (USNRC), the successor to the AEC in these aspects of the nation's nuclear activities, to initiate a full-scale application of what is now known as "probabilistic risk analysis" (PRA). This included accident consequence and uncertainty analysis and led to the publication of the *Reactor Safety Study* (USNRC, 1975). Due to controversies generated by the study, Congress created an independent panel to review it. They concluded that the uncertainties had been "greatly understated," and the report was withdrawn.

With the occurrence of the accident in 1979 at the Three Mile Island nuclear plant in Pennsylvania, new assessments were initiated that avoided some of the methodological defects of the *Reactor Safety Study*. This was followed by publication of the *Fault Tree Handbook* in 1981 and the *PRA Procedures Guide* in 1983. These two reports have provided a foundation for much of the PRA work that has followed (Cooke, 2009). Although the Three Mile Island accident was followed by the explosion of the Chernobyl plant in the U.S.S.R. in 1986, the defects in the design of this plant had been recognized years prior to its construction. In fact, a plant of this design would never have been licensed for operation in the United States or in any of the Western European or Asian nations. The next outcome in this series of studies was the issuance of NUREG-1150, which was based on a detailed assessment of the accident risks in five U.S. nuclear power plants. A key outcome was to establish safety goals for the average probabilities for an individual early fatality, or average latent cancer death, per reactor per year (USNRC, 1991). Since 2006, the major effort has been on analyses based on what are described as state-of-the-art reactor consequence analyses. That is to say, all safety analyses must incorporate all the analytical improvements that have been developed based on the lessons that have been learned.

Other Applications and Ramifications of Risk Assessments

Despite the progress that has been made, applications of risk assessment are not without difficulties and controversies. The public, in particular, finds the concept of risk difficult to understand. For many people, the concept of a "safe" dose is far more acceptable. Individuals also fail to realize that the risks associated with the hazards of everyday life vary widely and that the risks most feared are not necessarily those that are most important. People also generally voice far more concern over risks that they are asked to accept on an involuntary basis and that derive from man-made sources. They have far less concern about risks that they accept voluntarily and that derive from natural sources or risks that they have commonly accepted for years. An excellent example of the latter is the risk of lung cancer due to the presence of naturally occurring radon in the home (Chapter 5). When college students were asked to indicate their perception of the risks from pesticides and nuclear power plants, their rankings were much higher than those of a group of experts. When asked to express their perception of the risks of motor-vehicle accidents and coal-fueled electricity-generating stations, their rankings were well below those of the experts (Wilson and Crouch, 2001). Although individual members of the public may be concerned about contaminants (such as chloroform) in their drinking water or toxic substances (such as aflatoxin) in their food, the data show that many other risks in their daily lives should be of far more concern. For example, on a quantitative basis, the risk of death from motor-vehicle accidents is over 40 times as high, the risk of death due to violence (as exemplified by homicide) is over 100 times as high, and the risk from cigarette smoking is over 250 times as high as those associated with contaminants in drinking water and food.

Although initial applications of risk assessments were primarily directed to the effects on people, experience has demonstrated that these techniques are equally applicable to assessments of ecological and environmental impacts. Recognizing this fact, the National Council on Radiation Protection and Measurements (NCRP, 1991) has issued recommendations for the protection of aquatic species, and the International Commission on Radiological Protection (ICRP, 2008) has developed a framework that can be a practical tool to provide high-level advice and guidance and help regulators and operators to demonstrate compliance with existing environmental regulations. These include a small set of "representative animals and plants," comparable in a sense to the "representative person or individual" (ICRP, 2006), that can serve as a basis for a more fundamental understanding and interpretation of the relationship between exposure and dose and between dose and certain categories of effects.

Much earlier, the Environmental Protection Agency (EPA) initiated a program to develop ecological risk-assessment guidelines for protection against a full line of environmental impacts. The thought was that this would enable regulatory agencies to make better policy decisions on proposed developmental activities in geographical areas that might be particularly sensitive. The types of problems needing to be addressed range from regional issues, such as the possible impact of the drainage of wetlands on the local ecology, to global issues, such as the impacts on the environment of chemicals that can destroy the ozone layer and/or lead to worldwide warming (EPA, 1998).

As implied above, the goal of most risk assessments is to estimate the potential health impacts on the *representative individual.* In contrast, evaluations of the effects of toxic agents are generally directed to population groups. This occurs due to the fact that most measurements of air pollution, for example, are designed to determine the average concentrations of specific

Table 15.2   Estimates of comparative risks of various potential sources of death, 2009

Risk	Annual total deaths in the United States[a]	Degree of uncertainty
Cigarettes and other uses of tobacco	443,000[b]	Low
Poor diet, obesity, and physical inactivity	435,000[b]	Low
Alcohol abuse and motor-vehicular accidents	85,000[c]	Medium
Highway travel	45,000[c]	Low
Homicide	18,500[c]	Low
Airline travel	500[c]	Low
Outdoor air particles ($<PM_{2.5}$)	60,000[d]	Medium
Indoor air (radon)	21,000[e]	High
Pesticide residues in foods	3,000[f]	High

a. Values are rounded off.

b. American Cancer Society. *Cancer Facts & Figures.* Atlanta, GA.

c. National Safety Council. 2009. *Injury Facts, 2009 Edition.* Itasca, Il.

d. Pope, C. A., III, Burnett, R. T., Thun, M. J., Calle, E. E., Krewski, D., Ito, K., and Thurston, G. D. 2002. *Lung Cancer, Cardiopulmonary Mortality, and Long-term Exposure to Fine Particulate Air Pollution. Journal of the American Medical Association* 287, no. 9, 1132–1141.

e. EPA. 2003. *EPA's Assessment of Risks of Radon in Homes.* Report EPA 402-R-03-003, Environmental Protection Agency, Washington, DC.

f. Graham, John D. 1991. *Annual Report 1990,* unnumbered table, 3. Center for Risk Analysis. Harvard School of Public Heath, Boston, MA.

contaminants in the ambient environment (Chapter 5). Consequently, the resulting data are primarily limited to assessing the exposures of members of the public who spend large amounts of time outdoors. Equally important is that the accuracy and/or applicability of estimates of the risks either to individuals or to population groups depends on the nature of the toxic agent, the extent and availability of exposure measurements, the range and duration of the exposures, and other factors such as the physical characteristics and lifestyles of those exposed. Surveys show, for example, that the average member of the U.S. public spends 90 percent of his/her time indoors. Nonetheless, this does not mean that he/she is better off. Data show, for example, that air within homes and other buildings can be more seriously polluted than the outdoor air in even the most industrialized cities (EPA, 2009). Additional perspective can be gained by comparing various major sources of deaths in the United States, their associated uncertainties, and degree to which each person is vulnerable. This emphasizes the importance of the accuracy of the data on which such estimates were based, the methods that were utilized, and the underlying assumptions (Table 15.2).

Qualitative Risk Assessment

As the above examples illustrate, it is very difficult to quantify the risk from many commonly known sources. Exacerbating the situation are the wide variety and sources of risk (i.e., the large numbers of facilities having the potential for releasing toxic chemicals). For this reason, the common approach is to develop, as an initial step, some type of qualitative or semi-quantitative assessment. Possibilities include (a) qualitative characterizations where health risks are identified but not quantified, (b) qualitative risk estimations where the chemicals present are ranked or classified by broad categories of risk, and (c) semi-quantitative approaches where effect levels (for example, "no observable effect") are used in combination with uncertainty factors to establish "safe" exposure levels.

Perhaps the best example of a qualitative approach is the public health assessment methodology that has been developed by the Agency for Toxic Substances and Disease Registry, examples being their reports on benzene and hydrogen sulfide (ATSDR, 2006, 2007). This process includes a review and evaluation of data and information about hazardous substances at, for example, a Superfund site (Chapter 9), and a characterization of the nature and extent of the associated risk to human health. Frequently, the assessment includes recommendations on actions needed to prevent or mitigate any potential health effects and any additional studies that may be needed (ATSDR, 2003). Although fundamentally qualitative in nature, such assessments are based on an evaluation of a large amount of scientific data.

These data are supplemented in the case of the higher-risk Superfund sites by in-depth reviews, evaluations, and estimates of the frequency of occurrence, toxicity, and potential for human exposure for each of the substances present. Based on this information, ATSDR scientists maintain a priority list of the ten most hazardous substances known to exist at Superfund sites (Table 15.3a). The characteristics of these substances are, in turn, used to identify the priority human health effects that should receive specific attention in the evaluation of the potential range of health impacts of the associated releases (Table 15.3b). On the basis of the resulting public health assessment, each Superfund site on the priorities list is placed in one of five categories in terms of its overall significance to public health and the

Table 15.3a   Top ten substances on the ATSDR priority list

Rank	Substance
1	Arsenic
2	Lead
3	Mercury
4	Vinyl Chloride
5	Polychlorinated biphenyls
6	Benzene
7	Cadmium
8	Benzo(a)pyrene
9	Polyaromatic hydrocarbons
10	Benzo(b)fluoranthene

Table 15.3b   Priority health conditions to be considered

Birth defects and reproductive disorders
Cancers
Immune function disorders
Kidney dysfunction
Liver dysfunction
Lung and respiratory diseases
Neurotoxic disorders

associated requirements for follow-up action. These categories are (1) urgent public health hazard, (2) public health hazard, (3) indeterminate public health hazard, (4) no apparent public health hazard, and (5) no public health hazard (Johnson, 1992).

Quantitative Risk Assessment

In contrast, the EPA concentrates on quantitative risk assessments, its goal being to characterize in numerical terms the potential adverse health effects of human exposures to toxic agents. Such assessments involve four primary steps and serve as one of the principal elements of risk assessment and risk management, with the former being an essential preparatory step to the latter. Fundamental to these two steps is the conduct of the laboratory and field research to provide the necessary input data. Each of the primary risk assessment steps is described below (Wilson and Crouch, 2001).

HAZARD IDENTIFICATION

Hazard identification is a qualitative determination of whether human exposure to a specific agent has the potential to cause adverse health effects. This generally requires information on its identity and the outcomes of related mutagenesis and cell transformation studies, animal research, and human epidemiological studies. Information on the physical and chemical properties of the agent is also required for assessing the degree to which it can become airborne and be inhaled and absorbed into the body and to evaluate its solubility in water and availability for transport through the food chain.

DOSE-RESPONSE ASSESSMENT

Dose-response assessment involves evaluation of the hazard potency (power to produce adverse effects) inherent in receiving a dose from a specific toxic agent. If available, dose-response estimates based on human data are preferred. In their absence, information from studies of other animal species that respond like humans may be used. As previously discussed (Chapter 2), however, the use of animal data introduces multiple uncertainties into the accompanying risk estimates.

EXPOSURE ASSESSMENT

Exposure assessment is an estimate of the extent of exposure to the agent and the accompanying dose to people and the environment. The assessment may be directed to normal releases from a facility, or it may, in the case of accident assessments, involve estimating both the probability of the event and the magnitude of the accompanying releases. Factors considered in performing such assessments include the following:

- The chemical and physical characteristics of the agent, for reasons similar to those identified in the hazard identification step. Key parameters include partition coefficients, retardation factors, bioaccumulation factors, and degradation rates. To the extent possible, the values assigned to such parameters should be specific for the system and site being analyzed.

- Identification and characterization of the representative person to be protected. As noted earlier, this is especially important in cases where it is necessary to protect special groups, such as children and pregnant women, who may be more susceptible to a given agent.

- Recognition of the difference in the exposure measured and the dose that will actually be received. Although several people may be exposed to the same agent, the accompanying dose depends on a number of factors. In the case of airborne materials, this includes the age and breathing rate of the person exposed and whether he/she breathes through the mouth or nose.

RISK CHARACTERIZATION

Risk characterization involves estimating the dose and accompanying risk to people who have been exposed to a specific agent. Such a process requires integrating the results of the previously discussed hazard identification and exposure and dose-response assessments, to produce quantitative estimates of the associated health and environmental risks. Because risk estimates have significant limitations, the EPA requires that, in addition to the estimate itself (usually expressed as a number), the risk characterization contain (1) a discussion of the "weight of the evidence" for human carcinogenicity (for instance, the EPA carcinogen classification), (2) a summary of the various sources of uncertainty in the estimate, including those arising from hazard identification, dose-response evaluation, and exposure assessment, and (3) a report on the range of risks, using the EPA-based risk estimate as the upper limit and zero as the lower limit. Due to the magnitude of the accompanying uncertainties, it is important that those performing these exercises not be overly conservative in their assumptions. Otherwise, the associated risk estimates may be far in excess of what

Table 15.4 Primary components of the risk-assessment process

Component	Description
Source term	Characterization and quantification of releases into the environment
Environmental transport	Assessments of where within the environment the contaminants deposit and/or are absorbed, and their resulting concentrations in air, water, soil, agricultural products, etc.
Exposure factors	The time, location, and rate of their transport, and the traits of the individuals subject to exposure
Conversion to dose	This is accomplished by multiplying the intakes (via ingestion and inhalation) by the dose coefficients applicable to the contaminants and groups subject to exposure
Conversion of dose to risk	This is accomplished by multiplying the doses by the appropriate risk coefficients for the contaminants and age groups subject to exposure
Uncertainty in analyses	There are associated uncertainties in all of the components described above; these need to be evaluated and quantified
Validation	This is a very important ancillary requirement, namely, to verify (i.e., document the accuracy of) the estimates that have been made

will be experienced in the real world. For this reason, many public health officials urge that such assessments be designed so that the outcomes are as realistic as possible. A comprehensive summary of the primary components of the radiological risk-assessment process is provided in Table 15.4 (Till, 2008).

Temporal and Other Variations in Dose and Risk

Dose and risk may vary among different population groups. As has been emphasized, decision makers may also elect to manage risks differently depending on the estimated vulnerability of the affected group (Schmidt, 1999). As might be anticipated, the evaluation and management of risks to various groups is an ongoing challenge and is often influenced by socio-economic factors and political considerations.

One frequently neglected consideration is that the relationship between a given radiation dose and its accompanying risk (e.g., fatal cancer) varies with time, as well as other factors. This can be illustrated by the following three examples, in which a population group is exposed to different contamination sources:

Single or multiple radiation sources at a given time. This is the situation that is encountered under most circumstances. Under these conditions, the risk can be estimated by multiplying the dose for each source by the appropriate risk (e.g., fatal cancer) coefficient. Although it is more complicated, the total detriment can also be estimated.

Sources that contain mixtures of toxic chemicals with other carcinogens (i.e., radionuclides). This is exemplified by the assessment of the health impacts of cigarettes that contain both chemical and radioactive carcinogens. Under these circumstances, the sum of the risks can readily exceed that of them acting independently, that is, they are synergistic (ATSDR, 2001; NCRP, 2002; Moeller and Sun, 2010).

Sources that release radionuclides over a long time. Assessment of the associated risks in this case requires accounting for the relatively rapid progress being made in developing methods for preventing and medical therapies for curing the more common cancers afflicting humankind today. The latest prediction is that this goal could be achieved within the next 20 to 30 years (Healy, 2008). This is documented by the fact that between 1993 and 2001 the overall annual death rate from cancer among members of the U.S. population was reduced by an average of 1.1 percent. Between 2002 and 2004, the rate of reduction had been increased to 2.1 percent (Espey et al., 2007), and this trend is continuing (Associated Press, 2009). This is vividly reflected in the reductions in the lethality factors for cancers in various human body organs during the last 17 years (Table 15.5). Interestingly, it also means that it will not be possible to estimate future risks per unit dose due to the inability to predict the rate of progress in preventing and curing cancers in specific body organs, or the habits of the population (percent who will continue to smoke cigarettes in terms of lung cancer), or the racial composition of the population (percent who will be African Americans and not subject to skin melanomas). These advances are of special importance relative to assessments of the risks from the Yucca Mountain high-level radioactive waste repository (Moeller, D. W., 2010).

Table 15.5   Reductions in lethality factors for cancers in selected body organs (1990–2007)

Organ	Publication 60 (1991)	Publication 103 (2007)	Factor of reduction
Bladder	0.50	0.29	1.7
Bone	0.72	0.45	1.6
Bone marrow	0.99	0.67	1.5
Breast	0.50	0.29	1.7
Ovary	0.70	0.57	1.2
Thyroid	0.10	0.07	1.4

It is also important to realize that progress in the prevention and cure of various cancers will lead to corresponding reductions in the risks per unit dose for all types of carcinogenic agents, including chemicals such as arsenic, asbestos, beryllium, and nickel. In fact, it has been suggested that in many cases the existing limits on the release and human uptake of such agents "may no longer matter" (Karam, 2008). Karam cautions, however, that this does not mean that "reasonable" control of such materials should not be continued.

Risk Management

The distinction between risk assessment and risk management is very important in regulatory decision making. As noted earlier (Figure 15.2), risk management is the process of integrating the risk-assessment results with other information (engineering data, socioeconomic and political concerns), weighing the alternatives and the potential negative impacts of the proposed management approaches, and selecting the most appropriate action for reducing and/or eliminating the risk. In fact, these factors often play a more significant role in risk-management decisions than the assessment (i.e., the nature and magnitude of the risks themselves).

A successful program in risk assessment and risk management requires that certain aspects of the two processes be separated (Wiener and Graham, 1995). Otherwise, it is difficult not to confuse the scientific conclusions about the nature of a risk with the social, political, and economic concerns over how the risk should be managed. This does not mean, however, that there should be no communication between the two groups. Once an assessment

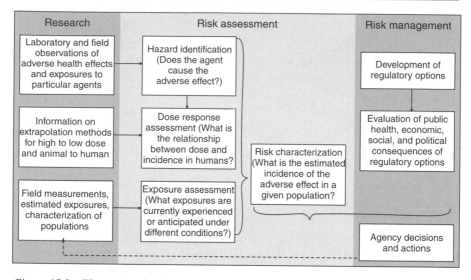

Figure 15.2   Elements of risk assessment and risk management

has been made, those responsible for managing the risk must be provided with both the risk estimate and the context under which it was developed. The primary areas of uncertainty must be defined, along with the degree to which they may influence the accuracy of the risk estimate. The techniques of risk management can then be used to aid in setting priorities for action and in analyzing alternative control strategies. A summary of the principles of risk assessment and management is presented in Table 15.6.

ACCEPTABLE LEVELS OF RISK

One of the primary considerations in risk management is the level at which controls should be applied. Such a level is required, for example, in determining when cleanup operations at a contaminated site can be terminated or whether a product is sufficiently safe for consumption or use by the public. Such a level has also been proposed in deciding the probability at which hypothesized future events, such as seismic events, are so "unlikely" that they need not be considered in evaluating the risks of a given facility. If scientific evidence were the only basis for establishment of these levels, such decisions would be relatively easy. As will be noted below, public input and the need for flexibility have played a major role, the result being that acceptable levels are generally established on a case-by-case basis.

Table 15.6   Principles of the risk assessment and management process

1	Estimates of attributable health risk should make use of the best available science.
2	Since reputable scientists often do not agree about how to assess risk, scientific disputes should be acknowledged.
3	When confirmed data are lacking, risk assessments should be explicit about any assumptions and should indicate the degree of sensitivity of results to plausible changes in assumptions.
4	Meaningful risk assessments usually develop a central estimate of risk, as well as upper and lower bounds on risk that acknowledge the extent of scientific uncertainty.
5	Public policy decisions about acceptable risk require public participation and application of democratic principles.
6	No quantitative level of risk exists that is universally acceptable or unacceptable; acceptable depends on the circumstances, the people affected, and the decision context.
7	Valid decisions about health risk require consideration of other cherished values such as quality of life, equity, ecological health, personal choice, and economic welfare.
8	Programs to reduce risk should be designed to avoid unintended side effects that may increase risk.
9	When risk reduction is desired, economic incentives and information should be considered in addition to conventional command-and-control regulation.
10	The context in which risk occurs (e.g., voluntary versus involuntary risk) may influence public reaction to risk as much as the magnitude of the risk in question.

One example of the application of risk levels is the determination of the degree of cleanup required for Superfund toxic-waste sites. In its initial efforts, the EPA ruled that if an existing site did not impose a lifetime cancer risk in excess of 1 in 1,000,000, no cleanup was required. In some situations, the same limit served as the goal for the degree of cleanup for sites that did not initially comply with this level. More recently, the EPA has defined acceptable excess cancer risk for the sites being evaluated under the Superfund program as a range from 1 in 10,000 to 1 in 1,000,000. This approach was adopted to provide risk managers the flexibility to consider site-specific factors, such as the number of people being exposed, and the feasibility and cost-effectiveness of cleanup. In other cases, they have adopted the following limits: (1) for state water-quality standards, 1 in 100,000 to 1 in 10 million; and (2) for air-quality standards, 1 in 1,000,000 for as many people as possible, and 1 in 10,000 for the maximally exposed individual (Graham, 1993).

With respect to seismic events, the USNRC has based its siting requirements for commercial nuclear power plants on the probability of a "capable" fault. In so doing, they have defined a "capable" fault as (1) one in which "movement at or near the ground surface has occurred at least once within the past 35,000 years, or (2) that has undergone movement of a recurring nature within the past 500,000 years" (USNRC, 1962). In a similar manner, the EPA ruled, in establishing standards for the proposed high-level radioactive waste repository at Yucca Mountain, Nevada, that performance assessments need not include consideration of very unlikely events (i.e., "those that are estimated to have less than one chance in 10,000 of occurring within 10,000 years of disposal" of the waste) (EPA, 2001).

APPLYING REGULATORY LIMITS

Although the establishment of regulatory limits is important, that is only an initial step in the process of risk management. Equally important is the development and implementation of the system through which the limits are to be applied. In fact, experience has shown that the techniques of application can influence not only the amount of protection provided but also the manner (favorable or unfavorable) in which the public views the limits. These aspects are vividly illustrated by previously discussed differences in the regulatory and risk-management policies of the EPA and the USNRC, namely, the "bottom-up" versus "top-down" approach (Chapter 13).

Both of these management policies have appealing features. One of the most important, in terms of the bottom-up approach, is the incorporation of community-based right-to-know programs, which mandate that the identification of the problem, the associated risks, and the options available for reducing the risks be communicated to, and shared with, the public in plain, easy-to-understand language. This also means that throughout the decision-making process, the public must be informed as well as involved. In a similar manner, the top-down approach has many good features, including the fact that the dose limits are risk based, the data supporting them were derived from human studies, and they are applied using a safety culture where simple compliance with the standard is not considered sufficient. In concert with this approach, the USNRC requires (Chapter 12) that licensees must not only meet the regulatory requirements but, once having done so, must demonstrate that all readily available means have been made to ensure that any releases to the environment and/or doses to their workers and members of the public are as low as reasonably achievable (ALARA; Gage, 2001).

There are, nonetheless, situations where confirmation of compliance with the regulations can be extremely difficult, a prominent example being the steps required in terms of the adequacy of the long-term disposal of high-level radioactive wastes. In the case of the proposed repository at Yucca Mountain, Nevada (Chapter 9), the regulations require that the facility be designed to contain these wastes for 1 million years. This is both impossible and unnecessary. Reviews based on detailed assessments document that not one of the eight radionuclides for which limits have been stipulated has any possibility of causing harm to any member of the public or the environment (Moeller, D. W., 2009). Nonetheless, in January 2009, the Secretary of the U.S. Department of Energy stopped work on the Yucca Mountain Repository that had been constructed at a cost of more than $12 billion and was ready to receive high-level radioactive waste. It is estimated that the cost to restore the site to its original state will require an equal amount (Moeller, 2010). In the meantime, an accumulated total of 60,000 metric tons (~66,000 U.S. tons) of spent fuel, removed from the 104 U.S. commercial nuclear power plants, remains in storage at their sites (USNRC, 2009).

PRECAUTIONARY PRINCIPLE

One of the principles that has served as a foundation for ensuring the protection of human health and the environment is that there need not be scientific certainty that an agent or activity causes harm before taking prudent action to limit or avoid exposure to it. Although this principle had been enunciated decades earlier, international attention was drawn to it through Principle 15 of the 1992 Rio Declaration on Environment and Development. Now identified as the "Precautionary Principle," it was expressed as follows:

> In order to protect the environment, the precautionary approach shall be widely applied by states according to their capabilities. Where there are threats of serious or irreversible damage, lack of full scientific certainty shall not be used as a reason for postponing cost-effective measures to prevent environmental degradation (United Nations, 1992).

Various expressions of this principle have subsequently been incorporated into international trade and environmental agreements, particularly in those of the European Community and the national laws of some of the countries in Europe. In many instances, however, applications of the principle have proved to be controversial. The primary reason is the widespread disagreement on what it calls for risk managers to do and the degree to which its application should supersede a fully developed risk analysis. Supporting groups believe that it means that no environmental releases should be permitted until they can be unconditionally proved safe, even though such a goal is not attainable. Other supporters, seeking to ensure adequate margins of safety, have called for the use of worst-case assumptions in the conduct of the associated risk assessments. In rebuttal, opponents emphasize that any precautionary actions must be cost-effective, appropriate, and proportionate to the threat that is to be avoided. They also point out that the Precautionary Principle in no way was intended to obviate the need to perform an analysis of the risk and that any precautionary actions should be viewed as provisional until such time that the necessary research is conducted to confirm whether the assumed risks are real (Rhomberg, 2001).

Regardless of these "flaws," there are cases where applications of this principle may clearly be warranted. Prominent among these are situations that involve what might be called "creeping" environmental change, for example, the possible long-term impacts of greenhouse gases on global climate (Chapter 18), and the "low-grade" cumulative environmental problems such as acid precipitation, air and water pollution, soil erosion, tropical deforestation, and habitat destruction. Another example is the debate on genetically engineered foods. In each of these cases, there is no clear threshold that distinguishes negligible harm from serious harm. There is also a risk that without action, the problem will languish without adequate attention until it is too late to prevent serious damage (Graham, 1999). At the same time, experience has shown that applications of the principle can lead to unanticipated and unfavorable results. A classic example was the addition of methyl tertiary-butyl-ether (MTBE) to gasoline, the intention being to reduce air pollution. Subsequent observations of MTBE contamination in groundwater supplies (Chapter 7) led to a retraction of this action (Graham, 1999).

Upgrading Methodologies for Risk Assessment and Management

As noted earlier, risk management involves the integration of the outputs of risk assessments with a variety of other types of information, coupled with the weighing of a host of contributing factors, to reach a decision on the most appropriate and/or most effective way for managing a specific risk. The types of topics to be considered, the decisions that enter into such evaluations, and the development of improved methodologies for reaching definitive conclusions are illustrated in the following example.

ENVIRONMENTAL RISK ASSESSMENT OF PHARMACEUTICALS

From 2004 to 2007, an international study was initiated to evaluate the potential impacts of the release of human and veterinary pharmaceuticals on the environment. The overall objective was to advance existing knowledge and procedures for performing related environmental risk assessments (ERAs). Specific topics included how to improve efforts to (1) assess, model, and account for the partitioning and persistence of pharmaceuticals in the environment, (2) identify those most likely to pose a significant risk below current action limits, (3) apply this information to establish alternative end points in ERAs, and (4) identify transformation products and assess their accompanying exposures and effects. The ensuing investigations involved the behavior of pharmaceuticals in water, freshwater and marine sediments, soil, dung, and sludge. A full range of environmental conditions were considered.

It was suggested that adoption of this approach would minimize the necessity for the generation of new data, reduce the uncertainties of risk characterization, and permit the selection of tests on a cost-effective basis. The outcomes of this effort subsequently enabled them to design a program to conduct related ERAs in a timely and effective manner and to provide useful guidance in conducting future assessments. Other benefits were that the studies provided information to address specific gaps in current guidelines and cope with similar issues in the future. One major outgrowth was the development of a Web-based tool, PharmaEcoBase, that provides risk analysts with ready access to the most-up-to-date scientific data and information (Knacker and Coors, 2007).

Risk Communications

Unless the outcomes of risk assessments are properly communicated to the public, many of the benefits of these efforts will not be realized. To ensure success, there are certain basic principles that need to be implemented. These are especially important when toxic chemical and/or radioactive materials are actually or potentially involved (Till, 2008).

- Environmental data related to the facility should be provided to all members of the potentially affected population group (i.e., the stakeholders). This information should include what materials are being or will be released; how these releases compare to the environmental regulations and to existing quantities of these materials already present in the potentially affected environment; and any other benchmarks that will provide perspective.

- Whenever possible, *human health risk* should be used as an endpoint in the assessment; impacts expressed in terms of dose will have little, if any, meaning. This will enable the potentially affected stakeholders to assess the risk in terms of other risks they face in their daily activities. One example would be to share with them the health effects on the Japanese survivors of the World War II nuclear detonations in Japan (Table 15.7; Clarke, 2008). As will be noted, the increase in cancer deaths was far less than one would have estimated on the basis of media reports.

- Readily accessible tools should be provided to assist the stakeholders in following and understanding the process through which the risk was estimated and in interpreting the results. These tools can be made available in multiple formats, ranging from software to reports specifically prepared for the public or decision makers.

- The software tools should provide (1) *transparency,* so users know how the calculations are being made, and the sources and nature of the input values that are being used, and (2) *flexibility,* so users can change the input values and gain perspective on the resulting changes in the risk estimates.

Table 15.7   Radiation-induced deaths among survivors of World War II nuclear weapon detonations in Japan (1945–2000)

Population	Number
Survivors (1945)	86,572 (cohort)
Alive (2000)	~38,000 (~44%)
Total death (cancer and leukemia)	~10,000
Cancer deaths (radiation-induced)	
Leukemia	~100
Solid cancers	~400
Cancer deaths due to radiation exposure	~5%

Table 15.8   Lessons learned about risk assessments and management

Experience documents that technological advances and the application of new computer models that permit the consideration of a wider range of influencing factors are yielding assessments that are far more reliable. For example, advances in analytical chemistry have enabled biological monitoring to serve as an excellent tool for validating or confirming the predicted degree of human exposure. These advances also permit estimates to be made of exposures that occurred 60 or more years ago.

In seeking to be prudent, assessors have placed too much emphasis on the so-called maximally exposed individual (now the *representative individual*); increased attention, however, needs to be directed to special groups who, because of their location and/or living habits, may receive unusually high exposures. At the same time, it is important to recognize that children and their exposure patterns are not the same as those of adults.

In most cases, the most significant risks due to exposures to chemicals occur in the workplace; for most persons, exposures to chemicals and bacteria in the home pose a higher risk than to those in the ambient air or through ingestion of water.

Advances in analytical chemistry have enabled biological monitoring to serve as an excellent tool for validating or confirming the predicted degree of human exposure; other improvements include estimating indirect exposure pathways such as the uptake via inhalation of volatile contaminants released from water while a person is taking a shower.

Care should be exercised to ensure data from samples that have no detectable amount of contamination are properly considered; otherwise, the impact of a few samples with detectable contamination can lead to improper conclusions about the actual level of risk.

There is a need to expand the scope of such assessments to include a wider range of impacts on the environment, including those on plants and animals. In all such efforts, the repeated use of conservative assumptions should not dictate the outcomes.

Both risk managers and the public want to understand the statistical confidence in estimates of risk; application of sensitivity analyses can yield important information about and help identify the critical exposure variables.

Risk assessors need to recognize the difficulty of this commitment, develop a well-defined schedule and end product, prepare a plan for receiving and responding to stakeholder input while clearly declaring their role and authority, and recognize that stakeholder involvement cannot be retracted once a commitment is made.

- All the risk estimates, and the tools that were used in preparing them, should be submitted for independent peer review by well-recognized independent specialists in the field of the toxic materials being evaluated. This is essential to ensure their scientific accuracy and credibility.

- Multiple opportunities should be provided for members of the potentially affected population group to provide comments and feedback, for example, through public meetings or through the Internet.

- A group of stakeholders should be actively engaged to participate in a meaningful and direct manner in conducting the assessment. Also critical to the success of both the decision makers and the potentially affected members of the public is that there is an ongoing dialogue throughout this entire process. A summary of the lessons that have been learned during the past several decades in this component of the risk-assessment process is presented in Table 15.8.

The General Outlook

Ideally, the risk-assessment process can provide structure to the process of setting standards and serve as useful input into public debates on the acceptability of proposed industrial facilities and other types of activities that may have both positive and/or negative impacts on the public and the environment. At the same time, however, regulators must be careful not to give disproportionate weight to the estimates generated through this process. In particular, they must not let quantitative risk assessments relegate human values and perceptions to a secondary role. Of key importance is that the efforts be conducted in an atmosphere (as just described) that increases public perception of the legitimacy of the regulatory process.

For optimal benefit from the risk-management perspective, it would appear to be appropriate to maximize the number of toxic agents regulated rather than the stringency with which they are controlled on an individual basis. It would likewise be wise to concentrate on controlling the principal uses of such agents. A pesticide, for example, may have a dozen applications, but only two of these may account for 98 percent of the amounts actually used. In the face of limited resources, regulators should concentrate on these applications. Concurrently, it is important to examine the total system to which a given risk-management procedure is being applied. Unless care is exercised and all interacting factors are considered, risk assessments directed at single issues and followed by ill-conceived management strategies can create problems worse than those they were designed to correct, a prime example being the previously cited addition of MTBE to gasoline. The single-issue approach can also create public myopia by excluding the totality of alternatives and consequences needed for an informed public choice.

Despite these caveats, our understanding of many components of risk assessment and risk management has been considerably enhanced in recent years. This is especially true with respect to our ability to analyze, quantify, and interpret exposures.

16

ENERGY RESOURCES
AND CONSERVATION

MEETING ENERGY NEEDS, conserving it, and protecting the environment are inseparable goals. Strip mining of coal can degrade the environment; drilling, acquisition, and transportation of oil can lead to spills that contaminate vast areas of land and water; and the consumption of gasoline in motor vehicles leads to air pollution, acidic deposition, and releases of greenhouse gases. Internationally, the production and consumption of oil can be a source of conflicts. The generation of electricity can also produce air pollution, problems of waste disposal, and other environmental impacts. Concurrently, an ever increasing portion of the use of energy in the United States is in the form of electricity, and overall energy consumption continues to increase more rapidly than the population (Figure 16.1). A comparison of the available fossil and nuclear fuels in the United States is illustrated in Figure 16.2.

After years of consuming energy resources as if they were unlimited, policymakers throughout the world now recognize that continued health and safety and environmental protection will be possible only if these resources, particularly the supplies of nonrenewable fossil fuels, are carefully managed and conserved. The urgency of coping with U.S. energy consumption is illustrated by the fact that this country, with 5 percent of the world's population, consumes about 25 percent of the global energy. Concurrently, our per capita annual consumption is increasing at a rate of 1 percent (*Wikipedia*, 2010).

The negative aspects of this situation have been vividly illustrated during the past four decades. In 1968, this country was exporting oil. By 1978, half of the oil being consumed was being imported. One year later, the ensuing situation culminated in a dramatic increase in gasoline prices that

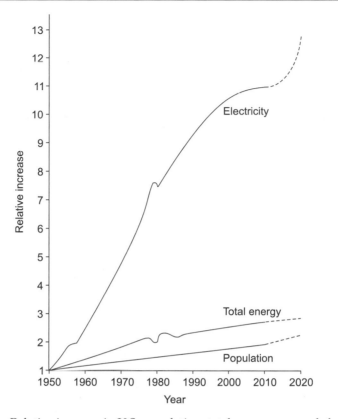

Figure 16.1   Relative increase in U.S. population, total energy use, and electricity consumption, 1950–2010 (data normalized to 1950)

created a financial disaster. Recognizing the situation, President Jimmy Carter instituted checks on oil prices and, in an infamous televised address (i.e., his "malaise speech") warned the nation about the long-term impacts on their lives of excessive consumption and the accompanying neglect of the environment. In so doing, he called for an increase in the price of fossil fuels and encouraged energy conservation and the use of alternative sources of energy, both by the nation and by individual members of the public (Johnson, 2009). Being neither what the public nor members of Congress wanted to hear, he was ridiculed and his message was rejected. After he left office, the checks on oil prices he had initiated were abolished. The ensuing energy policies ultimately led the United States to the situation that exists today (Mattson, 2009).

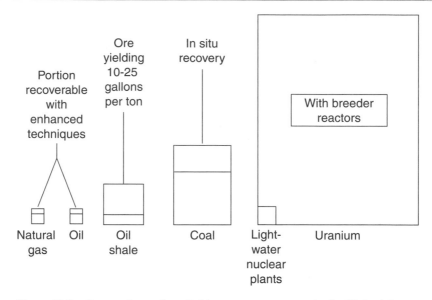

Figure 16.2   Comparison of available energy resources in the United States

Energy Resources

The fuels available to meet global energy needs fall into two broad catego-ries: (1) nonrenewable sources, such as fossil fuels, which include coal, oil, natural gas, and certain metals (i.e., lithium) used in batteries; and (2) renew-able sources, which include those of solar origin (i.e., wind and hydropower), as well as the forests (trees). Another such source is geothermal energy. Nu-clear fuels, as the discussion that follows will indicate, have the characteris-tics of both nonrenewable and renewable energy resources, depending on how they are used and their source.

NONRENEWABLE SOURCES

In considering the information that follows, it is important to recognize that the estimated reserves of each of these sources *are not static quantities.* Also important are the associated negative environmental impacts of ex-tracting these resources and the release of contaminants due to their use.

   Coal. Geologists estimate that economically recoverable U.S. resources
      will be adequate for several centuries. Globally, the estimated
      supply is a century or more. In reality, however, estimates such as
      these are less important than might be anticipated. Long before

existing supplies have been depleted, cleaner, less costly, sources of energy, such as solar power, should be available in abundance (Kerr, 2009).

Oil (petroleum). Globally, the confirmed reserves are sufficient to meet demands for three to five decades. This could, however, rapidly change. New methodologies, such as the use of bacteria, are being applied on an experimental basis to break up the complex carbon molecules in the two-thirds of the oil that is estimated to remain. In addition, there are enormous reserves in the oil sands of the Alberta province of Canada that could total as much as 1.7 trillion barrels. If so, this would place Canada second only to Saudi Arabia in proven oil reserves and could dramatically alter the situation. As is often the case, the associated environmental impacts of the recovery of these reserves are proving to be substantial. To date, these operations have created 50 square miles (130 square kilometers) of tailings ponds that contain multiple toxic chemicals. Although regulations require that these lands be restored, this has not been done (Kunzig, 2009).

Natural gas. The estimated recoverable U.S. reserves equal 150×10^{15} cubic feet (4×10^{15} cubic meters). These should be sufficient for more than two centuries. Estimated global reserves are more than $6,000 \times 10^{15}$ cubic feet ($>170 \times 10^{15}$ cubic meters) and are perhaps sufficient for five to six decades. Some experts predict that supplies yet to be discovered could prove adequate for as long as two centuries (OGJ Newsletter, 2008).

Metals (lithium). Although not generally considered an energy resource, lithium has enormous potential for such applications. It is being used to manufacture lithium-ion batteries that are lighter and longer lasting than any others. They serve, for example, as power sources for cell phones, laptop computers, and on a limited basis in motor vehicles. Although the last application has been hampered due to the limited availability of lithium, this problem now appears to have been solved. This is due to the discovery in 2009 that the Salar de Uyuni, a remote 5,000-square-mile area in the mountains of Bolivia, contains an estimated 50 to 70 percent of the world's lithium reserves (Kofman, 2009). This could reduce the cost of an all-electric automotive drive train system by an estimated 50 percent (Sagoff, 2009).

A comparison of the available fossil and nuclear fuels in the United States and the world, and estimates of how long they will meet demand, is provided in Table 16.1.

RENEWABLE RESOURCES

The advantages of solar power are obvious. Every minute of every day the sun transfers to the Earth more energy than the entire world uses in a year. From the standpoint of the United States, the dry, sun-drenched desert areas of the Southwest hold an enormous potential for the large-scale deployment of solar energy facilities and systems (U.S. Department of Energy, 2009a). At the same time, solar sources are versatile, being capable of being harnessed to generate electricity, monitor ecosystem conditions, pump water for livestock and irrigation, and provide lighting and communications in remote areas. Nonetheless, there is still the problem that on days without sun, reserves must be available to meet the needs. Other renewable sources include energy generated from forests and trees, and from geothermal sources that are released from inside the Earth, as well as from a variety of solar sources, including the sun and the wind. These are discussed below.

> *Forests and Trees.* Until the 1880s, when it was replaced by coal, wood supplied more U.S. energy than fossil fuels. Although this transition occurred earlier in Europe, its use has globally expanded in recent years. This has stimulated efforts to manage forests on a sustainable basis and to develop and apply improved technologies, such as "advanced wood combustion," that yields thermal efficiencies approaching 90 percent, combined with significantly reduced airborne emissions. More than 1,000 such facilities, supplying heat, cooling, and power, have been constructed in Austria, one of the latest being a 65-megawatt (MW) unit in Vienna that is providing both electricity to the grid and heat to the city's district energy system. Similar facilities are serving as sources of energy in France, Germany, and central and eastern Europe. The financial income could support restoration and improvements in forest management (Titus et al., 2009). Another important consideration is that the CO_2 released into the atmosphere, due to the combustion of wood, was already present in the carbon cycle (Chapter 18). Because of this, these releases do not increase atmospheric concentrations (Richter et al., 2009).

Table 16.1 Estimated U.S. and global nonrenewable energy resources, and adequacy for meeting needs[a]

Resource	United States		World-wide	
	Quantity	Duration of supply	Quantity	Duration of supply
Coal	$\sim$244$\times$10^9 tons	Several centuries	$\sim$900$\times$10^9 tons	More than one century
Oil (petroleum)	1.7$\times$10^{15} barrels[b]	Currently imports oil	4$\times$10^{15} barrels	Three to five decades
Natural gas	150$\times$10^{15} ft^3	Two or more centuries	>6,000$\times$10^{15} ft^3	Five to six decades
Nuclear fuels (uranium)	340$\times$10^3 tons of ore	5 or more decades[c]	3$\times$10^6 tons of ore	Several decades

a. Estimates are not static; they depend on resource exploration and development and improvements in extraction methodologies.
b. Includes potential contributions from the oil sands in the Alberta province of Canada.
c. Does not include potential contributions from breeder reactors.

Geothermal energy. Homeowners are increasingly using renewable energy through the application of geothermal heat pumps (GHPs) to heat and cool their homes. The ability of GHPs to perform this function is made possible by the fact that the soil has an average temperature of 50 to 60°F (10 to ~16°C) at a depth of 10 feet (3 meters). As a result, the underground zones are warmer during the winter and colder during the summer than ambient conditions above the Earth's surface. During the winter, a GHP uses water as a medium to transfer heat from the soil to a heat exchanger that warms the air that is circulated within the house. During the summer, it removes heat from the house to warm the soil (U.S. Department of Energy, 2009b).

A more extensive application of geothermal energy is exemplified by the ongoing operations in Iceland, where there are major sources of geothermal energy. These sources are being harnessed to serve a population of 300,000 people. This has been accomplished through the construction of five large geothermal power plants, which meet about 26 percent of their needs for electricity, supported by hydropower units that supply essentially all the remainder. Geothermal energy is also used to heat and provide hot water for almost 90 percent of their buildings (*Wikipedia*, 2009a). Unfortunately, the potential for the use of geothermal energy in the United States on this scale is possible only in localized areas.

Applications of Energy Derived from Solar Sources

Despite its advantages, by 2030 solar energy is projected to supply just slightly more than 1 percent of the electricity being consumed in the United States. Nonetheless, even with the application of existing technology, sufficient solar energy could be harnessed to supply a factor of 10 to 30 times this nation's demand. The major drawback is that the development of the accompanying infrastructure would cost much more than continuing to burn fossil fuels. Although this is admittedly a shortsighted view, there are factors that could reverse this situation: (1) the costs of generating electricity with solar power are rapidly being reduced, and (2) if the U.S. electric utilities are required to generate 25 percent of their power from renewable resources, which is an often-discussed probability, this could increase the contribution from solar power to 4 percent and that for wind to more than

10 percent by 2030. Nonetheless, if current policies continue and the demand for electricity increases, so will the consumption of fossil fuels and the emissions of carbon dioxide (Carroll, 2009; Johnson, 2009).

The same factors affect the use of solar energy on a global basis. Even so, certain other countries are making impressive progress. One is China. Although the achievement of their goal is subject to confirmation of a price of 7.6 cents per kilowatt-hour, if successful this could result in the available wind energy, which as noted earlier is one form of solar energy, providing their entire electricity demand by 2030, double the consumption in 2009. At present, coal serves as the energy source for the production of ~80 percent of their electricity, with wind producing about 0.4 percent (McElroy et al., 2009).

Summarized below are the primary methods for applying solar power. They are separated into two groups: those in which (1) solar energy is directly converted into electricity, and (2) the heat from the sun is harnessed for other types of applications.

PRODUCTION OF ELECTRICITY

Several examples of the applications of solar power to produce electricity, along with their advantages and disadvantages, are described below.

Hydroelectric power. To accomplish this on a major scale involves the construction of a dam on a river or major stream. The potential energy in the water behind the resulting lake is used to turn one or more turbines and generate electricity. The world's first such dam was the Hoover Dam, which was built on the Colorado River in the southwestern United States and began operating in 1935. This was followed by the enormously successful series of comparable facilities that were constructed by the Tennessee Valley Authority on major waterways in the south central United States during the mid-twentieth century. Another example is the Aswan High Dam located on the Nile River in Egypt. It was last modified in 1970 and generates about 2 million watts of electricity. The most recent is the Three Gorges Dam constructed on the Yangtze River in the People's Republic of China. It was completed in 2009 and generates about 18 million watts of electricity (Bosshard, 2009).

Worldwide, an estimated 45,000 large dams: (1) hold back about 14 percent of the runoff from precipitation and, in so doing, provide water for irrigating an estimated 40 percent of the world's agricultural lands; (2) enable large numbers of people to inhabit previously uninhabitable arid regions; and (3) protect plains from periodic flooding and permit the construction of cities that otherwise could not exist. In fact, hydroelectric systems provide more than half of the electricity needs for some 65 countries.

Wind power. Significant progress is being made in harnessing the wind, a source of energy that has been used for centuries. Globally, it is the most rapidly expanding energy source. Serving as a major stimulus for these developments is the dramatic reduction (by a factor of more than 10) during the past 20 years in the costs of producing electricity by this means. As a result, its cost today is about one-quarter of that for electricity generated using natural gas. New developments could also significantly increase the efficiency of wind turbines (Pacella, 2010). Another advantage is that whereas the costs of fossil fuels fluctuate, the costs of wind power, once a plant has been built and placed in operation, is steady. By 2020, the estimated wind electric generating capacity in the United States will provide 6 percent of the nation's needs, up from 0.3 percent in 2002. The overall wind generating capacity for the United States, however, pales in comparison to that in Europe. The new electric-generating capacity installed in Europe in 2009 was 23 percent above that in 2008. In fact, more wind power capacity was installed in 2009 than for any other generating technology, yielding a total of almost 26,000 MW of new capacity (Wilkes, 2010).

Tidal power. This is a form of hydropower that can be used to convert the energy of the tides into electricity. Because the energy is produced through the relative motion of the Earth, the sun, and the moon and their gravitational impacts on water levels, its availability is more predictable than that from the wind. The world's first commercial system for harnessing tidal power was installed in 2000 on the Scottish island of Islay. It includes a large, partially submerged concrete chamber built into the shoreline. The incoming water rotates the blades of a turbine that is connected to a generator. When the waves recede, the vacuum created sucks the water back through the turbine, permitting electricity to be generated both when the waves come in and recede (Gabraith, 2008; Staedter, 2002). Buoyed by this success, engineers in many countries are developing similar plans. These include explorations of San Francisco Bay as a potential resource for electricity generation. Another U.S. site deemed suitable is the East River in New York City (Electric Power Research Institute, 2006).

New more advanced projects are under construction. The most prominent are three 1.2 MW turbines under construction on the Campbell River and surrounding coastline of British Columbia, and the world's first tidal

energy farm off the coast of Anglesey in Wales. The target date for operation of these units is 2013 to 2014 (BBC, 2010). As in all such activities, there
will be associated ecological impacts. These include the alteration of the flow
of saltwater in and out of estuaries, a shift in the diversity of the species
in the affected waters, and up to 5 percent mortality of the fish that pass
through the turbines.

Another form of tidal power is represented by offshore structures that
generate electricity using energy created by ocean waves. Although these
have been deployed on a trial basis, multiple engineering challenges remain.
Even so, the potential for this resource is promising. The basic type of facility
under consideration would generate electricity through the use of an array of
buoyant floats on the ocean surface that, as the water rises and falls with the
waves, would push and pull electric coils up and down inside a cylinder of
stationary magnets mounted to spars anchored to the ocean floor. Although
developmental costs are expected to be high and additional assessments of
their environmental impacts and the associated permitting issues need to be
completed, analysts believe that this methodology represents a large and
viable source of renewable energy (Scruggs and Jacob, 2009).

Direct conversion of the heat of the sun into electricity. The discussion that follows describes the most common methods for the application of this approach.
The first two are, in reality, very similar to the use of fossil or nuclear fuels
to produce steam in an electric power plant. These, and the two that follow
them, involve various approaches for applying what is known as "concentrating solar power" (CSP) technologies.

- *Parabolic solar trough.* This, the most widely applied approach, involves
  the use of lenses or mirrors that condense a large area of sunlight into
  a small beam. A supporting tracking system continuously directs the
  beam onto a tube of mineral-oil transfer fluid that is heated to 390°C
  (734°F). The fluid then enters a heat exchanger and boils water, creating steam at high temperature and pressure that is passed through a
  turbine to generate electricity similar to the process used in a nuclear
  power plant (Hutchinson, 2008). One advantage of this method is that
  it requires the least land area per unit output of any solar technology.
  As of 2009, nine such plants were operating, some for as long as 25
  years, in California's Mojave Desert, providing electricity to about
  380,000 homes. (*Wikipedia*, 2009b).

- *Power towers.* These are an advanced version of the parabolic solar
  trough, and they operate on the same principles. At the same time,

they are more cost-effective and provide higher efficiencies. Units are in operation in Barstow, California, and in Sanlucar la Mayor, Spain (Johnson, 2009). Another version is what is called photovoltaic collectors. These have proved to be very popular in the United States. As of 2008, 15 gigawatts of electricity were being generated by facilities based on this design (Swanson, 2009).

OTHER APPLICATIONS OF SOLAR ENERGY

There are multiple other approaches for harnessing solar energy.

The Stirling engine. This is a heat engine that operates by cyclic compression and expansion of air or other gas, so that there is a net conversion of heat energy to mechanical work. This is accomplished by intermittently focusing the heat of the sun on the outside of an engine whose cylinders contain air. The heat causes the air to expand, moving a piston up and down much the same as the action of the steam in a steam engine (Walker, 1980). When the gas cools (which is relatively quickly), the piston returns to its original position and creates a vacuum that can be used to withdraw water from a lake or stream. Pumps based on this concept are in use in many of the developing countries of the world. Expanded commercial versions incorporate large mirrors that focus the heat of the sun onto multiple such units. In this mode, the engine can be used to generate electricity. One of its advantages is that it converts almost one-third of the incoming heat into electricity (*Wikipedia,* 2009c).

Methane production. The anaerobic digestion of animal and plant wastes in municipal sanitary landfills yields methane as a byproduct. Although for years this gas was vented to the atmosphere, recognition that it represented a valuable renewable energy source has changed this dramatically. As of 2007, there were an estimated 1,000 methane collection systems in operation globally. Serving as a stimulus in the United States is a directive from the EPA that requires operators of all U.S. landfills to have a leachate-collection system designed to accelerate the decomposition of the waste and the associated production of methane. As of 2007, such systems were annually collecting about 2.6 million tons of methane, 70 percent of which was being used to generate heat and electricity (Themelis and Ulloa, 2007).

Other direct applications. People in the developing countries use the heat of the sun to distill water, dry (preserve) agricultural crops, produce ethanol from sugarcane, methanol from wood materials, and methane through the anaerobic digestion of animal and plant wastes. Although the scale of each such activity is relatively small, the overall impact is large.

The Versatility of Nuclear Fuels

With modifications, commercial nuclear power plants can be operated such that the fuel can serve as either a nonrenewable or renewable source of energy, depending on the manner in which it is used.

Conventional applications. The fundamental source of nuclear energy is uranium, a naturally occurring radioactive material. To illustrate its abundance, it is estimated that, on average, there are about 30 tons of uranium within the first 5 feet of the ground for every square mile of soil on Earth (Evans, 1969). Although it contains three radioisotopes, two of these, ^{235}U and ^{238}U, represent 0.72 and 99.28 percent of the total, respectively.

Uranium. The basic nonrenewable approach is to fission ("burn") uranium, in which its ^{235}U isotope has previously been enriched to a concentration of 2 to 4 percent, in a nuclear power plant. The heat generated is used to convert water into steam, drive a turbine, and generate electricity (Figure 16.3). Seeking to reduce the threat of terrorist-related activities, the plutonium (^{239}Pu) resulting from the dismantlement of nuclear weapons in the former USSR is now being combined with uranium to form what is called mixed-oxide (MOX) fuel for use in conventional nuclear power plants. This has tremendous benefits since ^{239}Pu is comparable to ^{235}U as such a fuel. Following the lead of the European nations, newer U.S. plants are being designed to accommodate fuel containing from 50 to 100 percent MOX. As might be anticipated, security concerns require that sites for such facilities meet certain features (Sowder et al., 2009).

Interestingly, it had been demonstrated several decades earlier that a conventional nuclear power plant could be designed to produce more new ^{239}Pu than the amount of ^{235}U required to produce it. This was documented during a 10-year period of operation of the Shippingport nuclear power plant in Pennsylvania during the mid-1970s to the mid-1980s. The plant, which was used as a test bed for the design of nuclear power plants for ships in the U.S. Navy, was subsequently dismantled. Since such breeding (as this process is called) was not considered essential to the emerging U.S.

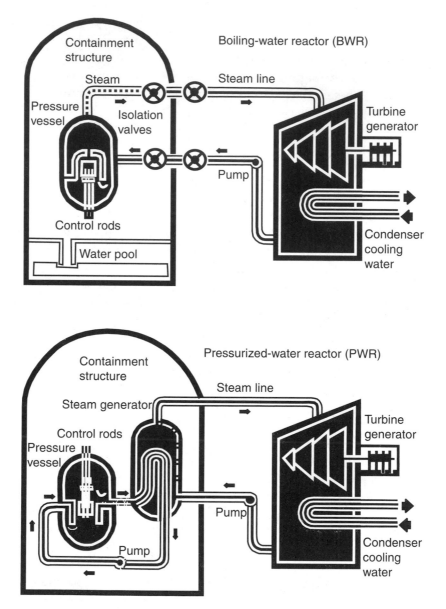

Figure 16.3 Schematic diagrams of boiling-water and pressurized-water nuclear power plants

commercial nuclear power plant industry, the concept was not further pursued (Adams, 1995). Nonetheless, in this mode uranium is a renewable source of energy.

Since the accident at the Three Mile Island nuclear power plant in 1979, no nuclear power plants have been constructed in the United States. In spite of the accident at the Chernobyl nuclear power plant (Chapter 17), other countries have continued to construct such plants. Nonetheless, this situation in the United States is now changing. The Nature Conservancy, for example, has confirmed that the production of electricity by nuclear power plants can reduce carbon emissions while consuming the least amount of land per unit of energy generated. For example, a typical nuclear power plant requires about 1 square mile of land. This compares to 15 square miles for a solar plant and 30 square miles for a wind power facility of an equivalent generating capacity (NEI, 2009b). Also of note is that the life-cycle airborne emissions from nuclear power plants are comparable to those from geothermal and wind power (Table 16.2). Interestingly, the EPA predicts that compliance with the carbon-emission provisions in the climate change bill, passed in the House of Representatives in June 2009, will require the construction of 187 new U.S. nuclear power plants by 2050 (NEI, 2009a).

Fusion Energy. In contrast to fission, fusion involves combining small atoms under conditions of extremely high temperature and pressure. This also yields enormous amounts of energy. In fact, it is the source of the heat being produced in the sun. Through a cooperative global effort initiated in 1985 by the Soviet Union, the United States, the European Union, and Japan, with the subsequent addition of China, France, India, and Korea, a major effort is under way to harness this source of energy for the generation of electricity. Global support is mandatory, for multiple reasons, one being the immensity of the challenges being faced. A major advance was the completion in 2009 of the National Ignition Facility (NIF) at the Lawrence Livermore National Laboratory in California. This facility is designed to deliver a power of 500 gigawatts, more than the capacity of all the power-generating units in the entire United States. The most recent tests were successful, and NIF researchers indicate that they will attempt a self-sustaining fusion reaction—one that produces excess energy—by the end of 2010 (Clery, 2010). If this test is successful, it will represent a historical step toward solving the energy problems of the world.

Another facility nearing operation is the Korean Superconducting Tokamak Reactor (KSTAR), which became operational in 2009 (Normile, 2009).

Table 16.2 Comparison of various sources of electricity in terms of life-cycle carbon dioxide emissions (2002) and quantities of emissions avoided[a] (2008)

	Coal	Natural gas	Solar power PV[b]	Hydroelectric dams	Nuclear	Geothermal power	Wind power
Life-cycle emissions[c]	1,041	622	39	18	17	15	14
Emissions avoided[d]	—[e]	—[e]	<1	206	689	13	44

a. Emissions avoided were estimated on the basis of regional and national fossil-fuel emission rates provided by the Environmental Protection Agency and the Energy Information Administration, U.S. Department of Energy (2009).
b. Photovoltaic cells.
c. Expressed in units of tons of CO_2 equivalent per gigawatt-hour.
d. Expressed in units of millions of metric tons.
e. Not applicable.

One of its primary advances is the use of high-performance superconducting magnets, composed of a niobium-tin alloy. It is expected to confine plasmas for up to 300 seconds, compared to about 20 seconds in the older models. The data generated by these two facilities will provide input for a third facility, the International Thermonuclear Experimental Reactor (ITER), located in France. Plans call for the initial production of a superhot hydrogen plasma in 2018 and a power-producing plasma of deuterium and tritium by the end of 2026 (Clery, 2009b). Because these efforts involve realms of chemistry, physics, and technologies never before explored, and will be accompanied by the possibility of multiple unanticipated difficulties, it is not possible to estimate when success will be achieved (Clery, 2009a). All such efforts are designed to achieve a common global goal, the harnessing of the fusion process to produce electricity. When success is achieved, this would meet the energy needs of the entire world.

Environmental Impacts

One concern that inevitably arises in any discussion of energy use is the accompanying environmental and public health impact of electricity-generating power plants. To provide perspective, a summary of the impacts for each of the methods of generating electricity are discussed and, in some cases quantified, in the sections that follow.

HYDROELECTRIC POWER

Large-scale dams can dramatically alter the environment and the lives of people. One example is the previously discussed Aswan High Dam on the Nile River in Egypt. As planned, this project had two objectives: (1) to produce electric power and (2) to irrigate the nearby desert. But many unforeseen impacts ensued. In its original state, the Nile River served as a mechanism for the downstream transport of tremendous quantities of silt and organic matter. During the annual spring floods, portions of this material were deposited on the banks of the river, where it served as fertilizer for agricultural crops. The material that remained ultimately reached the Mediterranean Sea, where it served as food for large numbers of fish. After the dam was completed, the river no longer flooded its banks and the fertility of the farming areas declined. Without the discharge of nutrients into the Mediterranean, the fishing industry was essentially destroyed. Construction of this facility also displaced 90,000 people who had lived in the areas being flooded. Compounding the problems, Lake Nasser, created by the dam, raised the water table and brought dissolved salts from the ground up into the desert topsoil, making it less usable for agriculture than anticipated. In addition, control of the flow rate in the river led to increased growths of algae and phytoplankton. This adversely affected water quality. The accompanying quiescence also promoted the growth of the snail population, which dramatically increased the incidence of schistosomiasis (Chapter 7) among nearby populations.

In a related manner, the lake created by the Three Gorges Dam in the People's Republic of China submerged hundreds of factories, mines, and waste dumps. In addition, the presence of massive industrial centers upstream is creating a festering bog of effluent silt, industrial pollutants, and rubbish in the reservoir. Erosion of the reservoir and downstream river banks also subsequently caused landslides and are threatening one of the world's largest fisheries in the East China Sea. Interestingly, recent evidence confirms that shifts in the weight of the water in the lake during wet and dry seasons were the probable source of several recent earthquakes. Even so, because of the need for electricity the Chinese government is replicating the Three Gorges model for additional domestic and international applications. Numerous other large hydroelectric projects have had related types of impacts. For example, the Grand Coulee Dam and its sister units on the Columbia River in Washington State have essentially eliminated salmon fishing in that region (Brink and Dvorak, 2009). Some biologists predict that without major changes, these fish will become extinct.

WIND POWER

This application also has its negative impacts. The blades on wind towers produce noise, interfere with television reception in nearby homes, and harm migratory birds. They have proved to be especially deadly for bats in Germany and Poland. One preventive measure, in this case, is to reduce operations during darkness when wind speeds are low. Analyses show that the financial impacts of reducing operations during those periods would be minimal (Curry, 2009).

GEOTHERMAL ENERGY

The pressure in geothermal hot-water reservoirs is often a result of the weight of the overlying land. For this reason, withdrawal of the water can lead to subsidence. Accidental spillage of the withdrawn water, which often has a high mineral content, can also lead to increased salinity of the soil and pollution of surface waters. Where water is injected to be converted into steam using heat from dry rocks in the vicinity, induced seismicity is a risk similar to the previously cited case in China. Significant problems can also arise from the release of radon and volatile gases such as hydrogen sulfide, which accompany the steam.

ELECTRIC POWER PLANTS

The generation of electricity in fossil-fuel and nuclear power plants also has a range of environmental impacts. For purposes of discussion, these have been separated into six categories.

Fuel acquisition. If coal is strip mined, the air is polluted with dust and the surface of the Earth is defaced unless the land is restored to its original state. Underground mining frequently produces "acid mine drainage"— sulfuric acid and iron salts that drain or seep from the mine during operation and for some years thereafter. When these materials flow into surface waters, they are toxic to most forms of aquatic life. When the coal is obtained from underground sources, the miners experience an array of occupational health problems. On average, the mining of sufficient coal to provide fuel for one 1,000 MW power station results in two to four accidental deaths and two to eight cases of black lung disease and other respiratory ailments among coal miners each year. Unfortunately, accidental deaths continue; 25 miners were killed in an accident in West Virginia in April 2010; 27 in Utah in 1984; and 38 in Kentucky in 1970. The total U.S. deaths during the past decade exceeded 550 (Associated Press, 2010).

Fuel transportation. A standard 1,000 MW coal-fired power plant requires about 8,000 tons of fuel per day, enough to fill 100 or more railroad cars. Transporting it to the power plant leads to an estimated 2–4 deaths and 25– 40 injuries each year, primarily as a result of accidents at railroad crossings. Similar impacts result from the shipment of oil in tankers. In 2002, the *Prestige* broke up off the western coasts of Spain and Portugal, resulting in the immediate release of several million gallons of fuel oil into the Atlantic Ocean. Another 20 million gallons sank to the bottom with the vessel and continues to leak out slowly and rise to the ocean's surface. The accompanying impacts forced officials to close ~100 kilometers (~60 miles) of the coastline to fishing. The worst such accident in recent years occurred in 1989, when the *Exxon Valdez* ran ashore and 11 million gallons of oil spilled into Prince William Sound in Alaska (Chapter 17). Shipments of natural gas pose similar problems. A modern tanker can transport about 125,000 cubic meters of liquefied natural gas at a temperature of –160 °C (~250°F). When the liquefied gas is being prepared for transfer to storage tanks, it is permitted to warm and gasify, increasing its volume by a factor of more than six. An accidental release of millions of cubic meters of natural gas (in expanded cloud form) could blanket a harbor city. Fortunately, no such events have occurred. In contrast, a 1,000 MWe nuclear plant requires only about 30–50 tons of fuel per year. Because the original uranium fuel is not a significant radiation source, its transportation does not present any unusual occupational or environmental health problems.

Power plant releases. Both fossil-fuel and nuclear plants release contaminants into the environment. Primary liquid releases are mercury in the case of the former, and tritium (^3H) in the case of the latter. Both types of plants also generate airborne releases. Average airborne releases for various types of fossil-fuel power plants and nuclear power plants are summarized in Table 16.3 and Table 16.4, respectively. To provide perspective, the relative amounts of air to dilute the releases to permissible limits are summarized in Table 16.5.

Steam condensation. If an electricity-generating plant is to be efficient, the steam leaving the turbines must be condensed to maximize the difference in the pressure of the incoming steam and that on the downstream side of the turbine (Figure 16.3). This is accomplished either (1) by using water from a nearby river or lake to condense the steam or (2) by constructing large cooling towers into which the water used for condensing the steam can be sprayed and release its excess heat into the atmosphere. In so doing, various gases and particles are released into the atmosphere, along with smaller quantities of liquid wastes.

Management of solid wastes. Power plants fueled by natural gas or oil have no spent-fuel disposal problems because combustion of these fuels produces no ash. In contrast, 12–25 percent of the fuel burned in coal-fired plants ends up as ash. Thus, a 1,000-MWe plant would require 12–25 railroad cars for the daily removal of ash. Where and how this ash is disposed of is important because it contains many toxic materials. Although a typical nuclear plant produces some 30–50 tons of intensely radioactive spent fuel each year, the volumes are much smaller. It does, however, pose significant problems from the standpoint of radiation protection. Once it has cooled (and decayed) it is sealed in thick casks (Figure 9.6, page 219) and placed in an underground facility, where it will not pose a threat to either the public or the environment (Moeller, 2009).

Accidents. Such events can occur during both the construction and operation of electric power plants. For example, a massive spill of coal ash produced in the operation of several fossil-fuel (i.e., coal) electricity-generating power plants operated by the Tennessee Valley Authority (TVA) occurred

Table 16.3 Annual airborne emissions from fossil-fuel electric power plants (assumed capacity 1,000 MWe)

Pollutant	Emissions (thousands of tons per year)[a]		
	Coal	Oil	Natural gas
Sulfur dioxide	43	29	7
Nitrogen oxide	25	11	7
Carbon monoxide	0.7	1.0	1.1
Particulates	120	0.6	<0.1

a. Emissions will vary depending on nature of fuel, plant design, and operating parameters.

Table 16.4 Annual airborne emissions from nuclear power plants (assumed capacity 1,000 MWe)

Pollutant	Pressurized-water reactor		Boiling-water reactor	
	Becquerels	Curies	Becquerels	Curies
Krypton-85[a]	7×10^{11}	20	—	—
Xenon 133	7×10^{13}	2,000	4×10^{12}	100
Carbon 14	4×10^{11}	10	4×10^{11}	10
Iodine 131	7×10^{7}	0.002	1×10^{8}	0.003
Tritium	6×10^{12}	150	4×10^{12}	100

a. Releases from boiling-water reactors are negligible.

Table 16.5 Annual dilution requirements for compliance with EPA and USNRC standards for airborne emissions from 1,000 MWe fossil-fuel and nuclear power plants

Type of plant	Limiting pollutant	Permissible concentration	Required dilution[a]	
			Cubic meters	Cubic miles
Coal	PM_{10}	$50 \, \mu g/m^3$	2×10^{15}	500,000
Oil	SO_2	$80 \, \mu g/m^3$	3.3×10^{14}	80,000
Natural gas	SO_2	$80 \, \mu g/m^3$	8×10^{13}	20,000
Nuclear (pressurized-water reactor)	^{133}Xe	$2 \times 10^4 \, Bq/m^3$	4×10^{9}	1

a. Approximate volume of air required to dilute the most critical pollutant from each type of plant to the permissible concentration as prescribed by federal standards for the ambient environment.

in late 2008. This was due to the failure of a large earthen impoundment in which the ash was being stored (Dewan, 2008, 2009). This was followed by a massive explosion due to improper purging of natural gas during tests of a plant under construction in Connecticut in February 2010, killing at least five people. This had been preceded by seven similar accidents in the United States during the past decade. Only 3 days prior to its occurrence, the U.S. Chemical Safety Board had issued urgent new recommendations on how to prevent such an explosion (Schauer and Clayton, 2010). Although other types of concerns have been expressed relative to U.S. commercial nuclear power plants, the only accident that has occurred in such a facility was at the Three Mile Island power plant in Pennsylvania in 1979. It caused no deaths, and detailed reviews documented that the associated impacts on the environment and public health were infinitesimal. As a result, most people cite the Chernobyl accident that occurred in the USSR in 1986 as justification for their fears of such events. As is discussed later (Chapter 17) the long-term health effects (cancer and leukemia) of this accident on adjoining population groups were significantly less than anticipated (UNSCEAR,

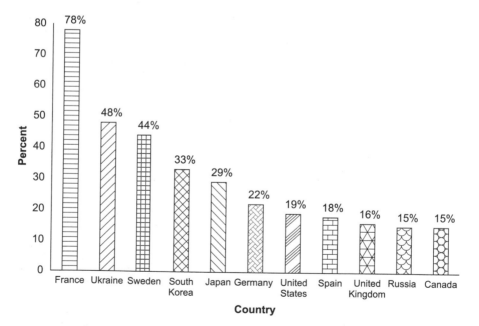

Figure 16.4   Percentage of electricity generated by nuclear power globally, 2007

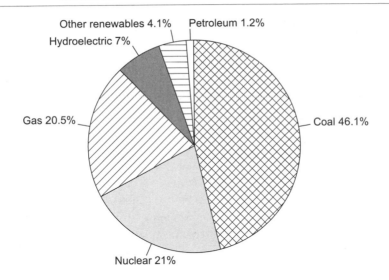

a. Other renewables include geothermal, wood and non-wood waste, wind, and solar energy.

b. Hydroelectric includes conventional hydroelectric and hydroelectric pumped storage.

c. Gas includes natural gas, blast furnace gas, and other manufactured and waste gases derived from fossil fuels.

d. Due to rounding, the total does not equal 100%.

Figure 16.5 Relative sources of energy used to generate electricity, United States, 2009

2000). Furthermore, the plant, as built, would never have been approved for use elsewhere in the world (Chapter 17).

The relative contribution of nuclear-generated electricity to the total for different countries on a global basis are shown in Figure 16.4 and the relative contributions of various sources of energy to the generation of electricity in the United States are shown in Figure 16.5. As noted, the global contributions from nuclear power are rapidly increasing (NEI, 2009a). The percentage contribution shown for the United States is less in Figure 16.4 than in Figure 16.5 due to the differences in the dates on which the data are based.

Energy Distribution and Conservation

There are multiple practical methods and technologies that could be applied to conserve energy. One that has widespread implications, particularly in terms of its distribution in the form of electricity, is the application of what is called *smart grid* digital technologies, combined with a national system of underground transmission cables. Although the primary purpose would be to transmit electricity, the system, as envisioned, would also (1) enable power companies to maintain a more accurate balance between electricity generation and demand on a grid-wide basis, (2) identify any user facility in which power is unnecessarily being wasted, (3) permit electricity generated by wind turbines, located far from consumers, to be delivered to consumers at the speed of light, and (4) reduce the threat of terrorist attacks on the nation's distribution systems (Amin, 2010; Schuler, 2010). Ancillary benefits would include a reduction in the current 12–20 percent losses in the transmission of electricity (Truly, 2002), the elimination of the environmental impacts of the consumption of up to 100 acres of land per mile as right-of-way for overhead power lines, and the loss of power due to various disasters of natural origin (i.e., winter storms, hurricanes, etc.).

Finally, such a system would permit the simultaneous two-way transmission of information. One of the many benefits of this feature is that it would not only eliminate the need to conduct onsite readings of meters to determine usage but also the consumption of fuel and associated airborne emissions of the vehicles operated by the meter readers. Although such a system would require a decade or more to plan and install, the benefits would be enormous. The economic savings alone would be an estimated $75 billion during the next 20 years (Kannberg et al., 2003).

Other approaches for conserving energy that are available and applicable to specific consuming sectors are discussed below. As will be observed, certain of these (e.g., light-emitting diodes) are applicable to several sectors.

INDUSTRIAL SECTOR

Relying on a mix of fuels, the industrial sector accounts for 33 percent of U.S. end-use energy consumption. Of this, 70 percent is used to provide heat and power for manufacturing; this, in turn, represents 25 percent of the U.S demand for petroleum. The chemical industry, recognizing the benefits of energy conservation, has been a leader in such efforts (Worrell et al., 2000).

Their analyses showed that energy productivity is just as important as labor productivity. Although their use of electricity increased by 20 percent between 1986 and 2000, production increased by more than 50 percent. This led to a 21 percent reduction in the consumption per unit of product, almost 60 percent higher than the industrial average. One of the outgrowths of this experience is that increasing numbers of industrial leaders are recognizing that in reducing their consumption of energy, they are saving money and reducing the release of contaminants into the environment.

COMMERCIAL SECTOR

A major portion of the energy used in the commercial sector is for lighting (e.g., for office buildings it represents more than 25 percent of the total electricity usage). To reduce such use, it is important to enhance the use of daylight and to incorporate features such as those that turn off the lights when the occupants leave a room. One of the most obvious energy-saving methods is to switch to more efficient lighting systems. Compact fluorescent units consume only 25 percent as much electricity as incandescent bulbs and can last 12 times as long. Another is to switch to light-emitting diodes (LEDs), now being used in traffic signals, pedestrian crossings, decorative holiday lights, and electric signs. Nationally, in 2007 this yielded an annual energy savings equivalent to more than the production of a large (i.e., 1,000 MW) electric power plant. Progress is being made to increase their output for use inside buildings as well as in homes. Once these new units become available, the savings could be considerable (U.S. Department of Energy, 2009c).

Other major consumers of energy are personal computers, printers, copiers, and facsimile machines. If the electricity required for additional ventilation and air conditioning is included, the combined load represents some 10 to 12 percent of the total electricity demand of a typical office. To reduce such usage, computers, their printers, and other desktop devices should be turned off at the end of each workday. Fortunately, the monitors on most computers do this automatically. Another factor that has reduced energy costs is the widespread adoption of a 4-day, 10-hour-per-day workweek. This has yielded significant reductions in energy usage, as well as other benefits (Walsh, 2009).

RESIDENTIAL SECTOR

Since more than 15 percent of U.S. energy is consumed in the home, this is obviously an important sector in which to practice conservation. Usage for heating and air conditioning can be significantly reduced by weather-stripping doors, caulking windows, and increasing the insulation on the attic floor. Cooling costs can be reduced by installing more efficient air-conditioning units. Improvements have been made at such a rapid pace that the cost of new units can be reclaimed within as short a time as 5 to 10 years. As was the case in the commercial sector, residents should switch to more efficient lighting systems. If every household in the U.S. replaced four 100-watt incandescent lightbulbs with compact fluorescent units, the savings would be equivalent to the output of about ten 1,000MW electric power plants. Residents should also insist on energy-efficient home appliances.

TRANSPORTATION SECTOR

This sector consumes the remaining 28 percent of energy in the United States, the major share of which is in the form of fuel for the 250 million U.S. automobiles. Nonetheless, the fuel economy of cars and light trucks improved from 18.0 miles per gallon in 1978 to 27.5 in 2009 (American Wind Energy Association, 2008). This should continue to improve as a result of the revised national standards for automobile emissions and fuel efficiency

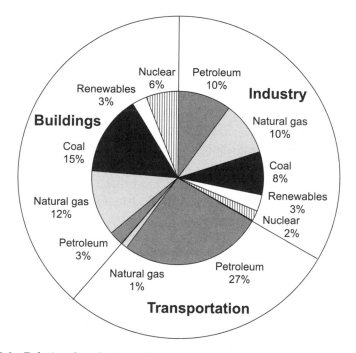

Figure 16.6   Relative distribution of energy usage in the United States

established in 2009. These will require that all new cars and trucks have an average fuel corporate fuel efficiency (CAFE) of 35.5 miles per gallon (6.6 liters per 100 kilometers) by 2016 (National Highway Transportation Safety Administration, 2009). If achieved, this will represent an almost 40 percent improvement. The standards also require reductions in greenhouse gas (GHG) emissions. Playing a somewhat unexpected role was the "Cash for Clunkers" program, initiated in late 2009, that led to the removal of almost 700,000 old cars and trucks from the highways. The new vehicles that replaced them averaged more than 9 miles per gallon more and had far fewer emissions (*Consumer Reports,* 2009). The degree to which hydrogen-fueled cars and hybrids will contribute to energy conservation is yet to be determined.

The relative distribution of energy usage in the United States is summarized in Figure 16.6.

The General Outlook

In any evaluation of the challenges faced in meeting energy needs, it is important to keep in mind its importance to society, especially in the form of electricity. Better lighting reduces accidents on highways and crime in cities. It is also necessary to clean the air, operate water-purification facilities and sewage treatment plants, dispose of old automobiles, and recycle other types of solid waste. Although conservation will continue to be important, the overall demand for electricity, and thus for more generating capacity, will not only continue but will also increase. The basic challenge is to educate people to use energy more efficiently and conservatively and to encourage the commercial sector to design, construct, and operate generating stations that function at maximum efficiency with minimal impact on public health and the environment.

On the positive side, progress is being achieved. Especially encouraging is the increasing recognition by major international energy corporations of the need to understand and incorporate the principles of sustainable development into their long-range plans. In a similar manner, regional electric utility companies in the United States are implementing positive environmental measures into their corporate policies (one example being to encourage customers to adopt more efficient lighting and install solar panels), their point being that conservation is preferable to the alternative—developing new electricity-generating capacity.

Advances in technology can also help. Electrochemical windows, which can be darkened or lightened to control heat from the sun, are now available; air circulation systems that include desiccant-based dehumidification systems will reduce cooling loads; and solar cells can be integrated into the roof and windows to reduce energy demand (Truly, 2002). Some of the more efficient buildings, which are equipped with solar collectors and energy-saving devices, are now producing the equivalent of almost the amount of energy they consume. Multiple colleges and universities are leading in the installation of such improvements. Hopefully, these concepts will later be adopted by their students.

Finally, there continues to be a need to define what a national energy plan should achieve, both in the near and long term, as well as on an international basis. On a national basis, there are four different goals that must be addressed (Sharp, 2009). The *first* is economic (i.e., to ensure that sufficient fuel is available for homes, schools, industries, and commercial activities). The *second* is to protect national security (this includes the stimulation of the production of oil, natural gas, and other sources of energy that are important to economic prosperity). The *third* is to protect the environment. And the *fourth* is to address issues of equity. This includes the need to address concerns for the poor and for regional impacts such as increasing fuel oil prices for home heating in northern regions and rising gasoline prices, especially for long-distance drivers, in the west, as well as meeting the basic nutritional needs of the poor in the developing nations of the world (Charles, 2009).

17

DISASTERS AND TERRORISM

IN A BROAD SENSE, disasters can be classified into two categories: (1) those of natural origin and (2) those of human origin. The former includes earthquakes, hurricanes, and cyclones; the latter includes industrial accidents, oil tanker spills, and terrorism. Few disasters, however, are totally of natural or human origin. The consequences of many are exacerbated by inadequacies on the part of humans—the failure to have a well-designed emergency response plan, buildings that are not designed to resist earthquakes, the lack of proper maintenance that leaves a ferry unable to resist heavy seas, or a single-hulled oil tanker that leaks oil because a double-hulled tanker was deemed too expensive. In a similar manner, the frequency and severity of events such as droughts and floods may be influenced by human destruction of forests. Selected examples of disasters of natural and human origin that have occurred, primarily during the last decade, are summarized in Table 17.1. Another approach is to classify disasters on the basis of the nature of their impacts on human health and the environment. In this sense, such events can be classified either as (1) those that are characterized by strong physical forces and impacts that affect both people and the environment (such as hurricanes, floods, and earthquakes), or (2) oil spills that involve the release of toxic materials and primarily affect the aquatic and marine environment and associated wildlife.

The globalization of terrorism, however, has dramatically changed the nature and potential magnitude of such events. This was vividly brought to the nation's attention by the September 11, 2001, attacks on the World Trade Center and the Pentagon in the United States, which caused the deaths of more than 3,000 people. These were followed by attacks on the transport systems in Spain, Japan, and London, and on embassies, nightclubs, and

Table 17.1 Selected examples of major disasters of natural and human origin, 1989–2010

Event	Location	Date of occurrence	Impact
Natural Disasters			
Earthquakes	Port-au-Prince, and adjoining regions, Haiti	January 12, 2010	An estimated 230,000 deaths; 300,000 injured, and 2 million left homeless; widespread destruction of homes, churches, and buildings, and accompanying devastation
	Mianyang, China	May 13, 2008	Upwards of 100,000 dead and missing, widespread flooding, landslides covered mountain villages, and blocked flow in the Qungzhu River; 5 million people left homeless
Hurricanes	Katrina, Gulf Coast	August 29, 2005	Largest natural disaster in the history of the United States; almost 2,000 deaths, and >$100 billion in property damage; flooded 80 percent of the city of New Orleans; the water lingered for weeks
Cyclones	Myanmar (formerly Burma)	May 2, 2008	Created 12-foot storm surge over large nearby land areas in the middle of the night, devastating villages and rice paddies, and leaving as many as 1,000,000 people homeless; >134,000 people dead or missing; widespread famine

Disasters of Human Origin

Wild fires	Australia	February 2009	Searing temperatures and winds created a firestorm that swept across Victoria state; 750 homes were destroyed and 850 square miles were burned; deaths exceeded 200
	California	July 2008	More than 1,400 wildfires choked the Sierra Nevada foothills, darkening a 100-mile stretch between Sacramento and Reno; almost 570 square miles of forest, grass, and brush were burned
Oil tanker spills	Exxon Valdez, Prince William Sound, Alaska	March 1989	Spilled 11 million gallons; extensive destruction of fishing and wildlife; impacts continued for a decade or more
Terrorism	New York, NY, Washington, DC, and rural area in PA	September 11, 2001	About 3,000 deaths; World Trade Center towers destroyed and Pentagon building extensively damaged by two hijacked airliners; a third hijacked plane (reportedly scheduled to destroy the White House) was commandeered by the passengers but crashed in a field near Shanksville, PA, killing about 40 people, including the crew and 4 hijackers

hotels (ICRP, 2005). Although acts of terrorism are also of human origin and, in some ways, the associated havoc and destruction are no different than those that result from other types of disasters, there are distinct differences in their nature and characteristics. For these reasons, they will be discussed separately.

Natural Disasters

Although natural disasters have been a problem for centuries, records show that the number of events of major consequence, that is, those that result in deaths or losses so high that outside assistance is required, has increased dramatically in recent years. This is due to many factors, including the destruction of forests and the construction of buildings along coastlines, on unprotected riverbanks, and on steep hillsides (Abramovitz, 2001). At present, an estimated 60 percent of the world's population lives within 100 kilometers (62 miles) of a coastline. Also increasing is the number of deaths from earthquakes and volcanic eruptions. Mexico City, for example, has nearly 20 million people living in a region subject to earthquakes and volcanoes. At the same time, the migration of the U.S. population to California, an active seismic area, continues unabated (Parfit, 1998). The probabilities of occurrence of a range of impacts caused by various types of natural disasters are summarized in Table 17.2.

For purposes of analyses, natural disasters can also be subdivided into two groups: (1) those related to geological factors, such as earthquakes, volcanic eruptions, and tsunamis, and (2) those that are related to climatological factors, such as floods and hurricanes. The latter occur more frequently and often have larger impacts, especially in terms of the size of the areas affected. Other sources of natural disasters include landslides and avalanches, droughts, and tornadoes. The characteristics of each of these are discussed below.

FLOODS AND HURRICANES

Floods can occur following heavy rains, a rapid snow melt, or as the aftermath of a hurricane. In some situations, the floods that accompany hurricanes can be more destructive than the winds. On a worldwide basis, floods account for about 40 percent of all natural disasters, and the impacts of hurricanes and floods have the largest impacts of all natural disasters in terms of the numbers of people affected. Overall, they annually cause about 165 deaths in the United States. In addition, floods often damage and destroy

Table 17.2 Major impacts of natural disasters

Impacts	Earthquakes	Hurricanes, high winds	Volcanic eruptions	Floods	Tidal waves, flash floods
Deaths	Many	Few	Variable	Few	Many
Severe injuries (requiring extensive medical care)	Overwhelming				
Increased risk of infectious disease	A potential problem in all major disasters; probability increases with overcrowding and deteriorating sanitation				
Food shortages	Rare (may occur as a result of factors other than food shortages)	Rare	Common	Common	Common
Major population shifts	Rare (may occur in heavily damaged areas)	Rare	Common	Common	Common

homes, displacing the occupants, and lead to crowded living conditions, contaminated drinking water, and the disruption of sewer systems and the collection and disposal of solid waste (WVU, 2004). Among all of these impacts, the most immediate need is to restore the drinking water supply (Servat, 2010).

LANDSLIDES AND AVALANCHES

These types of events occur when imbalances of rock, snow, and earth on steep slopes suddenly give way and fall to lower elevations. Historic examples include the estimated 90 million tons of limestone that roared down Turtle Mountain in Alberta, Canada, in 1903, crushing 70 people, and the rain-soaked soils that slid off hills into Minatitian, Mexico, in 1959 and buried 800 villagers (Parfit, 1998). Similar events, of a lesser magnitude in terms of fatalities, occurred in California in 2010.

EARTHQUAKES AND TSUNAMIS

Earthquakes can have major impacts. One such event in Mianyang, China, in 2008 caused upwards of 100,000 deaths and widespread flooding and landslides. A similar event in Port-au-Prince, Haiti, and surrounding areas in January 2010 killed an estimated 230,000 people, injured almost 150,000, left an estimated 2 million homeless, and caused widespread destruction of homes, churches, and buildings (Talmazan, 2010). This was followed by a cholera epidemic in October 2010, in the rural Artigonite Department, about 100 kilometers (62 miles) north of Port-au-Prince due to the inadequate treatment of the drinking water supply. This resulted in the deaths of 1,800 people and the hospitalization of thousands more (*Wikipedia*, 2010a). Earthquakes can also be of human origin. This was illustrated (Chapter 16) when an earthquake in China was attributed to the withdrawal of water from a major hydroelectric dam. Earthquakes were also initiated in Switzerland in 2010 when water, under high pressure, was injected into rocks 3 miles below the surface of the Earth to generate steam, bring it to the surface, and generate electricity. This caused 100 minor earthquakes within 1 week and an accompanying $9 million in damages to nearby buildings. For obvious reasons, the operations were immediately discontinued (Haring, 2010).

WILDFIRES

Wildfires have been especially destructive in recent years, leading to the devastation of entire towns in the southeast sections of Australia in February 2009. Searing temperatures and wind blasts created a firestorm that swept across a swath of the country's Victoria state. The accompanying airborne release rained ash, and trees exploded in the inferno as temperatures reached upwards of 117°F. At least 750 homes were destroyed, and the number of deaths exceeded 200 (Smith, 2009a, b). This led to a detailed evaluation of the accepted survival strategy—"to get out early or hunker down and fight"— which resulted in the deaths of many people who had delayed evacuating and succumbed to the fire as they attempted to escape (Associated Press, 2009).

VOLCANOES

Historically, the violent eruption of Mount Krakatau on a small island in Indonesia in 1883 stands as a classic example of volcanic activity. People throughout the world observed the clouds created by the accompanying dust and the reductions in temperatures that followed. Almost equally spectacular was the eruption of Mount Pinatubo in the Philippines in 1991 that produced a mushroom-shaped cloud 500 kilometers (310 miles) in diameter, the crown of which reached an altitude of 35 to 40 kilometers (22 to 25 miles). This led to a 2°C (3.6°F) reduction in surface air temperatures over the Northern Hemisphere during the summer of 1992, and up to a 3°C (5.4°F) increase during the winters of 1991–92 and 1992–93 (Robock, 2002). A volcanic eruption in Iceland in April 2010 released thousands of tons of ash and sulfur dioxide into the atmosphere (Associated Press, 2010a).

TORNADOES

On average, about 1,000 tornadoes occur annually in the United States. Many predominantly occur in a region stretching from Texas to Nebraska. On average, they annually cause about 50 deaths and 1,000 injuries. On March 1, 1997, a group of approximately nine tornadoes, originating from two separate thunderstorms, swept a distance of some 160 miles across the state of Arkansas. These caused 26 deaths and more than $100 million in property damage (CDC, 1997a). In April 2010, another series killed 10 people in southeast Mississippi (Associated Press, 2010b). In comparison to other natural disasters, however, tornadoes are a relatively minor source of deaths (CDC, 1997b).

WINTER STORMS

The buildup of ice due to a major winter storm in Canada and the northeastern United States in 1998 caused trees and utility towers to fall, roofs to collapse, and airports to close. A similar event occurred in Arkansas in late

Table 17.3   Examples of infrastructure failures and consequences in disasters

Event	Location	Infrastructure failure and consequences
1993 Great Midwest floods	Des Moines, IA	Businesses suffered higher economic losses from infrastructure outages (water, electric power, and wastewater services) than from physical flooding of their facilities.
1994 Northridge earthquake (M_w—6.7)	Los Angeles, CA	Damage to bridges, which closed portions of four major freeway routes, accounted for $1.5 billion in losses from business interruption (a quarter of the total).
1999 Great Hanshin-Awaji earthquake (M_w—6.9)	Kobe, Japan	Extensive infrastructure failures, including outages of electric power and telecommunications (1 week), water and natural gas (2–3 months), commuter railway (up to 7 months), and highway systems and poor infrastructure (~2 years). It required 10 years for the city population to recover. Economic activity, especially at the port, has still not fully recovered.
September 11, 2001 World Trade Center terrorist attack	New York, NY	Widespread disruption in lower Manhattan to emergency service facilities, transportation (including subways), telecommunications, electric power, and water.
August 14, 2003 blackout	Portions of Midwest and Northeast United States, and Southern Ontario	Power outages began in northern Ohio and cascaded through the electric power grid to cause the largest blackout in North American history (affecting 50 million people). Losses accounted for an estimated $10 billion. Water supply, telecommunications, transportation, hospitals, and other dependent infrastructures were disrupted.
2004 Hurricanes Charley, Frances, And Jeanne	Central Florida	Port closures disrupted delivery of fuel and emergency materials. Electric power outages lasted for more than a week. The supply of emergency generators was not sufficient to meet demand.

2000. Hailstorms, which often occur in winter, can also cause major damage. One such event in Fort Worth, Texas, in 1995 rained hailstones as large as softballs, causing injuries due to people being pelted with ice, shattering glass in buildings and cars, killing livestock, and destroying agricultural crops. The economic damage approached $2 billion (Parfit, 1998). One of the lessons that has been learned is that the impacts on the associated infrastructures can often be more disruptive than the initial physical damage (Chang, 1999). This is exemplified by the February 2010 blizzard across eastern sections of the United States that left hundreds of thousands of people without power and brought transportation to a halt (CBS Evening News, 2010). Other examples are illustrated in Table 17.3.

DROUGHT

About 10 percent of the Earth experiences drought in any given year. Although such events require a long time to develop, they spread farther, last longer, and touch more lives than any other type of natural disaster. Such conditions are common occurrences in the Prairie provinces of Canada, the western and central portions of the United States, and north and central Mexico. The impacts include the loss of agricultural crops and livestock and increases in the frequency and severity of wildfires (Parfit, 1998). A drought and heat wave in France in 2003 led to the deaths of up to 10,000 people (Cais et al., 2003). Similar events annually cause thousands of deaths, particularly in the less developed countries.

Disasters of Human Origin

Disasters of human origin are also diverse in nature. They range from acts of terrorism, such as the previously cited September 11th event, to various types of accidents.

INDUSTRIAL PLANT ACCIDENTS

One such accident occurred in Bhopal, India, in 1984 and caused the deaths of an estimated 2,500 to 7,000 people. Its impacts were exacerbated by multiple factors: (1) it occurred at night, (2) people were housed nearby, (3) there were delays in alerting the public, (4) the public had little knowledge of the potential toxicity of the emissions, and (5) local medical facilities were inadequate. Even today, few countries provide the public with adequate information about the location of chemical-manufacturing plants or the nature and quantity of chemicals being manufactured.

NUCLEAR POWER PLANT ACCIDENTS

Just as the Bhopal accident hopefully represents the most devastating industrial accident that will ever occur, the Chernobyl nuclear power plant accident in the USSR hopefully represents the worst possible nuclear-related accident. Immediate deaths due to acute radiation exposures, injuries, and efforts to bring the plant under control approached 50. In addition, there was widespread distribution of radioactive materials throughout neighboring countries. Fortunately, however, surveys showed that the more than 100,000 members of the public who were evacuated from contaminated areas received an average dose of only 30 mSv, 60 percent of the U.S. annual occupational dose-rate limit.

Soviet officials compounded the problem by seeking to conceal that the accident had occurred. As a result, many children consumed fresh milk containing a variety of radioactive contaminants, such as radioactive iodine, that concentrate in the thyroid. This led to about 5,000 cases of thyroid cancer among children. Subsequent epidemiological studies, however, showed that, 20 years later, the number of fatalities among this group was limited to 15. In addition, long-term follow-up revealed that there is no strong evidence of an increase in radiation-induced leukemia or solid cancers among the older population groups who were exposed (National Research Council, 2006). This coincides with the dose estimates. In contrast, due to the dire predictions by the media of serious radiation effects on the fetuses of babies conceived during the months immediately following the accident, several thousand mothers in Denmark, Greece, Italy, Norway, and West Germany, elected to have abortions (Trichopoulos et al., 1987). Most radiobiologists believe that this was not necessary. In fact, no hereditary effects have been observed among the children of mothers exposed to much higher doses during the detonations of nuclear weapons in Japan at the close of World War II.

Also of note, a plant of the Chernobyl design would not have been approved for construction and operation in the United States or anywhere else outside the USSR. One major difference was that it was not enclosed in a containment vessel. Detailed guidance for responding to nuclear power plant accidents has been provided by the U.S. Nuclear Regulatory Commission (USNRC, 2007, 2009b).

OIL TANKER SPILLS

Few disasters are more dramatic, particularly in terms of their impacts on the environment and associated ecological systems, than major oil spills. A classic example was the event involving the *Exxon Valdez* in Prince William Sound, Alaska, in 1989. The accompanying release impacted some 500 miles of the coastline of the Gulf of Alaska and had a devastating effect on the surrounding ecology. Although major efforts were initiated to clean it up, less than 15 percent was actually recovered, and what remained proved to be tenacious (Chapter 15).

While many of the wildlife species (such as harbor seals, herring, harlequin ducks, marbled murrelets, and pigeon guillemots) slowly recovered, others did not. In some cases, this may have been due to the lack of recovery of the marine life that served as their food supply (Mitchell, 1999). More recently (2002), another tanker, the *Prestige*, broke up off the western coast of Spain and Portugal. This resulted in the release of several million gallons of fuel oil into the Atlantic Ocean. During subsequent recovery operations, the vessel was towed out about 240 kilometers (150 miles) off the coast, where it sank (Wikipedia, 2010b). Exacerbating the impacts was that the oil contained polyaromatic hydrocarbons that are extremely toxic to plankton, fish eggs, and crustaceans and could have possible carcinogenic effects in fish and other animals. To resolve the problem, engineers removed the remaining oil in the sunken vessel by remotely pumping it into large aluminum shuttles that were floated to the surface. These activities, which were completed 2 years later, cost about $3 billion (Chapter 16).

Similar oil releases have occurred due to explosions in offshore oil drilling rigs. An event in April 2010 off the coast of Louisiana injured 17 and caused the deaths of 11 workers. Oil from the well leaked into the Gulf of Mexico at a rate of up to 70,000 barrels per day. Since the well is located about 1 mile beneath the ocean surface, control of the leaks proved to be difficult. Ultimately, the leak caused an enormous oil slick, with tar-like balls reaching the beaches and adjoining wetlands. Exacerbating the situation was the lack of a prompt response by the oilwell operator and what appeared to be false reports on the magnitude of the spill. In addition, multiple systems designed and installed to avert such an event failed. Marine experts predicted that the disaster would be the worst ever, threatening hundreds of species of fish, birds, and other wildlife along the Gulf Coast, as well as one of the world's richest seafood grounds, teeming with shrimp, oysters, and other marine life. In fact, the release was such that its volume matched that of the *Exxon Valdez* every 4 to 7 days (CBS Evening News, 2010). Based on an inquiry by the Departments of Homeland Security and the Interior that revealed multiple violations of safety requirements, investigators filed criminal charges against several of the companies involved. This also raised the prospects of significantly higher penalties than the current $75 million cap on civil liability (McClatchy Newspapers, 2010). Unfortunately, a larger sister facility in the Gulf of Mexico, with a similar lack of safeguards, holds the potential for an accident of even larger magnitude.

ACCIDENTS IN UNDERGROUND MINES

Although the number of workers being killed in underground mining operations is being reduced, a major reason is the shift from underground to surface mining. Nonetheless, many underground mines continue in operation, as do the associated accidents (National Safety Council, 2009). Two of the more recent U.S. accidents occurred in Hyden, Kentucky, and Montcoal, West Virginia, in 1970 and 2010, respectively. The Hyden incident resulted in 38 deaths, 29 deaths occurred in the Montcoal accident. In the most recent case, records revealed that the mine operator had been cited for multiple violations during recent years. As might be expected, this led to renewed calls for improvements in the safety of mining operations.

Many much more severe disasters have occurred in other parts of the world. An explosion in an underground mine in the People's Republic of China in July 2001 is reported to have killed 92 workers, and a flood in a mine a month later is reported to have killed 70 workers, with as many as 200 more missing. The total deaths for the year 2000 in coal mine disasters in that country apparently numbered in the thousands (U.S. Mine Rescue Association, 2009).

Terrorism

As noted earlier, there are basic differences in the nature and characteristics of terrorist attacks and those due to natural causes. Although the targets of natural disasters are random in nature, terrorists seek to identify and target a nation's vulnerabilities by identifying weak spots in its social infrastructures. While natural disasters can be predicted only statistically and not prevented at all, terrorist attacks can, through the effective application of crisis management (i.e., science, intelligence gathering, and improved investigative techniques) hopefully be predicted and prevented, or at least mitigated. In terms of specifics, terrorist attacks can range from subtle acts, such as the clandestine dispersal of an infectious agent within a populated area, to clearly visible acts such as the destruction of buildings and the deaths of the occupants.

POTENTIAL TOXIC AGENTS AND DISPERSAL MECHANISMS

The primary toxic agents that might be used are generally chemical, biological, and/or radiological in nature. The selection of the specific agent will probably depend on (1) the ease with which it can be disseminated, (2) the extent to which it can be expected to cause mortality, (3) its potential for inciting fear and panic, and (4) the difficulty in instituting countermeasures.

Chemical Agents. Because the effects of chemical agents that are absorbed either through inhalation, the skin, or mucous membranes will normally be immediate and obvious, attacks involving these agents are likely to be overt in nature. One example was the dispersal of *sarin* gas in a subway in Tokyo in March 1995 (CDC, 2000). Examples of rapidly acting potential chemical agents are presented in Table 17.4. Drinking water and food supplies represent two other specific dispersal pathways. The key points of vulnerability in drinking water systems include (1) the open reservoirs that hold the raw water prior to treatment, (2) the treatment plant itself, and (3) the elevated tanks in which the treated water is stored prior to distribution (Chapter 7).

Table 17.4 Examples of rapidly acting potential chemical agents

Category	Examples
Nerve and incapacitating agents	Sarin (isopropyl methylphosphanofluoridate) Soman (pinacolyl methylphosphonofluoridate) Tabun (ethyl N,N-dimethylphosphoramidocyanidate) BZ (3-quinuclidinyl benzilate) GF (cyclohexylmethylphosphonofluoridate) VX (o-ethyl-[S]-[2-diisopropylaminoethyl]-methyl-phosphonothiolate)
Pulmonary agents	Benzene Chlorine Chloroform Phosgene Trihalomethanes
Blood agents	Hydrogen cyanide Cyanogen chloride
Cutaneous (blister) agents	Lewisite (an aliphatic arsenic compound, 2-chloro-vinyldichloroarsine) Nitrogen and sulfur mustards Phosgene oxime
Poisonous and/or corrosive industrial compounds	Cyanide Nitriles Nitric acid Sulfuric acid
Nitro compounds and oxidizers	Ammonium nitrate combined with fuel oil

Biological Agents. One of the unique characteristics of biological agents is that they can be carried and transmitted by rodents and insects. Due to the delay between the time of exposure and the onset of illness, the dissemination of such agents will likely be covert in nature. If, after dispersing such an agent, the responsible parties make known their actions, this would be extremely effective in creating fear and panic among those who were either exposed or were thought to be exposed. Biological agents that could potentially be used to contaminate food supplies are included among those listed in Table 17.5. Of particular importance are *Clostridium botulinum toxin* (botulism), and *Staphylococcus enterotoxin B* (Hymann, 2008). Interestingly, viruses such as Ebola are not likely candidates since they kill their human victims rapidly and thus provide little opportunity for effective transmission of the disease from person to person. To enhance the capabilities of physicians in responding to such events, the American Medical Association has developed and distributed a CD-ROM that provides state-of-the-art medical and clinical information for dealing with both biological and chemical attacks (AMA, 2002). Similarly, the National Institute for Occupational Safety and Health (NIOSH, 2003) has issued a guidance document on filtration and air-cleaning systems for protecting the inhabitants of buildings within which airborne toxic materials have been dispersed.

Table 17.5   Characteristics of three groups of biological agents

Agents	Characteristics
Variola major (smallpox), *Bacillus anthracis* (anthrax), *Yersinia pestis* (plague), *Clostridium botulinum* toxin (botulism), *Francisella tularensis* (tularemia)	Easily disseminated or transmitted, cause high mortality, could cause public panic and social disruption; require enhanced epidemiology, vaccines and drugs, and diagnostic tests
Hantaviruses, tickborne hemorrhagic fever viruses, tickborne encephalitis fever, yellow fever, multi-drug-resistant tuberculosis	Available, easy to produce and disseminate, and have potential for high morbidity and mortality with major health impacts
Coxiella burnetti (Q fever), *Brucella* species (brucellosis), *Burkholderia mallei* (glanders), ricin toxin from *Ricinus communis* (castor beans), epsilon toxin of *Clostridium perfringens*, *Staphylococcus enterotoxin B*	Relatively easy to disseminate, cause moderate morbidity and low mortality, require specific diagnostic capacity and disease surveillance

Radiological Agents. One possible use of these agents would be to disperse specific radionuclides among a population group. These could be accomplished through the detonation of a so-called dirty bomb, using conventional explosives. The advantage to the terrorists, in this case, is that they could select a specifically hazardous radionuclide to disperse (Ring, 2004). Another possibility is the detonation of a nuclear device itself, or, in an extreme case, the detonation of an improvised nuclear weapon. This would also provide a mechanism for major physical destruction. Of these two options, the first would probably be the most logical since it would be difficult for terrorists to assemble an effective device/weapon on their own. Even so, depending on how many lives a terrorist organization is willing to sacrifice, the probability that they could obtain or construct a nuclear weapon cannot be ignored (Fleming, 2002). The identities, characteristics, and potential health effects of a range of radiological agents are summarized in Table 17.6.

Prominent among specific targets is a commercial nuclear power plant. Foreseeing such a possibility, such plants are protected by a reinforced concrete containment vessel that surrounds a heavy steel dome. In addition, the reactor fuel is enclosed within a steel pressure vessel that is 30 to 35 centimeters (12 to 14 inches) thick (NCRP, 2001). To confirm that the containment vessel will perform as anticipated, a variety of tests have been conducted. These included remotely flying a military fighter aircraft into a test wall that simulated a containment vessel. Although the engines of the plane penetrated approximately 5 centimeters (2 inches) into the wall, the test results showed that the reactor would not have been damaged (Chapin et al., 2002). Follow-up analyses show that even a large passenger plane fully loaded with jet fuel would not penetrate the containment structure. A list of relevant ICRP publications and NCRP reports is provided in Table 17.7.

Mitigating and Preventing the Impacts of Natural Disasters

One of the characteristics of many types of natural disasters is that people can be warned of their impending occurrence. In the case of hurricanes and floods, this can be as much as several days to a week. In the case of tornadoes, however, it may be only minutes, and for earthquakes there is generally no warning at all. Obviously, any type of a warning can significantly reduce losses of both lives and property. Although progress is being made, much work remains in developing methods both to increase the timeliness of the warnings and minimize the impacts of such events.

Table 17.6   Identities, characteristics, and health effects of a range of radiological
agents

Radionuclide	Characteristics	Potential effects
^{60}Co	Reasonably long-lived, concentrates in liver, would be difficult for terrorists to handle	Represents both an internal and external hazard; high risk on local basis
^{90}Sr	Long-lived, deposits in bone, retained by body	Possible latent leukemia among exposed people; taken up by dairy and agricultural products
^{99}Tc	Extremely long-lived, but not retained in the body, not very mobile within the environment	Contaminated land might not be usable for years
^{131}I	Short-lived, would require detonation of nuclear device to produce	Latent thyroid cancers in exposed children, but mortality is low
^{137}Cs	Long-lived, readily taken up by the body, exposes whole body	Represents both external and internal hazard; can be taken up by dairy and agricultural products
^{226}Ra	Naturally occurring, long-lived, retained by the body	Contaminated land might not be usable for years; can produce latent bone carcinomas, but required intake is high
^{238}U	Extremely long-lived	Primarily a chemical, not a radiological poison; highly contaminated land might not be usable for years
^{239}Pu	Very long-lived, retained by the body, feared by the public	Contaminated land might not be usable for years; might produce bone carcinomas, but required intake is high

HURRICANES

Hurricanes have three primary impacts: (1) loss of life, (2) direct property destruction, and (3) associated impacts on the infrastructure. Measures for mitigating the impacts include (1) improved technological forecasting, (2) reduction in the time required for evacuation, and (3) the provision of facilities for refuge. With respect to the first item, recent studies hold promise

Table 17.7 ICRP publications and NCRP reports related to nuclear accidents and terrorism

Organization	Title	Number: year
ICRP[a]	The principles and general procedures for handling emergency of accidental exposure of workers	28: 1978
	Protection of the public in the event of major radiation accidents: Principles for planning	40: 1984
	Principles for intervention for protection of the public in a radiological emergency	63: 1993
	Protecting people against radiation exposure in the event of a radiological attack	96: 2005
	Application of the Commission's recommendations for the protection of people in emergency exposure situations	109: 2009
NCRP[b]	Protection of the thyroid gland in the event of releases of radionuclides	55: 1977
	Management of persons accidentally contaminated with radionuclides	65: 1980
	Developing radiation emergency plans for academic, medical, or industrial facilities	111: 1991
	Recommended screening limits for contaminated surface soil and review of factors relevant to site-specific studies	129: 1999
	Management of terrorist events involving radioactive material	128: 2001

a. International Commission on Radiological Protection.
b. National Council on Radiation Protection and Measurements.

for achieving significant improvements. Based on analyses of the materials deposited at the bottom of swampy areas, scientists have obtained detailed information on weather patterns during the past several thousand years. These data, in turn, are being used to develop models that could have the capability for predicting hurricane patterns for a decade or more (Morrison, 2010). Combined with controlled residential and commercial development, better building practices, and safe in-place shelters for people who might otherwise have to leave potentially impacted areas, this could revolutionize the approaches in preparing for, and responding to, hurricanes.

Other measures that can be used include restricting development in high-risk areas, enforcing hurricane-resistant building codes, and educating the public on the successful implementation of steps to reduce losses of life.

FLOODS CAUSED BY HURRICANES

The 2005 Atlantic hurricane season was the costliest, as well as one of the five deadliest, in U.S. history. The primary example was Hurricane Katrina, which, after passing over Florida during August 2005, encountered land-fall again in southeastern Louisiana. Due to the accompanying storm surge, it caused severe destruction from central Florida to Texas. The most severe loss of life and property damage, however, occurred in New Orleans, Louisiana, due to the catastrophic failure of the levee system designed to protect the city from flooding (Swenson and Marshall, 2005). In retrospect, the fundamental cause of the failure was that dams that had been constructed over past decades on the upstream headwaters of the Mississippi River had caused the solids in the water to settle out and not be deposited on the wetlands and floodplains in Louisiana, the Mississippi delta, and in the Gulf of Mexico. This, in turn, had led to the loss of 100,000 square miles of floodplains. Interestingly, the sole response was to build higher levees around the city of New Orleans. As has been stressed throughout this book, it is mandatory that officials who are responsible for addressing such problems apply a systems approach. Without restoration of the floodplains and their ability to retain the excess water, another major storm will hit that area within the next decade or so, and an event similar to Katrina will be repeated (Lehrer, 2010).

EARTHQUAKES

Severe earthquakes occurred in Haiti, Chile, and China in early 2010. In all cases, the impacts were magnified by (1) the density of the population, (2) the magnitude of the event, and (3) the inadequacy of the building construction codes. At the same time, had the United States military and multiple other organizations from throughout the world not launched a herculean effort, the impacts in Haiti would have been far more destructive, especially on the welfare of the victims.

TORNADOES

Of all the tornado-related deaths that occurred in the United States in 1997, more than half of them were people who lived in mobile homes. Such facilities represent a form of housing for many who economically have few alternatives, and their use has dramatically increased in recent years. For example, the percentage of the U.S. population living in mobile homes increased from 1.3 percent in 1960 to more than 10 percent in 2008 (U.S. Census Bureau, 2008). Because of this, it is imperative that arrangements be made to enable these people to have access to underground shelters. For those living in permanent types of homes, comparable protection can be provided if they go to the basement, if one is available, or seek shelter in hallways and interior rooms (CDC, 1997a).

Mitigating and Preventing the Impacts of Disasters of Human Origin (Including Terrorist Events)

As is the case with natural disasters, it is important to initiate steps to minimize the impacts of disasters of human origin. In most cases, such efforts can be applied with equal effectiveness.

INDUSTRIAL PLANT ACCIDENTS

As noted earlier, the accident at the plant in Bhopal, India, showed that such events could be devastating. Stimulated by this event and a subsequent much smaller release from a similar plant in West Virginia, the U.S. Congress (1986) passed the Emergency Planning and Community Right-To-Know Act as part of the Superfund Amendments and Reauthorization Act. Industries subject to this act were required to evaluate the impacts of accidental toxic chemical releases, using worst-case scenarios, and develop plans for preventing them (Neville, 2001). Adding support, the U.S. chemical industry has independently initiated programs in support of these activities. These have led to reductions both in the amounts of hazardous materials present within various plant systems at any given time and the amount produced as intermediate products. Both industry and the U.S. Congress are directing efforts to improvements in the transportation of hazardous materials—for example, in 2009 Congress passed the Hazardous Material Transportation Act (U.S. Congress, 2009).

NUCLEAR POWER PLANT ACCIDENTS

Much the same as the Three Mile Island accident provided the stimulus for improvements in the safety of U.S. nuclear power plants, the Chernobyl accident in the USSR led to similar actions on a global basis. These included the development of (1) new designs to provide increased reliability and maintainability of the control of the reactor, (2) increased resistance of the plant structure to seismic events, and (3) the incorporation of three separate and independent systems for cooling the reactor core in case of an accident. Following the September 11 attacks, efforts specifically related to terrorism increased, including evaluations of the integrity of containment vessels and force-on-force training exercises of the armed guards stationed at all U.S. nuclear plants (USNRC, 2009a).

OIL TANKER SPILLS

As noted earlier, the grounding of the *Exxon Valdez* in Alaska's Prince William Sound in 1989 caused enormous destruction of marine life. As a result, nations throughout the world implemented stricter standards for cargo vessels, culminating in a stipulation by the United Nations that required the use of single-hulled tankers be phased out by 2012 (Wikipedia, 2010). Other advancements included the adoption of improved radar and global navigational satellite systems (National Research Council, 2001). Even so, oil tanker spills represent only about 8 percent of an estimated 1.3 million metric tons (1.4 U.S. tons) of oil that are released into the ocean annually. Almost half of this is due to seepage from natural deposits in the Earth (Gwin, 2009).

ACCIDENTS IN UNDERGROUND MINES

Improvements are being made in the safety of mining operations conducted underground, and the probability of events that result in worker deaths are being reduced. A significant contributor to these reductions, however, is the shift from underground to surface mining. While in 1950 more than 75 percent of the coal mined in the United States was obtained through underground mining, today less than 38 percent is being obtained through this approach. During the same time period, the total production of coal in the United States has doubled (CEQ, 1998). As is often the case in such changes, the new approach has major problems of its own. Strip-mining, for example, can have devastating impacts on the environment. In most cases, however, regulations require that the ground surface be restored by replacing the soil and replanting the trees and vegetation (Chapter 16).

NUCLEAR-RELATED TERRORIST EVENTS

Small amounts of highly enriched uranium (HEU) are located at civil, military, and space power facilities throughout the world. In addition, more than 40 countries—including Russia, other former Soviet states, and Pakistan—possess HEU. Although it would be difficult, it is not inconceivable that terrorists could acquire a sufficient amount, smuggle it into the United States, and assemble a bomb. Substantiating this possibility is that tests of monitors, including advanced spectroscopic machines, for detecting HEU inside U.S.-bound shipping failed to disclose its presence. Exacerbating the situation is that almost 300 million shipping containers are transported worldwide annually, about 42 million of which enter U.S. ports. If these challenges are to be successfully addressed, more resources should be directed to (1) identifying and arresting nuclear smugglers, (2), securing HEU that is being stored at overseas locations, and (3) reducing the enrichment by blending it with natural or depleted uranium so that it cannot be effectively used in a bomb (Cochran and McKinzie, 2009). On a global basis, this problem is being addressed by the Center for Arms Control and Non-Proliferation, which is based in Washington, DC (2009).

In developing plans to cope with emergencies following a radiological attack, radiation dosimetry experts are developing methods for assessing doses to individual members of an exposed group in the range of 2,000 to 10,000 mGy. These are based on the science of biodosimetry, which provides the capability to quantify external exposures through the assessment of residual signals in the body (Schwartz et al., 2010).

Emergency Preparedness Plan

Responding to the impacts of any disaster requires a carefully designed emergency preparedness plan. Essential components, in the case of industrial facilities, include an assessment of the internal, external, and natural phenomena that can cause accidental releases of toxic materials. Segments of such a plan include the need to (1) consider the full spectrum of possible accidents, (2) estimate their likelihood of occurrence, (3) analyze their potential consequences, (3) assess the effectiveness of the systems that have been incorporated to prevent and/or minimize releases; (4) identify significant structures, systems, and components designed to mitigate the consequences, and (5) identify a selected subset of accidents and related scenarios that need to be formally considered. Estimates of the accompanying potential doses to the public should include evaluations of both "realistic" and "conservative" assumptions. Finally, the plans should include the conduct of field surveys as soon as possible following any unanticipated releases, not only to confirm that the theoretical estimates of exposures/doses are as anticipated but also that adequate responses have been initiated.

Emergency Response Plan

Regardless of the type of disaster, a well-designed and well-executed emergency response plan is essential for the rapid mobilization of the resources necessary to respond in an effective manner to the immediate health-care needs of the people affected and to restore disrupted services. The plan should be clear, concise, and complete. It should also be dynamic, flexible, and subject to frequent evaluation and update. In addition, it should designate precisely who does what and when, and everyone involved should be thoroughly familiar with its contents. The priority should be to provide an immediate response to the event by locating and providing emergency medical services to the victims, controlling fires, removing downed power lines, and controlling leaks of natural gas. On a longer-range basis, the goals should be to provide health care and shelter for victims and, as noted earlier, to restore important services such as a safe water supply and basic sanitation. Next in importance are arrangements to provide a safe food supply.

TYPES AND CHARACTERISTICS

In general, there are two types of emergency plans. The *first* is national or regional in scope and defines the responsibilities and mobilization procedures of personnel in relevant public and environmental health departments and emergency preparedness agencies. Planning at this level frequently includes coordination of civil defense and military services. The *second,* which is local in scope, must be much more definitive and include detailed listings of the personnel involved, their individual responsibilities, and the range of countermeasures available for implementation. Properly coordinated, the two plans can provide a cadre of well-trained personnel to cope with disasters of natural and human origin of almost any magnitude.

PHASES

Experience demonstrates that disasters can generally be divided into four phases: (1) the pre-event phase, (2) the warning or alerting phase, (3) the response phase, and (4) the recovery phase. Although many aspects of the required responses are similar, there are sufficient differences to require different approaches in responding (ICRP, 2005).

> *Pre-event phase.* The objectives during this phase are to anticipate that accidents and disasters will occur and to plan a response. This includes initiating a program to (1) identify all available organizational resources, (2) inventory the types and locations of available

supplies and equipment, including hardware and medical supplies, (3) identify private-sector contractors and distributors who can provide otherwise-scarce specialized personnel and equipment, (4) review essential community and industrial facilities to identify those that may be vulnerable to a disaster, and (5) define the responsibilities of each agency or group and establish lines of communication.

Warning or alerting phase. In the case of hurricanes, tornadoes, and floods, where advance warning will be available, there will be an opportunity to alert emergency planning personnel and to have them move, where appropriate, to the emergency operations center. Hopefully, as discussed earlier, it will be possible to provide similar warnings for certain types of terrorist-related events, particularly those of a major magnitude. Prominent among activities during this phase should be the provision of timely and accurate information to the media and the public on what to anticipate. This should include specific details on preparations that should be made.

Response phase. Under normal circumstances, fire, emergency medical, and police personnel will be the first to arrive at the site of a major disaster. One of their first objectives will be to provide security to the affected area to ensure the safety of both the victims and the workers. The most experienced senior person should take charge, immediately surveying the area and assessing the scene, the number of victims, and their injuries. Officials at the center must then determine what additional support is needed.

Recovery phase. During this phase, substantial numbers of injured people may need follow-up care. All survivors will require food, water, shelter, clothing, and sanitation facilities. Floods, in particular, promote unsanitary conditions through the buildup of debris and blockage of sewer systems.

Countermeasures

The application of countermeasures is critical in mitigating the impacts of a disaster. To be acceptable, a countermeasure must be (1) effective (i.e., substantially reduce population exposures below those that would otherwise have occurred), (2) safe (i.e., it should introduce no health risks with potentials worse than those presented by the releases), (3) practical (i.e., it must be capable of being administered at a reasonable cost and without creating legal problems), and (4) defined (i.e., with no jurisdictional confusion about responsibility and authority for applying the measure). In addition, it must be recognized that most any countermeasure will carry with it health risks and social and economic disruption, depending on when and where it is applied. Based on these criteria, the following countermeasures have proved most useful.

Evacuation. This method of protective action can be effectively applied in many types of industrial accidents, especially those that result in releases of a toxic material into the atmosphere. The feasibility of evacuation, however, depends on (1) the magnitude and likely duration of the release, the weather and time of day, (2) the time interval between the accident and the order to evacuate, (3) the availability of transportation and suitable shelter for those evacuated, (4) the potential for vehicle accidents and personal injuries, and (5) the potential for increased uptake of releases as a result of the exertions involved. Evacuation is particularly effective for natural events involving airborne releases of toxic materials. As might be expected, the protective value increases with the rapidity with which evacuation is executed, the distance between the airborne release and the evacuated population, and the length of time the airborne material remains in the area (Figure 17.1).

Sheltering and respiratory protection. The simplest and least disruptive of all proposed countermeasures is to instruct people to remain indoors. The value of sheltering can be enhanced by encouraging the potentially exposed population to use common household materials for respiratory protection. In the Bhopal accident, for example, those exposed could have protected themselves simply by placing a wet cloth (e.g., a handkerchief, washcloth, or bath towel) over their face. Sheltering and respiratory protection can be especially effective for people who are close to an accident site and would need to move through the airborne cloud during evacuation.

Protective prophylaxis. If the nature of the release can be identified and a known antidote is available, it is possible that the antidote can be administered to counteract or negate the effects of the exposures. One prominent example is stable iodine, which will reduce the uptake of radioactive iodine by the thyroid, its effectiveness depending on how soon after exposure the prophylaxis is administered (Figure 17.2).

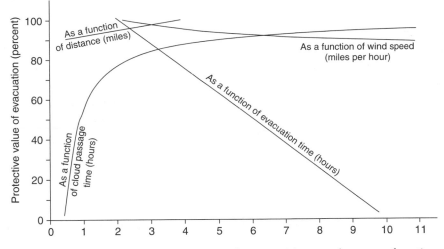

Figure 17.1 Protective value of evacuation from an airborne release as a function of the time required for evacuation, wind speed, distance, and cloud passage

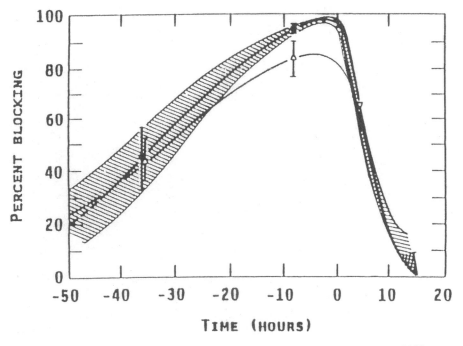

Figure 17.2 Percentage of thyroid blocking afforded by administration of 100 milligrams of stable iodine as a function of time before and after an assumed intake of 3.7×10^4 bequerels of ^{131}I

OTHER COUNTERMEASURES

It is important to recognize that airborne toxic materials can be transported many miles from an accident site, leading to the contamination of agricultural products, including milk, over large geographical areas. Due to their wide distribution, those produced in the more highly contaminated areas may, within a short period of time, be consumed by people some distance away. Adding to these concerns is that the movement of contaminants through the environment can be extremely rapid. Radionuclides, for example, can appear in milk within hours after being deposited on pasture grass (Figure 17.3).

Since agricultural products can be directly contaminated by airborne deposition, those of immediate concern are those that grow above ground,

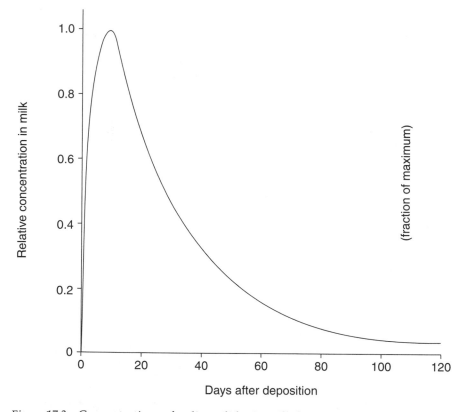

Figure 17.3   Concentrations of radionuclides in milk from cows as a function of time following a single deposition on pasture grass

such as broccoli, cabbage, celery and lettuce. Harvesting them immediately and removing their outer membranes may provide a simple, low-risk method of averting the intake. For crops that can be washed, such as tomatoes and squash, this may provide similar protection. Similarly, removal of the peel from fruits, such as oranges and bananas, should make them acceptable for consumption. Crops that grow underground, such as carrots, beets, and potatoes, should pose no problem until sufficient time has passed for the contaminants to gain access via root uptake through the soil.

Fortunately, the quantities of toxic agents that terrorists might add to public drinking-water supplies, especially if they desire to cause acute effects, are large and would be difficult to disperse effectively. In addition, the reservoirs and holding tanks of most large water-treatment plants are equipped with surveillance systems and physical barriers to deter access. To strengthen the resistance of these and related facilities, the Pan American Health Association recommends what is described as a vulnerability analysis, the basic objective being to identify measures that can be taken prior to a disaster to strengthen such systems (Wisner, Blaikie, and Cannon, 2004). Additional guidance has been provided by the Commissioned Officers Association of the U.S. Public Health Service (Carmona et al., 2010).

RESPONSE COORDINATION AND FOLLOW-UP

Recognizing that responses to all types of disasters are similar, it is now common practice to combine such activities within one organization. The objective is to help ensure that response activities are better organized, staffed, and focused, and that attempts to mitigate the impacts do not exacerbate the negative consequences. Unfortunately, this is often not the case, especially when governments and organizations outside the affected area attempt to assist. For many disasters—the 2010 earthquake in Haiti was an exception—it is unlikely that medical personnel from outside the affected country will be required. Another frequent problem is the influx of volunteers who neither speak the local language nor have adequate disaster-relief experience. Unless these situations are avoided, they can lead to what some have called the *secondary* disaster. One of the more common fears is that hurricanes and floods will be followed by outbreaks, and even epidemics, of infectious disease. In reality, experience shows that such outbreaks are rare. A summary of the positive actions that should be taken, and the negative actions that should be avoided, in response to a disaster is presented in Table 17.8.

Table 17.8   Responses to disasters: positive and negative actions

Positive actions	Avoiding negative actions
Recognize that every disaster is unique. Its effects on health depend on the degree of development within the affected country.	Do not overreact to media reports for urgently needed international assistance. Take care to ensure that donations meet real needs.
It is incumbent on the part of the affected people to inform donors about what types of assistance are needed. This is just as important as specifying what is not needed.	Recognize that the quality and appropriateness of the assistance are more important than its amount or its monetary value.
Emergency assistance should be designed to complement, not duplicate, what is being done by those in the affected country.	Do not promote shipments of medical or paramedical personnel, medical equipment, or field hospitals, unless the need has been documented.
For disasters restricted to small geographic areas, cash donations may be preferable. These will enable the affected groups to purchase what they need at the local level. This will also save time, storage, and transportation costs.	Do not promote the shipment of used items, such as clothes and shoes. Never donate medicines that have expired or are about to expire.
Recognize that the affected populations may continue to need assistance during long-term rehabilitation and reconstruction.	If a product is not acceptable in the donor country, it is also not acceptable as a donation.
Because of their importance, every effort should be made to reestablish communication and transportation systems as rapidly as possible	Donors should not compete with each other.

In addition to addressing drinking-water and wastewater disposal, recovery efforts need to be focused on two objectives: (1) ensuring the safety of the people and (2) maintaining critical fuel supplies (Cunningham, 2009; Qureshi et al., 2005). Other key objectives are to ensure that (1) the response is well coordinated, rapid, and effective, (2) communication channels with local, state, and federal agencies are maintained, (3) health-care workers are qualified and willing to serve in their assigned capacities, and (4) the health and safety of workers who are involved in rescue and recovery operations will be protected. Although it may not be possible to attain all these objectives, every effort should be made to do so. Also to be considered are the impacts of urbanization as a factor in exacerbating the impacts of disasters. This was vividly illustrated by the earthquakes in Haiti. They include the need to assess the role of housing for low-income families. Although it will be difficult to address these issues, they should certainly be a part of our long-range planning (PAHO, 2010).

Benefits of Advances in Technology

The application of advances in technology can be especially valuable in supporting the early response to disasters. One example is the use of the capabilities of geographic information systems (GIS) to provide a broader and more comprehensive picture of the extent and nature of the damage, as well as to identify the location of victims who might otherwise be missed. This technology can also facilitate the identification of areas of flooding that can, in turn, foster the breeding of mosquitoes and the potential spread of malaria. This has led the United Nations (UN) to establish an Office for the Coordination of Humanitarian Assistance through which international space agencies can collaborate and make satellite imagery available during disasters. In a similar move, the Internet-based Relief Web is providing a system through which maps and reports can be shared and funding appeals from dozens of international aid agencies can be coordinated. Another advance is the establishment of the UN Humanitarian Early Warning Service (HEWS). This is designed to provide continually updated information on existing and predicted droughts, floods, storms, locust invasions, and the like, the objective being to enable the initiation of interventions before a nascent crisis worsens (Fink, 2007).

The General Outlook

Although more work is needed, there have been significant advances in the methods for addressing disasters. For natural events, these include methods for forecasting their occurrence. For events of human origin, they include applications of engineering technology to enable industrial plant personnel to respond in a manner so as to mitigate a potential triggering event. In addition, steps are being implemented to limit the releases of harmful substances in case they occur. Models are also now available through which real-time monitoring data can be used to update predictions of the magnitude of the consequences and the geographical areas that will be affected. Another advance has been the development of methods for factoring into accident prevention and mitigation the possible contributions from malevolent acts such as terrorism (National Research Council, 2002).

In the meantime, experience has demonstrated the importance of informing the public on what to expect and protective actions to avoid harm. These include providing advance information about the possible effects of and countermeasures for exposures to chemicals, as well as timely warnings so that people can seek shelter or evacuate. At the same time, it is important to avoid providing information that will interfere with recovery operations (Table 17.8). Concurrently, hospital staffs and emergency department personnel should be provided guidance in responding to the potential violence within the affected populations (Ketterlinus, 2008). In all cases, the shareholders (i.e., members of the public) should be invited to participate in the development of disaster response plans. Not only does this ensure their cooperation and support, but it also leads to the development of plans that are superior to those that would otherwise be produced. One reason is that, in most cases, the first people to respond in a disaster will be those in the affected community.

From the standpoint of prevention, analyses show that most industrial accidents occur through a combination of several factors, the most important of which are hidden or apparently minor design errors in a piece of equipment, poor maintenance, taking shortcuts to maintain schedules, and bad communications. In fact, disasters at the human-machine interface rarely have one cause. Sometimes it is a technical "blind spot," such as the absence of coolant-level indicators and confusing control panel alarms in the Three Mile Island nuclear power plant (Okrent and Moeller, 1981). Other times, it is the perceived need to complete a mission, such as in the case of the disaster that followed the launch of the space shuttle *Challenger* in 1986 (Moeller, 2005).

Adding support to such developments is the increasing recognition that disaster preparedness represents sound economic investments. In fact, data show that every dollar spent on disaster preparedness saves seven dollars in economic losses. If the averted social and ecological costs were considered, the estimated rate of return would be even higher (Abramovitz, 2001).

A GLOBAL VIEW

MANY OF OUR environmental problems—air and water pollution, solid waste, and food contamination—are consequences of large-scale cultural patterns. Some are the result of millions of people making individual decisions; others are triggered by decisions of a small number of key decision makers in industry, government, and academia. Although many of the problems are local, others, such as ozone depletion, acidic deposition, ecosystems impacts, and the potential for climate change, have global implications. Solutions will require cutting across national jurisdictions and a shift in focus from protection and restoration to planning and prevention.

These problems also reflect three major trends. *First,* due to more than a fourfold increase in the world's population (U.S. Census Bureau, 2009) and more than a twentyfold increase in the values of goods produced since 1900, the quantity of pollutants being generated has significantly increased. *Second,* there has been a shift from the use of natural products to the production and use of synthetic chemicals. For example, a billion pounds of synthetic pesticides are used annually in the United States. In addition, some 70 percent of the antibiotics manufactured in the United States are used in agriculture and animal food production (Chapter 2). Many have proved to be highly toxic, and some persist and accumulate in biological systems and in the atmosphere. *Third,* expanded technological capabilities, and in some cases the export of hazardous technologies and toxic solid wastes (Chapter 9), have led to a situation in which the developing countries are now more polluted than the developed nations. Even though there is no consensus on how to solve these increasingly difficult problems, progress is being made at both the national and international levels. Several examples are discussed later in this chapter.

Examples of Current Challenges

URBANIZATION

Closely related to the rapid rate at which the world's population is increasing is the problem of urbanization. In fact, this is the most powerful and visible anthropogenic force on Earth, with the portion of the world's population that lives in cities having reached 50 percent in 2007 and projected to reach 60 percent by 2030 (Chapter 1). Ideally, cities exist as places where technology, population, culture, economics, and natural systems intersect and interact. Nonetheless, the space they occupy and the resources required to fulfill their needs absorb, transform, and/or consume either directly or indirectly ever-larger amounts of forests and arable land. Although cities are essential instruments of social advancement, wealth creation, globalization, creativity, psychic energy, and birthrate reduction, many continue to be dysfunctional (Bugliarello, 2001).

Recognizing these problems, multiple experts have concluded that the concept of a city must be rethought in terms of efficiency, manageability, and quality of life, including the emotional aspects. If these challenges are to be successfully addressed, there will be a need to learn how to organize and engineer the city as an organic whole, addressing its biological, social, and physical functions. The challenges that this poses are larger in scale, complexity, and involve more disciplines than any previously encountered. As such, they need to be addressed both nationally and globally.

OZONE DEPLETION

Ozone depletion is an instructive problem to discuss because it has been effectively addressed through cooperative efforts at the international level. The problem was created by the release of chlorofluorocarbons (CFCs) into the atmosphere. For some 80 years, these had been used as refrigerants in household appliances and air conditioners, as industrial solvents, and as propellants for aerosol sprays. Once in the atmosphere, they mix with other compounds and rise slowly into the stratosphere. The ultraviolet (UV) radiation from the sun then destroys the CFC molecules and releases highly reactive chlorine atoms. These, in turn, interact with ozone and convert it into normal oxygen. Although ozone is considered a pollutant when it is near the ground (Chapter 5), in the stratosphere it shields the Earth from UV radiation. Since destruction of this shield increases the amount of UV reaching the Earth, it leads to increases in skin cancers and cataracts, lower crop yields, and damage to materials such as vinyl plastics. The magnitude of the problem is illustrated by the fact that a single molecule of CFC can destroy tens of thousands of molecules of ozone.

Recognizing the problem, the international community took action that led to the *Montreal Protocol on Substances That Deplete the Ozone Layer,* an international treaty that was developed in 1987. Under the agreement, 47 nations agreed to reduce their production of CFCs by 50 percent by 1998. Subsequently, the developed nations agreed to eliminate all CFC production by 2000, a target that was later shortened to 1996. Studies conducted in 2003 showed that the hole in the ozone layer was being reduced. Ultimately, the production of such substances was eliminated and the problem resolved (WMO, 2007). Due to the more significant technical and economic challenges facing the developing nations, they were given an additional 10 years to accomplish this goal. To assist them, a fund was established to help them transition to replacement chemicals. Later, the Environmental Protection Agency (EPA) estimated that the $45 billion cost to rid the United States of CFC emissions was more than offset by the $32 trillion in crop damage, skin disease, and ecological problems that were averted.

ACIDIC DEPOSITION

As is well known, airborne contaminants have widespread effects and their movement does not respect national boundaries (Chapter 5). Exacerbating these problems is that, in a manner similar to the CFCs, pollutants such as oxides of nitrogen, once released into the atmosphere, can be converted into nitric acid, and oxides of sulfur can be converted into sulfuric acid. Subsequent deposition of these acids, either dry or as nitric or sulfuric acid in rain or snow, has been shown to impose an unprecedented burden on forests, streams, and lakes throughout the world. The Earth is truly a system, and impacts are readily transferred from one component to another.

The principal measures for controlling acidic deposition and its effects are (1) to reduce the discharges of oxides of nitrogen and sulfur into the atmosphere and (2) to treat sensitive ecosystems to make them less susceptible to damage. Recognizing that sulfur dioxide (SO_2) accounts for about two-thirds of the acidic deposition in the northeastern United States and eastern Canada, the U.S. Congress (1990) mandated a 50 percent reduction by the year 2000 in the releases of this contaminant from coal-fired plants in the Midwest. Similar mandates have been imposed for the control of nitrogen oxide releases from automobiles.

Although no formal remedial actions have been stipulated for treating terrestrial ecosystems damaged by acidic deposition, it is common practice to add lime to lakes to neutralize the acids that have been deposited in them. While emissions of sulfur dioxides were reduced by about 10 percent during the first half of the 1990s, emissions of nitrogen oxides increased by about 5 percent. Exacerbating the problem is that following the mandate by the EPA that required utilities operating coal-fired electric generating station to apply best available technology to control airborne releases of heavy metal contaminants, the selected alternative was to discharge them as liquid wastes. Now, the EPA is mandating that they remove them from these wastes (Associated Press, 2009b).

BIODIVERSITY

Biologists have identified and assigned formal classes to more than 15 million species of plants, animals, and microorganisms (Figure 18.1). The actual number is estimated to range from 3 to 100 million (Wilson, 2002). In fact, some form of life occupies essentially every available niche on Earth. These include photosynthetic bacteria, microscopic invertebrates, fungi, and mites that inhabit the cold and dry environment of Antarctica; specialized microbes that flourish in volcanic hydrothermal vents on the ocean floor with water temperatures near the boiling point; and marine organisms that survive at depths with pressures 1,000 times higher than those at the surface. Still other species prosper at altitudes equivalent to that of Mount Everest (Myers, 2002).

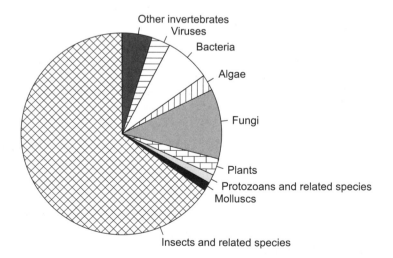

Figure 18.1   Relative sources of diversity in the biosphere

Many of these organisms, even though vast in numbers, are extremely fragile. As such, they are poorly equipped to withstand the relentless assault of humanity and other impacts on their habitats. Among those most affected at the present time are frogs and bees (discussed later) and bats (Chapter 10), which are disappearing at alarming rates (Debnam, 2009). In fact, the current rate of extinction of organisms, plants, and animals is estimated to be 100 to 1,000 times higher than before the coming of humanity. Concurrently, the birthrate of new species has declined (Wilson, 2000). As the negative effects of other factors such as climate change occur, the situation could worsen. Compounding the problem is that once a species is destroyed, it cannot be restored.

As scientists and the public are increasingly aware, biodiversity provides multiple benefits. This was first documented in the classic book *Origin of Species by Means of Natural Selection* (Darwin, 1859). In this and five subsequent editions, he revealed that all species of life have evolved over time from common ancestors. With respect to the benefits of resulting biodiversity, he shared this observation: "It has been experimentally proved that if a plot of ground is sown with one species of grass, and a similar plot be sown with several distinct genera of grasses, a greater number of plants and a greater weight of dry herbage can thus be raised" (Darwin, 1985). Experience shows that this is but one of multitudes of examples of such benefits that have been derived from animals, trees, plants, and inhabitants of aquatic and marine environments. These include blood thinners based on the venom of the deadly Asian saw-scaled viper and the European leech; digitalis, derived from foxglove and used to boost the pumping of the heart; quinine, obtained from the cinchona tree and used to treat malaria; and a blood-clotting agent from the horseshoe crab that is used to detect potentially fatal bacteria in vaccines, drugs, and medical devices. One of the more important examples is Taxol, a drug used to treat ovarian and breast cancers. It was discovered in the bark of the Pacific yew tree. Chemists later acknowledged that its chemical structure was too complex to have been developed by humans.

Losses of amphibians. Recent observations document that multiple factors, ranging from habitat loss, pollution, climate change, and an exotic fungus, are proving lethal to amphibians on a global basis. Although these include toads, salamanders, and newts, the discussion that follows will be restricted to frogs that are being eliminated due to *chytridiomycosis,* a fungus that causes a skin disease that prevents their skin, through which they both breathe and absorb water, to function. Even though this fungus is estimated to have existed for half a billion years, it has recently evolved into this more deadly form. Exacerbating the situation, it can survive at elevations from sea level to 20,000 feet and kill animals that are both aquatic and land-bound. Although thought to have originated in Africa, it was first observed in the wilds of Australia. In 2008, it was observed moving south from Costa Rica at a rate of up to 27 miles (43 kilometers) per year. After reaching the Panama Canal, it crossed and is continuing its march both southward and northward. By 2009, it had afflicted at least 200 species. Its initial victims in the United States were the yellow-legged frogs in California's Sierra Nevada. It has also been discovered in *Atelopus* frogs in Ecuador's Andes. In this case, the stimulus for the problem is believed to have been the interruption of the flow in the streams, where frogs breed, by construction debris. Other contributing factors include forest clearing, aridity, and infectious diseases, as well as the shipping of exotic frogs to various zoos throughout the world. It can be transmitted by the legs of a frog, the feathers of a bird, or the muddy boots of a hiker (Holland, 2009).

Fortunately, progress is being made. One program, under the auspices of the Amphibian Ark Network, is to establish captive breeding programs to prevent their extinction. Such programs, however, are incredibly expensive and can save only a fraction of the world's frog species. For this reason, studies are under way to discover bacteria that can prevent the fungus from adhering to the skin (Weise, 2009). All such efforts are important because frogs are a key link in the food chain and are crucial to a healthy ecosystem. This includes being an essential part of the food supply for snakes and birds and many other animals.

Losses of Honeybees. In a related series of events, honeybees in the United States began abandoning their hives in 2006. No bodies of the missing bees have been found, and no one knows what happened to them. Interestingly, other bees avoided the deserted hives, even though honey remained in them. Although this has been attributed to several sources, the basic cause appears to be an (as-yet-unidentified) virus. Nonetheless, multiple U.S. honeybee keepers have continued to be successful through (1) importing queen bees from countries, such as Italy, where the bees are resistant to disease, and (2) ensuring that sufficient honey is left in the hives in the fall to enable the bees to remain healthy through the winter and resist the virus (Krouse, 2009).

The introduction of nonnative species. Although biodiversity, if left alone, provides many benefits, the introduction of nonnative, foreign, or alien species can have negative effects equivalent to those caused by the loss of diversity. In some cases, such introductions were inadvertent; in other cases, they were intentional, the goal being to control an existing problem. A prime example is the Jackson's chameleon, a three-horned variety that can change color and is often kept as a pet. Introduced into Hawaii in the 1970s, they subsequently escaped or were released. They, in turn, challenged many of the native species that had evolved during the archipelago's eons of isolation. Coping with the problems of this and other alien invaders is costing the state millions of dollars in damage to forests, crops, and buildings (Leslie, 2002). Another example is an invasive strain of the tropical alga, *Caulerpa taxifolia*, that was introduced, perhaps accidentally, through the discharge of aquarium waste into a San Diego County lagoon in 2000. Since it has runners that grow several centimeters each day, it soon forms a dense green carpet that excludes all other plants. Herbivores that might otherwise check its growth are deterred by the toxins it produces. Attempts to eradicate the strain in the Mediterranean and southeast Australia, where it is also a problem, have failed (Withgott, 2002).

Such problems, however, do not end here. For example, over 100 different species of aquatic plants and animals are being raised in aquaculture farms in the United States. This has led to the introduction of nonnative aquatic plants, fish, invertebrates, parasites, and pathogens into many areas. If this continues, aquaculture may well become the leading vector of aquatic invasive species worldwide, with many such introductions leading to unpredictable and irreversible ecological impacts. The magnitude of this problem is exemplified by surveys that have shown that every U.S. river system contains at least 1 invasive species, and 60 percent have between 2 and 10 (Powell, 2002).

Preserving Biodiversity

Noting that a major share of the world's biodiversity is concentrated in a relatively small number of coral reefs, forests, savannas, and other habitats, environmentalists have initiated a vigorous international program to preserve these areas. Prominent examples are those in Madagascar, the Philippines, and the Mediterranean-climate coast of California. Less well known are specific areas in Ecuador, India, and South Africa. In all, some 25 such areas, which occupy only about 1.4 percent of the Earth's land surface, are believed to be the exclusive homes of 44 percent of the plant species and 35 percent of the birds, mammals, reptiles, and amphibians that exist (Wilson, 2002).

One of the largest such preservation efforts is a cooperative venture involving Mexico and seven countries in Central America. Called the Mesoamerican Biological Corridor, the goal is to link existing protected areas and to provide additional ones so that scores of corridors will be established that will ultimately enable animals to have safe passage from Chiapas, Mexico, southward to Darien Gap, Panama. Similar efforts are under way throughout the world (Kaiser, 2001). Even so, such projects do not guarantee that endangered species will receive the attention they require. In fact, scientists continue to report that elsewhere in the world the loss and degradation of natural wildlife habitats are continuing at essentially an unabated rate. One possible approach to halt this trend is to provide financial incentives, ranging from tax breaks to future development rights, to owners who choose to provide such habitats. Another is to develop programs to help people recognize the significant economic benefits that habitats generate. At a minimum, these are estimated to be 100 times as valuable as those that might be gained through continuing to destroy them for short-term gains (Balmford et al., 2002).

Unique Roles of Ecosystems

According to the Council on Environmental Quality (CEQ, 1997), there are four major types of ecosystems (Table 18.1). Recognizing their complexity, an alternate categorization has also been proposed (Table 18.2). In fact, humans might be added as another type of ecosystem since essentially every person, at one time or other, serves as a semipermanent host for organisms, such as mites, lice, bedbugs, and fungi, or as an intermediate host for other organisms, such as mosquitoes. This is exemplified by the information in Table 18.3, which provides the key characteristics of ecosystems. As noted, it includes plants and animals.

As scientists have explored ecosystems in more depth, they have increasingly recognized that they are "extraordinarily complex and dynamic, poorly

Table 18.1   Major types of ecosystems

Type	Examples
Terrestrial	Forests; agricultural lands
Aquatic	Freshwater rivers
Coastal and marine	Coral reefs, major bay and ocean fishing areas
Riparian	Floodplains and wetlands

Table 18.2   Alternate categorization of ecosystems

1.	Coasts and oceans
2.	Forests
3.	Farmlands
4.	Freshwaters
5.	Grasslands and shrublands
6.	Urban and suburban areas

Table 18.3   Key characteristics of ecosystems

1.	Ecosystem extent, that is, whether it's growing or shrinking
2.	Fragmentation
3.	Presence or absence of key chemicals needed for life
4.	Contaminants
5.	Physical conditions, including factors such as erosion or depth to groundwater
6.	Plants and animals
7.	Biological communities
8.	Plant growth and productivity
9.	Production of food and fiber and use of water
10.	Recreational use and other services provided by the ecosystem

understood, and prone to unforeseeable behavior that may alter their functionality" (Prugh and Assadourian, 2003). It is therefore not surprising that none of the major ecosystems has been immune to the impacts of humans. Also of note, these impacts are frequently due primarily to a lack of recognition and understanding of the interplay with and interdependence of each of the several subunits within such systems. For example, excessive harvesting of one species of fish from the ocean has ramifications far beyond the depletion of that species. Each species within such an ecosystem is linked to many others, either as a predator, as a scavenger, or as a source of food or shelter. If the species that is removed is peripheral to the system, it may be possible for the ecosystem to continue to function. If, however, it is a major or keystone player and is overly impacted, the ecosystem will be forced to establish a new equilibrium that inevitably will not be as functional as the original (Hayden, 2001). Some of the ways in which humans have impacted each of the four major ecosystems are discussed below.

TERRESTRIAL ECOSYSTEMS

Forests and trees. These constitute one of two major constituents of the terrestrial ecosystem. Among their better-known benefits is that they provide both ecological services and economic goods, ranging from soil and watershed protection, to timber and firewood, to wildlife habitat and recreation. They also moderate climate, capture and store precipitation, serve as sinks for carbon dioxide, and are home to two-thirds of all species. In fact, as much as half of the carbon in the world's biomass may be stored in forests. Any loss of this resource thus reduces the Earth's capacity to absorb carbon dioxide from the atmosphere. In spite of these attributes, an estimated 10 percent of the forests in the United States were destroyed by harvesting between 1960 and 1990. Such activities lead to the removal of essential nutrients and topsoil, the water and nutrient cycles are destabilized, and the soil itself is left without protection from flooding and erosion.

During more recent years, increased mortality has been observed among trees in the western United States (van Mantugun et al., 2009) and among the hardwood forests of New England due to infestations of longhorn beetles (Alsop, 2009). Spruce trees in the forests in the Yukon Territory are also being devastated by infestations of beetles, combined with fires (Associated Press, 2009a). Concern is likewise being expressed with respect to rain forests that are being severely impacted both by excessive harvesting and various insects. Some scientists estimate that these types of forests are vanishing at a rate of as much as a million square miles annually. In a like manner, prominent biologists have predicted that only 5 percent of the tropical old-growth forests will survive by 2050. This will yield an accompanying extinction of 75 percent of the species they contain. Adopting a more optimistic approach, other experts argue that a trend toward urbanization will reduce the number of people living in or near forests and the accompanying rates of loss (Stokstad, 2009). Obviously, the potential outcome warrants the careful attention of national and world leaders.

Agricultural lands. These represent a second major constituent of the terrestrial ecosystem. Nonetheless, farming techniques such as deep plowing lead to wind and water erosion and the accompanying runoff of billions of tons of U.S. topsoil every year. Its loss, in turn, causes turbidity, silting, and deterioration of aquatic habitats and reduces the storage capacity of lakes and reservoirs. Accompanying fertilizers, pesticides, and salts also reduce water quality. Furthermore, windblown particles exacerbate respiratory ailments, impair vision, and create dust storms that damage crops and buildings. Urban sprawl also contributes to these problems since it leads to thousands of acres of the world's richest farmland being covered every year by parking lots, housing developments, and shopping malls.

Nonetheless, techniques are available to correct these problems. These include contour plowing, maintenance of vegetative buffer strips between fields and along waterways, planting trees or grass cover on highly erodible soils, and maintaining a vegetative cover on idle land. One of the most successful is no-till farming, in which the residue from a previous crop is left on the field, with seeds and/or plants being placed in the ground the next season without prior plowing. In addition to reducing topsoil losses, such practices reduce the associated greenhouse-gas emissions (Johnston, 2009). Although less than 7 percent of the world's farmers now apply these methods, this is rapidly increasing, especially among those who grow organic crops (Huggins and Reganold, 2008; Johnston, 2009). Hopefully, this will continue as more farmers recognize that, once removed, topsoil requires hundreds to thousands of years to regenerate (Montgomery, 2008).

AQUATIC ECOSYSTEMS

Although the destruction of aquatic ecosystems continues to be a problem, the consequences are being recognized. This is illustrated by an assessment of the impacts due to the construction of a series of dams for hydropower and flood control on the Missouri and Columbia Rivers. While the discussion that follows will emphasize the negative impacts of these installations, it is important to keep in mind that the hydropower units that were constructed have served as major sources of low-priced electricity for people living in these areas. In addition, the reservoirs created by the dams have enhanced the recreational opportunities for millions of people and have enabled similar numbers to inhabit otherwise forbiddingly arid regions. Another benefit has been a major increase in food production through the water made available for irrigation (Reisner, 2000).

In the case of the Missouri River, the benefits were achieved through the construction of a system of multiple levees and six dams (Kearney, 2002). Experience has revealed, however, that these undertakings have had many negative impacts, the most prominent being the loss of habitats for dozens of native species of fish, birds, and other wildlife. The primary reason for these losses is that the absence of periods of quiescent water flow has kept fish hatchlings from drifting to the shore to develop. Another negative impact is that the sediment that formerly moved with the water now settles out in the reservoirs created by the dams. This, in turn, eliminates the creation of shoreline habitats and shallow shoals for fish and other aquatic life.

Likewise, the removal of plants and trees along the river for farms and developments has destroyed the habitats for other types of wildlife.

Restoration actions being considered include opening one or more of the dams. While this would resolve some of the problems, there would obviously be many negative impacts. As discussed earlier, the loss of the upriver reservoirs created by the dams would eliminate major fishing and recreational areas. In addition, the homes of many people who live on what were formerly the downstream floodplains would now be subject to flooding. For these and other reasons, any such action would be highly controversial. As such, it is doubtful that any of the proposed restoration actions will be implemented within the near future (Thigpen, 2002).

In the case of the Columbia River, the major negative impacts have been on the fishing industry. Prior to installation of the hydropower dams, some 10 to 15 million adult salmon annually swam from the Pacific Ocean to the upper river watershed that provided about 40 percent of their ancestral spawning habitat. Although salmon that spawn on tributaries to the Columbia can make it successfully up the fish ladders around the smaller dams, sooner or later they encounter a dam that is too imposing to pass. Logging, livestock grazing, and the diversion of water for irrigation also cause losses of the spawning habitat. Even though efforts have been made to restore the fisheries, there is evidence that these may not be adequate. This is confirmed by the fact that coho salmon that spawn in Lagunitas Creek, whose origin is in Mount Tamalpais north of San Francisco, California, are nearing extinction (Miller, 2010). The seriousness of this situation is also illustrated by the fact that, as of 2004, 44 percent of all freshwater rivers and streams, 64 percent of all lakes, ponds, and reservoirs, and 30 percent of all bays and estuaries in the United States were either "polluted or impaired" (EPA, 2009).

COASTAL AND MARINE ECOSYSTEMS

As noted earlier (Chapter 17), people throughout the world are migrating to coastline areas. This shift is most dramatically illustrated by the giant coastal cities in Africa and Asia. These and those in other parts of the world frequently lack adequate facilities for treating their domestic and industrial wastes (Tibbetts, 2002). As a result, the coastal zones of the world are being subjected to increased nutrient loading, toxic contamination, and habitat alteration. The resulting impacts are illustrated by conditions in the Chesapeake Bay, the largest, and the most complex, U.S. estuary. In this case, the clearing of the land and the establishment of plantations along its shores have led to increased loss of soil and organic matter via runoff. Early on, the oyster population was sufficient to filter and cleanse the water in the Bay in an estimated time of less than a week, the result being that the pollution was kept in balance. With the passage of time, however, the harvesting of oysters became a major industry. This ultimately reduced their population to the point where the estimated time for the oysters to cleanse the water had increased to more than a year. The increase in suspended matter soon thereafter made the water uninhabitable for the rockfish and blue crabs, which were, in turn, replaced by less-desirable sea nettle jellyfish and toxic algae (Hayden, 2001). To remedy the situation, the Chesapeake Bay Program Office of the EPA is leading a major effort, jointly supported by state and local agencies and environmental organizations, to address these issues. Although, as of 2010, they had met only 24 percent of their cleanup goal, this represented a 2 percent increase during the past year (EPA, 2010).

Related events occurred in the shallow bays of South Florida and the Caribbean. When settlers first arrived, green sea turtles were abundant. By the late 1600s and early 1700s, their population was almost depleted. Concurrently, turtle grass, which had served as the primary food for the turtles, began to flourish. As the plants died and decayed, the oxygen content of the water was reduced to such an extent that fish and shrimp could no longer survive. Ironically, by the 1980s the turtle grass was, in turn, destroyed by molds that grew in the decaying organic matter. Only by restoring the turtle population, and the grass they eat, will it be possible to revitalize what was once a major ecosystem. Another example was the depletion of marine fishing areas, such as Georges Bank off the coast of New England and Canada. One of the primary causes, in this case, is thought to be the use of trawlers and scallop dredges that compact and kill bottom dwellers, such as sponges, deep-sea corals, bryozoans, and other sedentary animals. Without the protection against predators provided by these types of sea life, baby fish are unable to survive (Hayden, 2001). Recognizing that all would suffer unless these problems were solved, fishermen, environmentalists, scientists, and the courts agreed to seek a solution. The answer came through the application of scientific principles, the basic approach being to evaluate the situation in terms of population dynamics. This led to the establishment of a schedule for opening and closing the region's scallop grounds to permit the marine life to recover. Within 3 years, the mass of scallops had increased sufficiently to support a stable and productive scallop industry (Greene, 2002).

Similar problems can be created by activities of a military and/or industrial nature. One example was the beaching and deaths of a group of whales following underwater tests of a sonar system being developed by the U.S. Navy. These occurred in the Atlantic Ocean along the Outer Banks of North Carolina. Similar events were observed near the Bahamas and Canary Islands. Autopsies revealed hemorrhages consistent with acoustic trauma in and around the ears of those that died (Pickrell, 2004). Continuation of the sonar tests was later challenged by the National Resources Defense Council. The Supreme Court, however, permitted them to continue for purposes of national security (NRDC, 2010). Other sources of potential stresses on marine life include offshore activities associated with seismic exploration for oil and drilling rigs and those associated with the operation of supertankers, icebreakers, cruise ships, tugboats, and ferries (Carpenter, 2002).

RIPARIAN ECOSYSTEMS (SAVANNAS AND WETLANDS)

Savannas. These ecosystems are commonly described as those that are influenced by the intermittent interaction of freshwater systems with low-lying land or are at the interface of freshwater and saltwater ecosystems. Prominent examples are the savannas on plains characterized by coarse grasses and scattered tree growth, for example, in the Sudan of Africa. Although covering only 20 percent of the Earth's surface, they contribute an estimated net primary production equivalent to that of the tropical forests. At the same time, however, they are being lost at a rate higher than that of the forests (Lehmann, 2010).

Wetlands. These include tidal marshes, swamp forests, peat bogs, prairie potholes, and wet meadows. Biologically, they are the most productive ecosystems in the world, serving as nurseries and feeding grounds for a range of commercial fish species, as nesting and feeding grounds for waterfowl and migratory birds, and as habitats for many other forms of life (otters, turtles, frogs, snakes, and insects). Wetlands also trap nutrients and sediments; purify water by removing coliform bacteria, heavy metals, and toxic chemicals; provide flood protection by slowing and storing water; and anchor shorelines and provide erosion protection (CEQ, 1997). Without the removal of nutrients, the resulting excessive growth of phytoplankton or algae can destroy coral reefs and other coastal environments that serve as habitat for fish, seabirds, and other animals. Later when the algae die and sink to the bottom, their decay can deplete the concentrations of oxygen in bottom waters, causing environmental hypoxia (Stegeman and Solow, 2002).

In spite of their well-known benefits, by the late 1980s less than half of the wetlands originally in the contiguous United States remained. Fortunately, major efforts are under way to protect these resources. Major stimuli were provided by the Clean Water Act and Amendments (U.S. Congress,

1977) and the Water Quality Act (U.S. Congress, 1987). These require that a permit be obtained from the Army Corps of Engineers to alter, fill, or otherwise change the characteristics of wetlands. In so doing, an applicant must first avoid, or at least minimize, any ensuing damage. If damage cannot be avoided, the Army Corps of Engineers requires the permit holder, or a third party paid by the permit holder, to restore or replace the impacted wetlands (Kearney, 2001). This can be accomplished by the creation of new wetlands and the enhancement and guaranteed preservation of existing wetlands, or the payment of a fee to compensate for their loss.

A classic example of the negative impacts of human activities on wetlands is the situation in the Florida Everglades, a river of grass that once covered almost 12,000 square kilometers (5,000 square miles). With the increasing urban sprawl, water within this area was gradually diverted to urban and agricultural uses. As part of this process, a major channel was dug to facilitate the flow of the water into the ocean. As a result, bird populations decreased, and 68 species such as the manatee and the panther that lived there are now endangered. In response, the Army Corps of Engineers initiated a program in 2001 to undo the earlier changes and restore the ecosystem. Although the program was designed to duplicate the successful efforts of the cooperative multination group that restored the vitality of the wetlands of the Romanian Danube delta (Schmidt, 2001), a subsequent evaluation revealed that scant progress has been made. This was due to a lack of the basic data on which to plan the effort and with which to evaluate whether it was succeeding (National Research Council, 2008). This example illustrates one of the basic tenets of this book, namely, that prior to initiating any major environmental project, it should be mandatory that all potential interactions and impacts be thoroughly evaluated and assessed.

Addressing Other Environmental Challenges

The challenges in the field of environmental and public health extend well beyond the previously discussed examples. This is illustrated by the topics identified by the National Research Council as deserving immediate attention from the standpoint of research on environmental health (Table 18.4). While certain topics can be addressed by the scientific community, essentially all of them will, during some phase or time, require cooperative efforts on the part of the leaders of the global community of nations. At the other extreme, certain challenges and problems will always remain that, of necessity, must be addressed at a local or regional level. Examples of several of these issues are discussed below.

Table 18.4 Research needs in environmental health

Subject area	Description of need
Deserving immediate attention	
Biodiversity and ecosystem functioning	Improved understanding of the factors, including human activities, that affect biodiversity and how biodiversity relates to the overall functioning of an ecosystem
Hydrologic forecasting	Capability to help predict changes in freshwater resources and changes in the environment caused by floods, droughts, sedimentation, and contamination
Infectious diseases and the environment	Better understanding of how pathogens, parasites, and disease-carrying species, as well as the humans and other species they infect, are affected by changes in the environment, the goal being to prevent outbreaks of infectious diseases in plants, animals, and humans
Land-use dynamics	Methods for applying recent advances in data collection and analysis to document and understand the causes and consequences of changes in land cover and use
Also important	
Biogeochemical cycles	Understanding of how changes in the balance of carbon, oxygen, hydrogen, sulfur, and phosphorus in soil, water, and air affect the functioning of ecosystems, atmospheric chemistry, and human health
Climate variability	More complete comprehension about how the Earth's climate varies over a wide range of time scales, from extreme storms that develop quickly to changes in weather patterns that occur over several decades
Institutions and resource use	More information about how the condition of natural resources is shaped by markets, governments, international treaties, laws, and informal rules that govern environmentally significant human activities
Reinventing the use of materials	Additional data on the forces driving human use of reusable metals such as copper and zinc, hazardous metals such as mercury and lead, reusable plastics and alloys, and ecologically dangerous compounds such as CFCs and pesticides

SOLVING PROBLEMS REGIONALLY AND LOCALLY

As noted above, many environmental problems are regional or local in nature. Their effective prevention and control will require input from all levels—industry, governmental agencies, and society.

Responsibilities of industry. Of primary importance is that industrialists and industrial engineers learn to incorporate sound environmental thinking into the initial selection and design of manufacturing processes and products. Pollution controls must be designed into industrial equipment, not added on later. Multinational corporations and financial institutions also have an obligation to set a new moral tone for the world. They must commit themselves to a sustainable future and be prepared to sacrifice a portion of their profits to do so. Close examination will demonstrate that such an approach is in their best interests, as well as essential to their survival.

One promising development is the increasing awareness of major industrial corporations of the need to be responsible for limiting the environmental impacts of their operations. Recently, the editors of *Newsweek* magazine evaluated the 500 largest U.S. companies based on their environmental impacts, as well as their related policies and reputation among their peers and environmental experts. The outcome of these efforts was based on an evaluation of 700 metrics, which included water use and acid-rain emissions, corporate policies and initiatives, and extensive surveys of the reputations of their chief executive officers and environmental directors. The result, a "Green Score" for each company, was a weighted average of these three components based on an allocation of 45 percent for the "Environmental Impact Score," 45 percent for the "Green Policies Score," and 10 percent for "Reputation Survey Score." The rankings highlighted greenhouse-gas emissions because, for many companies, these represented their most severe impact on the environment (McGinn, 2009).

The companies with the five highest rankings were Hewlett-Packard, Dell, Johnson & Johnson, Intel, and IBM. The review indicated that many other companies had initiated major programs to improve their rankings. These included Nike, Wells Fargo, McDonald's, Proctor & Gamble, Microsoft, Walt Disney, First Solar, Coca-Cola, Walmart, Whole Foods Market, General Electric, and Marathon Oil. Also noted were the increasing number of companies that promote their green initiatives in their advertisements (McGinn, 2009). Overall, this is a significant advance in ensuring the protection of public health and the environment, especially because it recognizes how such achievements can be accomplished and will hopefully provide an incentive to other companies to design and implement comparable efforts.

Responsibilities of governmental agencies. Government agencies likewise need to assume increased responsibility for protecting the environment and improving its two-way relationship with people. As enumerated by EPA's Science Advisory Board (Johnson, 1995), the principal needs are as follows:

- Develop programs that will provide continuing evaluation of key environmental areas, such as ecosystem sustainability, noncancer human health effects, nontraditional environmental stressors, and the health of the oceans.

- Emphasize the avoidance of future environmental problems as much as the control of those that exist.

- Stimulate coordinated efforts among federal, state, and local agencies and the private sector to develop the capability to anticipate and respond to environmental change; an integral part of this effort is the establishment of an early warning system to identify emerging environmental risks.

- Recognize that global environmental quality is a matter of strategic national interest and to adopt policies that link security, foreign relations, environmental quality, and economic growth.

Also noted was that if various governmental agencies, as well as the private sector, are to have the capabilities for anticipating and responding to environmental changes, it will be necessary for them to expand the database both in the scientific and policy sector. One step in this direction is the UN program related to biodiversity.

Responsibilities of society. Concurrently, society as a whole needs to develop a forward-looking attitude in dealing with environmental problems. Societal behavior will change only if enough people become aware of environmental problems and act, both as individuals and through their elected governmental representatives. The American Medical Association (AMA, 1989), in developing its policy statement *Stewardship of the Environment,* suggested that the United States play a leading role in effecting such change:

> The U.S. and the world at large appear to be facing environmental threats of unprecedented proportions, and scientists, environmental activists, health professionals, politicians and world leaders are beginning to realize the need for changes in societal behavior (i.e., human behavior as well as the conduct of business and industry) as a means of forestalling these potential threats. Societal changes must be initiated worldwide if they are to have any significant effect overall. However, their implementation will need a model, most suitably a national model . . . The United States could well become a model for environ-

mental stewardship if a grassroots movement were to develop to encourage and endorse a protective and nurturing philosophy towards the environment at both the personal and societal levels.

As is evident, the accomplishment of these objectives will require action on several fronts, most notably in education. As emphasized by Bruce Alberts (2010), "Why is it that students often become bored with education before their teenage years? How might the United States produce an education system that allows a child with a specific fascination to explore that interest in depth as an integral part of his or her early education?"

SOLVING PROBLEMS INTERNATIONALLY

One of the best examples of how a global environmental problem can be successfully controlled followed the realization during the late 1950s and early 1960s that radioactive materials released during atmospheric testing of nuclear weapons were being transported throughout the world. In fact, tests of the air, water, and soil in all countries in the Northern Hemisphere showed the presence of these materials. In a joint effort, the United States, the United Kingdom, and the Soviet Union in 1963 signed a treaty that banned nuclear tests in the atmosphere, outer space, and underwater (*Encyclopedia Britannica Online*, 2010). Later the ban was expanded to include underground nuclear tests.

Under the leadership of the United Nations, several other environmental problems have been similarly addressed. These include the successful efforts to halt the depletion of the ozone layer and the treaty that initiated a worldwide phaseout of the use of certain persistent organic chemicals (United Nations, 2002). With respect to the ozone layer, the developed nations were willing to take a lead role because the overall benefits outweighed the costs of eliminating CFC production. The developing nations also benefited through being granted extra time to reduce their CFC emissions as well as funds to help them pay for the transition. No country was granted big concessions, but neither did any country suffer a significant loss. The result was, and continues to be, that a significant problem was addressed and its long-term consequences averted. The effort to phase out the use of certain persistent organic chemicals also appears to be following a road to success. The chemicals in question, widely referred to as the "dirty dozen," include polychlorinated biphenyls (PCBs), DDT, dioxins, furans, and other pesticides that have been shown to contribute to developmental defect, cancer, and other health effects in humans and animals (Chapter 10). Because eliminating their production would significantly reduce potential health effects to billions of people, there was widespread support for the phaseout. Adding to the impetus for success was the immediate endorsement of the action by the United States and the availability of less persistent substitutes at comparable costs.

Climate Change

One of the most contentious and debated issues confronting the world today is climate change. As will be noted in the discussion that follows, this is due to the fact that its potential impacts will affect the entire world, and there are difficult and complex scientific questions that need to be answered. Seeking to address these issues, the United Nations convened an international meeting on climate change in Japan in 1997. This concluded with the issuance of the Kyoto Protocol, which outlined a plan for stabilizing the atmospheric concentrations of greenhouse gases. It was based on the fact that chemical compounds such as carbon dioxide (CO_2) methane, and CFCs, when present in the atmosphere, are transparent to incoming shortwave electromagnetic radiation reaching the Earth from the sun. Once the radiation scatters off the surface of the Earth, however, its energy is reduced such that it is readily absorbed by these compounds on its outward path. In essence, the heat from the incoming radiation is trapped in a manner similar to that of a greenhouse. For this reason, any increase in the concentrations of these gases in the atmosphere will increase the temperature near the surface of the Earth (Figure 18.2) and could contribute to global warming. The Kyoto Protocol, however, led to little action. This was due to the reticence of the developed countries to establish goals and to seek to meet them. At the same time, the developing countries feared the accompanying economic impacts.

In early December 2009, the United Nations convened a second conference in Copenhagen, Denmark (Brown, 2009). Little was accomplished, however, the primary weaknesses being the lack of agreement on credible emission targets, the unwillingness of many nations either to measure or quantify their emissions, and disputes between the goals for developing and developed nations. Another interpretation was that leaders in many of the countries still questioned the extent to which global warming is of human origin. Subsequent to the conference, many delegates challenged the degree to which politics influenced key statements in the wording of the final report. For this reason, one of the key objectives of the delegates at the 2010 follow-up conference in Cancun, Mexico, was that there would be no secret negotiations conducted behind closed doors (Associated Press, 2010).

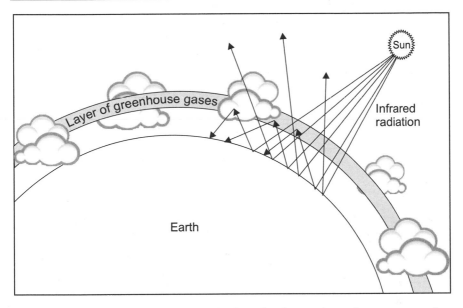

Figure 18.2   Schematic presentation of infrared radiation from the sun, some of which is absorbed by the Earth, some of which is scattered back into outer space, and some of the latter which interacts with greenhouse gases during the outward passage and is scattered back and absorbed by the Earth

THE FUNDAMENTAL QUESTIONS

Based on these observations, there are three fundamental questions that need to be addressed:

- Is the temperature of the planet increasing?
- To what extent do the greenhouse gases created by human activities contribute to this warming, and what can be done to reduce them?
- Warming of the planet will cause many other climate patterns to change at speeds unprecedented in modern times. This being the case, what can be done to enable the population of the world to survive an inevitable increase in the global temperature?

With regard to the *first* question, the planet is, beyond any doubt, warming. There are multiple sources of evidence that this is the case:

- During the last half century, the dates on which 385 species of British plants have flowered has advanced by an average of 4.5 days (Fitter and Fitter, 2002). Similarly, the leaves of most deciduous plant species in various Mediterranean ecosystems are unfolding an average of 16 days earlier and falling an average of 13 days later than they did 50 years ago. In Western Canada, *Populus tremuloides* is now blooming 26 days earlier than a century ago (Penuelas and Filella, 2001).

- During the past four decades, there have been measurable changes in the species composition of certain marine plankton in the North Atlantic. These include a northward extension, by more than 10 degrees in latitude, of warm-water species, accompanied by a decrease in the number of cold-water species (Beaugrand et al., 2002). There have also been significant advances in the dates for initial breeding in a number of bird species. For example, migratory birds that winter south of the Sahel are now arriving later in Europe (Penuelas and Filella, 2001). Interestingly, these types of changes can also cause disruptions in the relationships between blooming plants and their pollinators—for example, between the blooming of a particular flower and arrival of the butterfly that previously pollinated it. If there is no nectar for the insect, it could well be that the future could bring the world to a time when there will be no seeds to grow more plants (Kennedy, 2002). Also of interest is that a 1- to 2-degree increase in temperature could lead to insect-borne outbreaks such as malaria, dengue, and yellow fever in areas previously not affected. This has occurred in Hawaii, where mosquitoes, now able to survive at higher elevations, have bitten and killed multitudes of native birds that were not immune to the disease (Harvell et al., 2002).

- Concurrently, coral reefs are disappearing throughout the world. For example, the size of the Great Barrier Reef of Australia has declined by nearly 15 percent since 1990 (De'ath, Lough, and Frambricius, 2009). This is due to the disruption of certain basic characteristics and interactions of these ecosystems. In essence, corals are to coral reefs as trees are to a forest. The impacts of global warming and the accompanying environmental changes could affect both the nutritional value and structural formation of this ecosystem. Another example is the increasing acidification of the oceans due to the dissolution of larger quantities of atmospheric carbon dioxide (CO_2). A reduction in pH of only 0.2 unit during the next century could reduce their calcification rates, necessary to form new coral, by up to 50 percent. It will also promote the dissolution of the corals that exist (Weis and Allemand, 2009).

Another contributing factor is ocean warming, In fact, recent studies confirm that temperatures in the ocean are increasing more rapidly than those in the air (NASA, 2010). This causes corals to expel the algae from their bodies, turn white, and die. At the same time, if additional nutrients are discharged into the ocean, the growth of seaweed will be stimulated and the coral will be smothered and unable to feed or draw sunlight during the day (Chapple, 2009). Since coral reefs serve as the habitat of countless forms of marine life, including fish, lobsters, oysters, shrimp, and turtles, their existence could also be threatened.

• The loss of Greenland ice is accelerating, with the melting of its glaciers contributing to two-thirds of the loss (Figure 18.3). In fact, satellite data have revealed that the Antarctic ice sheets are shrinking, with projections that this could lead to a rise in sea level of 1 meter (3.3 feet) or more by 2100. This, in turn, could lead to flooding of coastal areas throughout the world. Although the complete disintegration of the Greenland ice sheet would require a 6°C (11°F) increase in temperature, a 15 percent loss would translate into a 1 meter rise in

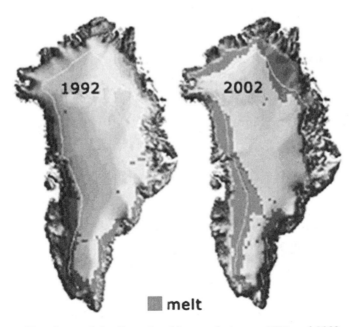

Figure 18.3   Shrinkage of the Greenland ice cap between 1992 and 2002

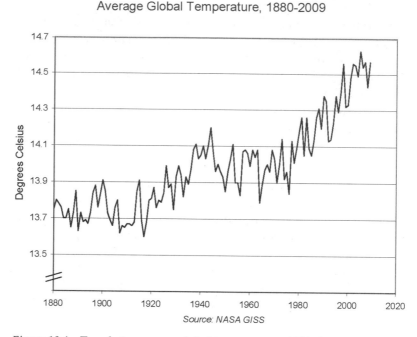

Figure 18.4   Trends in average global temperature, 1880–2009

sea level (Kintisch, 2009). The warming of the Earth is also documented by records of the trends in global temperatures during the
past 130 years (Figure 18.4).

The answer to the *second* question is that global warming is due both to
natural and human activities. The best example of the latter is the increased
concentrations of heat-trapping gases being released due to burning of fossil fuels and the destruction of CO_2-absorbing forests. This is the conclusion of the scientists who served on the Intergovernmental Panel on Climate
Change (Sills, 2010). Without question, every effort should be made to reduce
the production and release of CO_2 and other heat-trapping gases. This is true
not only because they are major contributors but also because they may serve
as triggers in initiating changes earlier than they might otherwise have occurred (Chu, 2009).

With regard to the *third* question, a carefully designed program should
be initiated to address the known impacts climate change will have. Topics
that should be addressed include how to (1) ameliorate its potential impacts

and (2) cope with the threats to coastal communities and cities, our food and water supplies, marine and freshwater ecosystems, forests, and high mountainous environments (Sills, 2010). Other challenges include how to cope with the potential massive migrations of the population toward the North and South Poles, and how to respond to the shifts in disease patterns that may occur due to the potential rise in the number and types of insects and other vectors. Other, now unknown, challenges will undoubtedly emerge if, and when, climate changes reveal the need.

Due to the extent of the threat, this must be a coordinated global effort with the full support of nations throughout the world. To the maximum possible extent, the developed countries should assist the developing countries in meeting these goals. The conclusions of the experts are clear. Delay is not an option.

The General Outlook

In spite of what may often seem to be an insurmountable task, progress is being made in protecting the environment. Although global surveys do not yield universal results, they confirm that public concern for the environment has generally been increasing on both cultural and national boundaries. This is attributed to the more effective and rapid dissemination of scientific knowledge on this subject (Wolinsky-Nahmias and Kim, 2008). This trend is especially true in the United States, where polls show that about two-thirds of the population, if asked to choose between protection of the environment and economic growth, would select the former. This has proved true even in times of economic downturns. Problems that were listed as of highest concern included pollution of drinking water, contamination of soil and water by toxic wastes, pollution of rivers, lakes, and reservoirs, and pollution of the atmosphere. Although farther down on the list of immediate priorities, it is interesting to note that those polled were also knowledgeable about essentially all of the global problems discussed above (Greenberg, 2001). The will is there; the knowledge and technology continue to evolve. The primary need is for leadership to marshal existing support and to set priorities for action.

REFERENCES • CREDITS
INDEX

REFERENCES

1. The Scope

American Cancer Society. 2008. *Cancer Facts & Figures 2008*. Atlanta, GA.

ATSDR. 2002. "A Pledge to Protect Children—The Bangkok Statement." *Public Health and the Environment* 1, no. 1/2, 21 (fall).

Brink, Susan. 2002. "Phys Ed Redux." *U.S. News and World Report* 132, no. 19 (3 June), 50–52.

Bugliarello, George, 2001. "Rethinking Urbanization." *The Bridge* 31, no. 1 (Spring), 5–12.

Carson, Rachel. 1962. *Silent Spring*. Houghton Mifflin. Boston, MA.

CDC. 1998. "Youth Risk Behavior Surveillance—United States, 1997." *Morbidity and Mortality Weekly Report* 47, no. SS-3 (14 August), 1–89. Centers for Disease Control and Prevention. Atlanta, GA.

———. 2002. "Guidelines for School Programs to Prevent Skin Cancer." *Morbidity and Mortality Weekly Report* 51, no. RR-4 (26 April), 1–21. Centers for Disease Control and Prevention. Atlanta, GA.

———. 2008. "Smoking-Attributable Mortality, Years of Potential Life Lost, and Productivity Losses—United States." *Morbidity and Mortality Weekly Report* 57 (14 November), 1226–1228. Centers for Disease Control and Prevention. Atlanta, GA.

———. 2009. "State-Specific Prevalence and Trends in Adult Cigarette Smoking—United States, 1998—2007." *Morbidity and Mortality Weekly Report* 58 (12 March), 221–226. Centers for Disease Control and Prevention. Atlanta, GA.

Chad, Chao, 2009. "Letter to the Editor: Effects of Increased Urbanization." *Science* 326, issue 5923 (3 April), 37.

Colditz, G. A., Samplin-Salgado, M., Ryan, C. T., Dart, Fisher, H. L., Tokuda, A., and Rockhill. B. 2002. "Fulfilling the Potential for Cancer Prevention: Policy Approaches." *Cancer Causes and Control* 13, no. (3 April), 199–212.

Davis, Brennan, and Carpenter, Christopher. 2009. "Proximity of Fast-Food Restaurants to Schools and Adolescent Obesity." *American Journal of Public Health* 99, no. 3 (March), 505–510.

DeKay, Mark, and O'Brien, Michael. 2001. "Gray City, Green City." *Forum for Applied Research and Public Policy* 16, no. 2 (Summer), 19–27.

———. 2009. "Health Effects of Secondhand Smoke." http://www.epa.gov/smoke free/healtheffects.html. Accessed 26 January 2010.

Eskenazi, Brenda, and Landrigan, Philip J. 2002. "Environmental Health Perspectives and Children's Environmental Health." *Environmental Health Perspectives* 110, no. 10 (October), A559–A560.

Fields, Scott, 2002. "Urban Issues: If a Tree Falls in the City." *Environmental Health Perspectives* 110, no. 7 (July), A 392.

Forastieri, Valentina, 1997. *Children at Work, Health and Safety Risks*. International Labour Office, Geneva, Switzerland.

Gerber, Joel, 2010. "Federal Government Takes on Childhood Obesity: September Recognized as Childhood Obesity Awareness Month." Freedom ENC, *Sun Journal* New Bern, NC (September 7, 2010), 7.

Gill, Thomas, M., Williams, Christianna S., Robinson, Julie T., and Tinetti, Mary E. 1999. "A Population-Based Study of Environmental Hazards in the Homes of Older Persons." *American Journal of Public Health* 89, no. 4 (April), 553–556.

Guyer, Ruth Levy, 2001. "Backpack = Back Pain." *American Journal of Public Health* 91, no. 1 (January), 16–19.

Hayden, Thomas, and Potts, Malcolm. 2010. "Make Birth Control, Not War." *Miller-McCune* (May–June), 48–57.

Holden, Constance. 2009. "Youth Bulge." *Science* 325, issue 5943 (21 August), 923.

Kaiser, Jocelyn, 2002. "Texas Surgeon Vows to Take Next Step in Beating Cancer." *Science* 296, issue 5572 (24 May), 1394–1395.

Kunzig, Robert. 2010. "Population Explosion." *National Geographic* 19, no. 1 (January), 42–69.

Moeller, Dade W., and L-SC Sun. 2010. "Chemical and Radioactive Carcinogens in Cigarettes: Associated Health Impacts and Responses of the Tobacco Industry, U.S. Congress, and Federal Regulatory Agencies." *Health Physics* 99, no. 5, 674–679.

Newswatch. 2009. "8 Million Americans Considering Suicide." *Sun Journal*, New Bern, NC (19 September), 10.

Pell, J. P., Haw, S., Cobbe, S., Newby, D. E., Pell, A. C. H., Fischbacher, C., McConnachie, Pringle, Murdoch, D., Dunn, F., Oldroyd, K., MacIntyre, P., O'Rourke, B., and Borland. W. 2008. "Smoke-free Legislation and Hospitalizations for Acute Coronary Syndrome." *New England Journal of Medicine* 359, 482–491.

Sheehan, Molly O., 2002. "What Will it Take to Halt Sprawl?" *World-Watch* 15, no. 1 (January/February), 12–23.

Suk, William A., 2002. "Beyond The Bangkok Statement: Research Needs to Address Environmental Threats to Children's Health." *Environmental Health Perspectives* 110, no. 6 (June), A 284–A 285.

United Nations. 2009. *World Population Growth Will Occur in Urban Areas of World, United Nations Report States*. Vienna, Austria. http://www.unis.univienna.org/ uris/pressrels/2000/pop29.html. Accessed 1 August 2009.

U.S. Congress. 1970. *National Environmental Policy Act of 1969*. Public Law 91-190, 42 USC (1 January), 4321–4347. Washington, DC.

———. 2009. *The Family Smoking Prevention and Tobacco Control Act*. HR 1256 (22 June).

U.S. Department of Agriculture. 2008. "Food Pyramid." http://www.mypyramid .gov/html. Accessed 6 October, 2009.

USDHHS. 2009. *Healthy People 2010: Understanding and Improving Health*, 2nd ed. (November). U.S. Department of Health and Human Services. Washington, DC.

U.S. Public Health Service. 2001. *Surgeon General's Call to Action to Prevent and Decrease Overweight and Obesity* (December). Washington, DC.

Wakefield, Julie. 2002. "Learning the Hard Way." *Environmental Health Perspectives*, 110, no. 6 (June), A 298–A 305.

Yen, Hope. 2009. " '100-Year-Olds' Club Rapidly Growing Members." *Sun Journal*, New Bern, NC (20 July), 8.

Zoroya, Greg. 2009. "Ban on Tobacco Urged in Military." *USA Today* (10–12 July), 1A.

2. Toxicology

Ames, B. N. 1971. "The Detection of Chemical Mutagens with Enteric Bacteria." In Hollander, A., ed., *Chemical Mutagens: Principles and Methods for Their Detection*, 1, 267–282. Plenum Press. New York, NY.

ATSDR. 2009. *Toxicological Profiles*. Agency for Toxic Substances and Disease Registry. Atlanta, GA. Available from the National Technical Information Service, 5285 Port Royal Road, Springfield, VA 22161.

Butterworth, B. E., Dorman, D. C., Gaido, K. W., Sumner, S. J., Corton, J. C., Borghoff, S. J., and Conolly, R. B. 1999. "Research at CIIT on the Risks to Human Health from Exposure to Chemicals." CIIT *Activities* 19, no. 10 (October), 1–8. Research Triangle Park, NC.

Calabrese, Edward J., Baldwin, Linda A., and Holland, Charles D. 1999. "Hormesis: A Highly Generalizable and Reproducible Phenomenon with Important Implications for Risk Assessment." *Risk Analysis* 19, no. 2, 261–281.

CDC. 2001. *Why Poison Ourselves? A Precautionary Approach to Synthetic Chemicals. National Report on Human Exposure to Environmental Chemicals*. Centers for Disease Control and Prevention. Atlanta, GA.

CDC. 2003. *Second National Report on Human Exposure to Environmental Chemicals*. Centers for Disease Control and Prevention. Atlanta, GA.

Croteau, M., Luoma, S. N., and Stewart, A. R. 2005. "Trophic Transfer of Metals along Freshwater Food Webs: Evidence of Cadmium Biomagnification in Nature." *Limnology and Oceanography* 50, no. 5, 511–519.

Doull, John. 1992. "Toxicology and Exposure Limits." *Applied Occupation Environmental Hygiene* 7, 583–585.

Doull, J., and Bruce, M. C. 1986. "Origin and Scope of Toxicology." In Klaassen, C. D., Amdur, Mary O., and Doull, J., eds., *Casarett and Doull's Toxicology: The Basic Science of Poisons*, 3rd ed., 3–10. Macmillan Publishing Company. New York, NY.

Elfarra, Adnan, ed. 2009. *Advances in Bioactivation Research*. Springer Publishing. London, England.

EPA. 1986. *Guidelines for the Health Risk Assessment of Chemical Mixtures*. Federal Register 51, 34014–34025. Washington, DC.

Gochfeld, Michael. 1998. "Principles of Toxicology." In Wallace, Robert B., ed., *Maxcy-Rosenau-Last Public Health and Preventive Medicine*, 415–427, Appleton & Lange. Stamford, CT.

Greenlee, W.F. 2002. "Message from the President." In *2001 Annual Report*, 7–8, Research Triangle Park, NC: Chemical Industry of Toxicology.

Henry, Carol J., and Bus, James S. 2000. "Long-Range Research Initiative of the American Chemistry Council." *CIIT Activities* 20, no. 7 (July), 1–5. Chemical Industry Institute of Toxicology. Research Triangle Park, NC.

Kaiser, Jocelyn. 2003. "A Healthful Dab of Radiation." *Science* 302, issue 644 (17 October), 378.

Kamrin, Michael. 2003. *Traces of Environmental Chemicals in the Human Body: Are They a Risk to Health?* New York: American Council on Science and Health (May), revised edition.

Klaassen, C. D. 1986. "Principles of Toxicology." In Klaassen, C. D., Amdur, Mary O., and Doull, J., eds. *Casarett and Doull's Toxicology: The Basic Science of Poisons*, 3rd ed., 11–32. Macmillan Publishing Company. New York, NY.

Lippmann, Morton. 1992. "Introduction and Background." In Lippmann, Morton, ed., *Environmental Toxicants: Human Exposures and Their Health Effects*, p. 129. Van Nostrand Reinhold. New York, NY.

Loomis, T. A. 1968. *Essentials of Toxicology*. 3rd ed. Lea & Febiger. Philadelphia, PA.

Los Alamos Scientific Laboratory. 1995. "Radium–The Benchmark for Alpha Emitting Radionuclides." *Los Alamos Science* 23, 224–233.

Lu, Frank C. 1991. *Basic Toxicology: Fundamentals, Target Organs, and Risk Assessment*, 2nd ed. Hemisphere Publishing Corporation. New York, NY.

Moriarty, F. 1988. *Ecotoxicology: The Study of Pollutants in Ecosystems*, 2nd ed. Academic Press. New York, NY.

NCRP. 1993. *Limitation of Exposure to Ionizing Radiation*. Report No. 116. Bethesda, MD.

NRC. 1983. *Drinking Water and Health. Vol. 5*. Board on Toxicology and Environmental Health Hazards. National Academy Press. Washington, DC.

———. 1986. *Drinking Water and Health. Vol. 6*. Board on Toxicology and Environmental Health Hazards. National Academy Press. Washington, DC.

———. 1995a. *Radiation Dose Reconstruction for Epidemiologic Uses*. Committee on an Assessment of CDC Radiation Studies, National Research Council, National Academy of Sciences. Washington, DC.

———. 1995b. *Prudent Practices in the Laboratory Handling and Disposal of Chemicals*. Washington, DC.

OSHA. 2007. *Laboratory Safety Standard*. 29 CFR 1910, 1450. Washington, DC.

Random House. 2009. *Dictionary of the English Language. The Unabridged Edition*. New York, NY.

Schmidt, Charles W. 2002. "Assessing Assays." *Environmental Health Perspectives*. 110, no. 5 (May), A248–A251.

Seeley, Mara. 2001. "Carcinogen Classification." *Trends in Risk and Remediation* 3 (Spring), 5. Gradient Corporation. Cambridge, MA.

Service, Robert F. 2002. "More on Drug Pollution." *Science* 296, issue 5567 (19 April), 463.

Smith, R. P. 1992. *A Primer of Environmental Toxicology.* Lea & Febiger. Philadelphia, PA.

Society of Toxicology. 2009. "Definition of Toxicology." Communique, Reston, VA.

State of California, 2009. *Notice of Proposed Rulemaking Title 27.* California Code of Regulations: Proposed Amendment of Section 25903, Notice of Violation (19 November).

Stone, R. 1993. "FCCSET Develops Neurotoxicology Primer." *Science* 261, issue 5124 (20 August), 975.

3. Epidemiology

Brain, J. D., Kavet, R., McCormick, C. Poole, Silverman, L. B., Smith, T. J., Valberg, P. A., Van Etten, R. A., and Weaver. J. C. 2003. "Childhood Leukemia: Electric and Magnetic Fields as Possible Risk Factors." *Environmental Health Perspectives* 111, no. 7 (June), 962–970.

Cantor, K. Blair, P. A., Everett, G., Gibson, R., Burmeister, L. F., Brown, L. M., Schuman, L., and Dick, F. R. 1992. "Pesticides and Other Agricultural Risk Factors for Non-Hodgkin's Lymphoma among Men in Iowa and Minnesota." *Cancer Research* 52, no. 9 (1 May), 2447–2455.

Clarke, Roger H. 2008. "The Risks from Exposure to Ionizing Radiation." In Till, John E., and Grogan, Helen A., eds., 2008, *Radiological Risk Assessment and Environmental Analysis*, chapter 12. Oxford University Press. Oxford, NY.

Darby, S., Hill, D., Auvinen, A., Hill, D., Barros, J. M., Dios, H., Baysson, H., Bochicchio, F., and Falk, R. 2005. "Radon in Homes and Risk of Lung Cancer: Collaborative Analysis of Individual Data from 13 European Case-Control Studies." *British Medical Journal* 330, 223–227.

Dockery, D., Pope, C. A., Xiping, X., Spengler, J., Ware, J., Fay, M., Ferris, B., and Speizer, F. 1993. "An Association Between Air Pollution and Mortality in Six U.S. Cities." *New England Journal of Medicine* 399, no. 24, 1753–1759.

Doll, Richard, and Hill, A. Bradford. 1950. "Smoking and Carcinoma of the Lung: Preliminary Report." *British Medical Journal* 2 (30 September), 739–748.

Doll, Richard, and Peto, Richard. 1976. "Mortality in Relation to Smoking: 20 Years' Observations on Male British Doctors." *British Medical Journal* 2 (25 December), 1525–1536.

EPA. 1992. *Respiratory Health Effects of Passive Smoking (Also Known as Exposure to Secondhand Smoke or Environmental Tobacco Smoke ETS).* Report EPA/600/6-90/006F. Office of Health and Environmental Assessment. Environmental Protection Agency. Washington, DC.

———. 2003. *Assessment of Risks from Radon in Homes.* Report EPA 402-R-03-003. Office of Health and Environmental Assessment. Environmental Protection Agency. Washington, DC.

Goldsmith, John R. 1986. *Environmental Epidemiology: Epidemiological Investigation of Community Environmental Health Problems.* CRC Press. Boca Raton, FL.

Hande, M. Prakash, Azizova, Tamara V., Geard, Charles R., Burak, Ludmilla E., Mitchell, Catherine R., Khokhryakov, Valentin F., Vasilenko, Evgeny K., and Brenner, David J. 2003. "Past Exposure to Densely Ionizing Radiation Leaves a

Unique Permanent Signature in the Genome." *American Journal of Human Genetics* 72, 1162–1170.

Hill, A. B. 1965. "The Environment and Disease: Association or Causation?" *Proceedings of the Royal Society of Medicine* 58, 259–300.

Monson, Richard. 1990. *Occupational Epidemiology,* 2nd ed. CRC Press. Boca Raton, FL.

Muirhead, Colin. 2001. "Radiation Risks in Kyoto." *Radiological Protection Bulletin,* no. 231 (September), 34–35. National Radiological Protection Board. Chilton, Didcot, United Kingdom.

NRC. 1991. *Environmental Epidemiology: Public Health and Hazardous Wastes.* National Research Council. National Academy Press, Washington, DC.

———. 1995. *Radiation Dose Reconstruction for Epidemiologic Uses.* Committee on an Assessment of CDC Radiation Studies, National Research Council, National Academy of Sciences. Washington, DC.

———. 2006. *Health Risks from Exposures to Low Levels of Ionizing Radiation: BEIR VII Phase 2,* Chapter 6. Committee to Assess Health Risks from Exposure to Low Levels of Ionizing Radiation. The National Academies Press, Washington, DC.

Pell, J. P., Haw, S., Cobbe, S., Newby, D. E., Pell, A. C. H., Fischbacher, C., McConnachie, A., Pringle, S., Murdoch, D., Dunn, F., Oldroyd, K., MacIntyre, P., O'Rourke, B., and Borland, W. 2008. "Smoke-Free Legislation and Hospitalizations for Acute Coronary Syndrome." *New England Journal of Medicine* 359 (31 July), 482–491.

Pope, C. A., III, Thun, M. J., Namboodiri, M., Dockery, D. W., Evans, J. S., Speizer, F. E., and Heath, C. W. 1995. "Particulate Air Pollution as a Predictor of Mortality in a Prospective Study of U.S. Adults." *American Journal of Respiratory and Critical Care Medicine* 151, 669–674.

Pope, C. A., III, Burnett, R. T., Thun, M. J., Calle, E. E., Krewski, D., Ito, K., and Thurston, G. D., 2002. "Lung Cancer, Cardiopulmonary Mortality, and Long-Term Exposure to Fine Particulate Air Pollution." *Journal of the American Medical Association* 287, no. 9, 1132–1141.

Terracini, B. 1992. "Environmental Epidemiology: A Historical Perspective." In Elliott, P., Cuzick, J., English, D., and Stern, R., eds. *Geographical and Environmental Epidemiology: Methods for Small-Area Studies,* 253–263. World Health Organization Regional Center for Europe: Oxford University Press.

Trichopoulos, Dimitrios. 1994. "Risk of Lung Cancer from Passive Smoking." *Principles and Practice of Oncology: Principles and Practices of Oncology Updates* 8, no. 8, 1–8.

U.S. Congress. 1987. *Clean Air Act Amendments of 1987.* 100th Congress. Washington, DC.

USPHS. 1964. *Smoking and Health: Report.* Surgeon General's Advisory Committee on Smoking and Health, Publication no. 1103, U.S. Public Health Service, Washington, DC.

WHO. 1983. *Guidelines on Studies in Environmental Epidemiology. Environmental Health Criteria,* Report 27. World Health Organization, Geneva, Switzerland.

4. The Workplace

ACGIH. 2008. *TLVs and BEIs—Based on the Documentation of the Threshold Limit Values for Chemical Substances and Physical Agents & Biological Exposure Indices.* American Conference of Governmental Industrial Hygienists, Cincinnati, OH.

ATSDR. 2007. *ToxProfiles: 2007.* Computer discs. Agency for Toxic Substances and Disease Registry. Atlanta, GA.

Bass, Carolle, 2009. "Tiny Troubles: How Nanoparticles Are Changing Everything from Our Sunscreens to our Supplements." *E Magazine* XX, no. 3 (July/August): 21–24, 26, 28–29.

Burgess, William A. 1995. *Recognition of Health Hazards in Industry.* 2nd ed. New York: Wiley-Interscience.

CDC. 2001a. "Nonfatal Occupational Injuries and Illnesses Treated in Hospital Emergency Departments—United States, 1998." *Morbidity and Mortality Weekly Report* 50, no. 16 (27 April) 313–317. Centers for Disease Control and Prevention. Atlanta, GA.

———. 2001b. "Pesticide Related Illnesses Associated with the Use of a Plant Growth Regulator—Italy, 2001." *Morbidity and Mortality Weekly Report* 50, no. 39 (5 October), 845–847. Centers for Disease Control and Prevention. Atlanta, GA.

Clayton, George D., and Clayton, Florence. 2001. *Patty's Industrial Hygiene and Toxicology,* 5th ed., Hoboken, NJ: John Wiley & Sons.

CommonDreams.org. 2009. *Workplace Exposures Rise as OSHA Health Inspections Fall.* http://www.commondreams.org/newswire/2009/09/24-2. Accessed 27 October 2009.

Daum, Kent. 2004. "New Study Links Computer Vision and Productivity." *Industrial Hygiene News* 27, no. 2 (March), 15.

Fine, Lawrence J. 1996. "The Psychological Work Environment and Heart Disease" (editorial). *American Journal of Public Health* 86, no. 3 (March), 301–303.

Franco, Giuliano. 2001. "Bernardino Ramazzini: The Father of Occupational Medicine." *American Journal of Public Health* 91, no. 9 (September), 1382.

Hamilton, Alice. 1943. *Exploring the Dangerous Trades.* Little Brown. Boston, MA.

Helman, Greg. 2009. "Preliminary Findings Show Image of Carbon Nanotubes in Mouse's Lung." *Bureau of National Affairs* (27 March). National Institute for Occupational Safety and Health. Cincinnati, OH.

Herrick, Robert F. 1998. "Industrial Hygiene." In Wallace, Robert W., ed. *Maxcy-Rosenau-Last Public Health and Preventive Medicine,* 14th ed., pp. 661–667. Appleton and Lange. Norwalk, CT.

Keyserling W. M., Sudarsan, S. P., Martin, B. J., Haig, A. J., and Armstrong, T. J. 2005. "Effects of low back disability status on lower back discomfort during sustained and cyclical trunk flexion." *Ergonomics* 48(3), no. 3; 219–233.

Levine, Steven P. 2001. "An Industry Safety and Health Forgot: Health Care." *Synergist* 12, no 4. (April), 33–34.

NIOSH. 2004. *NIOSH ALERT:Preventing Deaths, Injuries, and Illnesses of Young Workers.* Publication No. 2003-128. National Institute for Occupational Safety and Health. Cincinnati, OH.

———. 2007a. *NIOSH ALERT: Simple Solutions: Ergonomics for Construction Workers.* Publication No. 2007-122. National Institute for Occupational Safety and Health. Cincinnati, OH.

———. 2007b. *NIOSH ALERT: Waste Anesthetic Gases: Occupational Hazards in Hospitals.* Publication No. 2007-151. National Institute for Occupational Safety and Health. Cincinnati, OH.

———. 2007c. *Injuries to Youth on U.S. Farm Operations, 2004.* Publication No. 2007-161. National Institute for Occupational Safety and Health. Cincinnati, OH.

———. 2009. *Approaches to Safe Nanotechnology: Managing the Health and Safety Concerns Associated with Engineered Nanomaterials.* Publication No. 2009-125. National Institute for Occupational Safety and Health. Cincinnati, OH.

NSC. 2009. *Injury Facts, 2009 Edition.* National Safety Council. Itasca, IL.

———. 2006. *Respiratory Protection.* Code of Federal Regulations, Title 29, Part 1910, 134, Occupational Safety and Health Administration, U.S. Department of Labor. Washington, DC.

OSHA. 2006. *Respiratory Protection.* Code of Federal Regulations, Title 29, Part 1910, 34, Occupational Safety and Health Administration, U.S. Department of Labor. Washington D.C.

Schlecht, P. C., and O'Connor, P. F., eds. 2003. *NIOSH Manual of Analytical Methods.* Publication 2003-154, 4th ed. 3rd Supplement. National Institute for Occupational Safety and Health, Cincinnati, OH. (pfo1@cdc.gov).

Seeley, Patricia. 2008. "Injuries Getting You Down? Move Safety Upstream." *Synergist* 19, no. 5, 71–73 (May).

U.S. Congress. 1913. *An Act to Create a Department of Labor.* Public Law 426, 62nd Congress. Washington, DC.

———. 1970. *Occupational Safety and Health Act of 1970.* Public Law 91-596, 91st Congress. Washington, DC.

———. 1976. *Toxic Substances Control Act.* Public Law 94-469, 94th Congress. Washington, DC.

———. 1990. *Pollution Prevention Act of 1990.* Public Law 101-508, 101st Congress. Washington, DC.

———. 2000. *Energy Employees Occupational Illness Compensation Program Act of 2000.* Public Law 106-398, 106th Congress. Washington, DC.

Wassell, James T., Gardner, Lynn I., Landsittel, Douglas P., Johnston, Janet J., and Johnston, Janet M. 2000. "A Prospective Study of Back Belts for Prevention of Back Pain and Injury." *Journal of the American Medical Association* 284, no. 21 (6 December), 2727–2732.

5. Indoor and Outdoor Air

Architecture Week. 2007. "Saving the Taj Mahal," C2.2 (May). http://www.architectureweek.com/2007/0509/culture_2–2.html. Accessed 27 January 2007.

Associated Press. 2009. "National Lake Fish Tissue Study." www.epa.gov/waterscience/fishstudy. *Sun Journal,* New Bern, NC (17 November), 9.

Biswas, Pratim, and Wu, Chang-Yu. 2005. "Nanoparticles and the Environment." *Journal of the Air & Waste Management Association* 55, 708–746.

Bortnick, S. M., Coutant, B. W., and Hanley, T. 2002. "Public Reporting of an Air Quality Index Using Continuous PM$_{2.5}$ Monitoring Data." *EM* (March), 27–33.

Boyes, R. 2009. "World Agenda: 20 Years Later, Poland Can Lead Eastern Europe Once Again. *The Times* (London). http://www.timesonline.co.uk/tol/news/world/world_agenda/article6430833.ece. Accessed 4 June 2009.

CEQ. 1997. Environmental Quality—The Twenty–fifth Anniversary Report of the Council on Environmental Quality, Executive Office of the President. Washington DC.

Dooley, Erin E., ed. 2002. "Biodiesel Bulldozes Ahead." *Environmental Health Perspectives* 110, no. 453 (August).

Dumyahn, Thomas S., Spengler, John D., Burge, Harriet A., and Muilenburg, Michael. 2000. *Comparison of the Environments of Transportation Vehicles: Results of Two Surveys*. Standard Technical Publication 1393, American Society for Testing and Materials. West Conshohocken, PA.

EPA. 2000. *National Emission Trends, 1900 to 1998*. Report EPA 454-R-00–002 (March). Environmental Protection Agency, Research Triangle Park, NC.

———. 2009. *National Ambient Air Quality Standards (NAAQS)*. http://www.epa.gov/air/criteria.html. Accessed 8 January 2009.

Findley, Roger W., and Farber, Daniel A. 2000. *Environmental Law in a Nutshell*, 5th ed., West Group Publishing Company. St. Paul, MN.

Garth, David E. 2001. "Letter to Nebraska Senators from San Luis Obispo Chamber of Commerce in Favor of Smokefree Legislation" (29 January). Tobacco.org. http://www.tobacco.org/News/010129garth.html. Accessed 7 April 2007.

Gilman, Sander L., and Xun, Zhou. 2004. *A Global History of Smoking*. Reaktion Books Limited, London, UK (November).

Gruenspecht, Howard K., and Stavins, Robert N. 2002. "New Source Review Under the Clean Air Act: Ripe for Reform." *Resources*, issue 147 (Spring), 19–23.

Helfand, William H. 2001. "Donora, Pennsylvania: An Environmental Disaster of the 20th Century." *American Journal of Public Health*, 91, no. 4 (April), 553

Jezouit, Debra J., and Frank, Joshua B. 2002. "BART Sent Back to the Drawing Board." *EM* (August), 30–33.

Keady, Patricia, and Halvorsen, Tom. 2000. "A New Tool for Eliminating Indoor Air Quality Complaints." *Journal of Nanoparticle Research* 2, no. 2 (June), 205–208.

Krupnick, Alan J. 2002. "Does the Clean Air Act Measure Up?" *Resources*, issue 147, 2–3 (Spring). Resources for the Future, Washington, DC.

Larsen, Ralph E. 2002. Personal communication. Environmental Protection Agency, Research Triangle Park, NC (June).

Latko, Mary Ann. 2000. "Guidelines and Resources for Indoor Air Quality Professionals." *EM* (September), 15–17.

Lohn, Martga. Associated Press. 2007. "Minnesota Lawmakers Approve Smoking Ban." *Boston Globe* (13 May).

Long, Christopher M. 2002. "Overview: Indoor Air Quality." *TRENDS: Risk Science & Application* 1–2 (Spring). Gradient Corporation, Cambridge, MA.

Lydersen, Karl. 2009a. "Low-Emission Locomotives May Boost Public Health." *Washington Post* (17 November), 12.

———. 2009b. "A Glut of Mercury Raises Fears." *Washington Post* (17 November), E1 and E6.

Moeller, Dade W. 1990. "Some Facts on Radon." *Radiation Protection Management* 7, no. 1 (January/February), 40–46

NPO Staff. 2009. *Workplace Smoking Bans and Restrictions*, National Program Office, Temple University Beasley School of Law. Philadelphia, PA (7 December).

NRC. 2001. Evaluating Vehicle Emissions Inspection and Maintenance Programs, National Research Council, National Academy Press, Washington, DC.

Pope, C. A., III, Burnett, R. T., Thun, M. J., Calle, E. , Krewski, D., Ito, K., and Thurston, G. D. 2002. "Lung Cancer, Cardiopulmonary Mortality, and Long-term Exposure to Fine Particulate Air Pollution." *Journal of the American Medical Association* 287, no. 9 (6 March), 1132–1141.

Proctor, R. N. 1997. "The Nazi War on Tobacco: Ideology, Evidence, and Possible Cancer Consequences." *Bulletin of the History of Medicine* 71, no. 3 (Fall), 435–488.

Service, Robert F. 2002. "Cleaning Air While Sparing Water." *Science* 296, issue 5567 (19 April), 463.

Spengler, J. D., and Sexton. K. 1983. "Indoor Air Pollution: A Public Health Perspective." *Science* 221 (1 July), 9–17.

State News. 2010. "EPA Gets Input on Plans to Regulate Coal Ash." *Sun Journal*, New Bern, NC (15 September), 11.

Stone, Richard. 2002. "Air Pollution: Counting the Cost of London's Killer Smog." *Science* 298, issue 5599 (13 December), 2106–2107.

USA Today. 2005. "San Francisco Enacts Broad Smoking Ban" (28 January).

U.S. Congress. 1987. *Ban on Smoking of Flights Less Than 2 Hours Duration. Amendments to the Transportation Appropriations Bill*. HR 2890 (13 July). Washington, DC.

———. 1990. *Clean Air Act Amendments*, Code of Federal Regulations, Title 42, Chapter 85. Washington, DC.

———. 2009. *Consumer Assistance to Recycle and Save Act; the Cash for Clunkers Program*. H. R. 1550 (8 June). Washington, DC.

Weinhold, Bob. 2002. "Fuel for the Long Haul?" *Environmental Health Perspectives*, 110, no. 8 (August), A 458-A 464.

West's Encyclopedia of American Law. 2010. http://www.answers.com/topic/air-pollution. Accessed 27 January 2010.

Wilkening, Kenneth E., Barrie, Leonard A., and Engle, Marilyn. 2000. "Trans-Pacific Air Pollution." *Science* 290, no. 5489 (6 October), 65 and 67.

6. Food

American Council on Science and Health. 2004. *Assessment of Thanksgiving Dinner Menu*. Brochure, New York, NY. http://www.acsh.org/publications/pubID.103/pub_detail.asp. Accessed 5 January 2010.

Associated Press. 2007. "U.S. Food Imports Rarely Inspected." *Sun Journal*, New Bern, NC (16 April), A7.

———. 2009a. "Officials Want to Decrease Salt Use." *Sun Journal*, New Bern, NC (23 April), A6

———. 2009b. "FDA Says E. coli in Nestlé Sample." *Sun Journal*, New Bern, NC (30 June), 19.

————. 2009c. "Government Tightening Food Safety Standards." *Sun Journal*, New Bern, NC (8 July), 14.

ATSDR. 2001. *Toxicological Profile for Polychlorinated Biphenyls*. Agency for Toxic Substances & Disease Registry. Centers for Disease Control and Prevention. Atlanta, GA.

Begley, S. 2009. "Born to be Big: Early Exposure to Common Chemicals May be Programming Kids to be Fat." *Newsweek* (21 September), 56–58, 62.

Besser, T., and Sischo, B. 2009. *Safe to Eat? The Science, and Economics of Food from Farm to Fork*. News Release (December). Washington State University. Pulman, WA.

Calafat, A.M., Kuklenyik, Z., Reidy, J.A., Caudill, S.P., Ekong, J., and Needham, L.L. 2005. "Urinary Concentrations of Bisphenol A and 4-Nonylphenol in a Human Reference Population." *Environmental Health Perspectives* 113, no. 4, 391–395 (April).

CDC. 2001. "Botulism Outbreak Associated with Eating Fermented Food–Alaska, 2001." *Morbidity and Mortality Weekly Report* 50, no. 32 (17 August), 680–682.

————. 2002. "Norwalk-Like Viruses–Associated Gatroenteritis in a Large, High-Density Encampment—Virginia, July 2002." *Morbidity and Mortality Weekly Report* 51, no. 30 (2 August), 661–663.

————. 2006a. "Diagnosis and Management of Foodborne Illnesses—A Primer for Physicians." *Morbidity and Mortality Weekly Report* 53, no. RR4 (16 April), 1–33.

————. 2006b. "Botulism Associated With Commercial Carrot Juice in Georgia and Florida." *Morbidity and Mortality Weekly Report* 55 (September), 1–2.

————. 2007. "Shigella Infection." http://emedicine.medscape.com/article/968773-overview. Accessed 8 July 2007.

————. 2008. "Illness Associated with Red Tide—Nassau County, Florida, 2007." *Morbidity and Mortality Report* 57, no. 26 (4 July), 717–720.

————. 2009a. "Multistate Outbreaks of Salmonella Infections Associated with Live Poultry—United States, 2007." *Morbidity and Mortality Report* 58, no. 2 (23 January), 25–29.

————. 2009b. "Multistate Outbreak of Salmonella Infections Associated with Peanut Butter and Peanut Butter–Containing Products—United States, 2008–2009." *Morbidity and Mortality Weekly Report* 58, no. 18 (29 April), 1–6.

Codex Alimentarius Commission. 2009. http://www.codexalimentarius.net/web/index_en.jsp. Accessed 5 January 2010.

Consumer Reports. 2007. "Dirty Birds." http://www.consumerreports.org/cro/food/food-safety/chicken-safety/chicken-safety-1-07/overview/0107_chick_ov.htm. Accessed 29 October 2009.

Davis, B., and Carpenter, C. 2009. "Proximity of Fast-Food Restaurants to Schools and Adolescent Obesity." *American Journal of Public Health* 99, no. 3 (March), 505–510.

Downes, F.P., and Ito, K. 2001. *Compendium of Methods for the Microbiological Examination of Foods*. 4th ed. American Public Association, Washington, DC.

Eubanks, M. 2002. "Allergies a la Carte—Is There a Problem with Genetically Modified Foods?" *Environmental Health Perspectives* 110, no. 3 (March), A130–A131.

Falkow, S, and Kennedy, D. 2001. Antibiotics, Animals, and People—Again! (editorial). *Science* 291, issue 5503 (19 January), 397.

FDA. 2005. *Food Code*. Food and Drug Administration, National Technical Information Service. Springfield, VA.

————. 2009. "Generally Recognized as Safe (GRAS)." http://www.fda.gov/Food/FoodIngredientsPackaging/GenerallyRecognizedasSafeGRAS/default.htm. Accessed 2 November 2009.

Federated Mills, Inc. 2010. "Sorbic Acid." http://www.federatedmills.com/sorbic-acid-pr-6.html. Accessed 9 January 2011.

Heymann, D.L., ed. 2008. *Control of Communicable Diseases Manual*. 19th ed. American Public Health Association. Washington, DC.

Hoffmann, S.A. 2009. "Attributing U.S. Foodborne Illness to Food Consumption." *Resources* 172 (Summer), 14–18. Resources for the Future. Washington, DC.

HSUS. 2009. *HSUS Fact Sheet: Antibiotics in Animal Agriculture and Human Health*. The Humane Society of the United States. Washington, DC. http://www.hsus.org/farm/resources/research/enviro/fact_sheet_antibiotics.html. Accessed 8 July 2010.

Kava, R., Ross, G.L., and Whelan, E.M., eds. 2009. *Obesity and Its Health Effects*. American Council on Science and Health, New York, NY.

Kliebenstein, D.J., and Rowe, H.C. 2009. "Plant Science: Anti-Rust Antitrust." *Science* 323, issue 5919 (6 March), 1301–1302.

Krattinger, S.G., Lagudah, E.S., Spielmeyer, W., Singh, R.P., Huerta-Espina, J,, McFadden, H., Bossolini, E., Selter, L.L., and Kelter, B. 2009. "A Putative ABC Transporter Confers Durable Resistance to Multiple Fungal Pathogens in Wheat." *Science* 323, issue 5919 (6 March), 1360–1363.

Loaharanu, P. 2003. *Irradiated Food*. 5th ed. American Council on Science and Health. New York, NY.

Lynch, M.F., Tauxe, R.V., and Hedberg, C.W. 2009. "The Growing Burden of Foodborne Outbreaks Due to Contaminated Fresh Produce: Risks and Opportunities." *Epidemiology and Infection* 137, no. 3, 307–315.

Marchione, M. 2009. "Staph Germs Found on West Coast Beaches." *Sun Journal*, New Bern, NC (13 September), 18.

Marshall, D.L., and Dickson, J.S. 1998. *Ensuring Food Safety*. In Wallace, Robert B., ed., *Maxcy-Rosenau-Last Public Health and Preventive Medicine*, 14th ed., 723–736. Appleton & Lange. Norwalk, CT.

Metcalfe, D.D. 2003. "Introduction: What are the Issues in Addressing Allergenic Potential of Genetically Modified Foods?" *Environmental Health Perspectives* 111, no. 8 (June), 1110–1113.

NRC. 2008. *National Plant Genome Initiative and New Horizons in Plant Biology*. National Research Council. National Academies of Science. Washington, DC.

NRC. 2009. Through Genomics, New Rice Can Impact World Hunger. National Research Council. National Academies of Science. Washington, DC.

Pooladi, P. 2009. "Through Genomics, New Rice Can Impact World Hunger." National Academies of Science. Washington, DC. http://www.nationalacademies.org/headlines/20090312.html. Accessed 9 January 2011.

Reid, I. 2009. "You Could Reduce Your Health Risks by Eating Tomatoes." *Jones Post*, Kinston, NC (30 July), 4.

Scudder, B.C., Chaser, L.C., Wentz, D.A., Bauch, N.J., Bringham, M.E., Moran, P.W., and Krabbenhoft, D.P. 2009. *Mercury in Fish, Bed Sediment, and Water from Streams Across the United States, 1998–2005*. U.S. Geological Survey. Washington, DC. http://pubs.usgs.gov/sir/2009/5109/. Accessed 11 January 2011.

Taylor, M.R., and Hoffman, S.A. 2001. "Redesigning Food Safety— Using Risk Analysis to Build a Better Food Safety System." *Resources* 144 (Summer), 13–16. Resources for the Future. Washington, DC.

Ulger, F., Esen, S., Dilek, A., Yanik, K., Gunaydin, M., and Leblebicioglu, H. 2009. "Are We Aware How Contaminated our Mobile Phones Are With Nosocomial Pathogens?" *Annals of Clinical Microbiology and Antimicrobials* 8, no. 7 (6 March).

U.S. Congress. 1999. "U.S. Regulatory Requirements for Irradiating Foods." *Federal Register* 64 (17 February), 7834. Washington, DC.

U.S. Department of Health and Human Services. 2009. "HHS Rolls out New Food-Safety Website." *The Nation's Health* 7 (November). American Public Health Association. Washington, DC.

Webster, P. 2009. "Playing Chicken with Antibiotic Resistance." *Turning Research into Solutions* 2, no. 5 (September/October), 16–19. Sara Miller McCune. SAGE Publications USA. Thousand Oaks, CA.

Weise, E. 2009. On tiny plots, a new generation of farmers emerges. *USA Today,* 1A and 1B (14 July). McLean, VA.

Weiss, Giselle. 2002. Acrylamide in Food: Uncharted Territory. *Science,* 297, issue 5578, 27 (5 July).

Wikipedia. 2009a. "Bovine Spongiform Encephalopathy." http://wikipedia.org.wiki/Bovine_spongiform_encephalopathy. Accessed 7 July 2009.

———. 2009b. "Minamata Disease." http://en.wikipedia.org/wiki/Minamata_disease. Accessed 2 November 2010.

7. Drinking Water

Adham, S., Chiu, K., Howe, K., Lehman, G., Mysore, C., and Clouet, J. 2006. *Optimization of Membrane Treatment for Direct and Clarified Water Filtration.* American Water Works Association Water Research Foundation. Denver, CO.

Agricultural Water Conservation Clearinghouse. 2009. "Water Supply, Sources, and Agricultural Use." http://www.agwaterconservation.colostate.edu/FAQs _WATER%20SUPPLYSOURCESAGRICULTURALUSE.acpx. Accessed 9 September 2009.

Babin, Steven, Burkom, Howard S., Mnatsakamyan, Zaruhi R., Ramac-Thomas, Liane C., Thompson, Michael W., Wojcik, Richard A., Lewis, Sheri Happel, and Yund, Cynthia. 2008. *Drinking Water Security and Public Health Disease Outbreak Surveillance. Technical Digest* 27, no. 4, 403–411. The Johns Hopkins University Applied Physics Laboratory. Howard County, MD.

Burt, Charles, and Styles, Stuart W. 2007. *Drip and Micro Irrigation Design and Management for Trees, Vines, and Field Crops,* 3rd ed. Irrigation Training and Research Center. California Polytechnic State University, San Luis Obispo, CA.

Canby, Thomas Y. 1980. *Water—Our Most Precious Resource. National Geographic* 158, no. 2 (August), 144–179.

CDC. 1994. *Cryptosporidium Infections Associated with Swimming Pools—Dane County, Wisconsin, 1993. Morbidity and Mortality Weekly Report* 43, no. 32 (12 August), 561–563.

————. 2001. *Prevalence of Parasites in Fecal Material from Chlorinated Swimming Pools—United States, 1999. Morbidity and Mortality Weekly Report* 50, no. 20 (25 May), 410–412.

————. 2009. *Drinking Water and Water-Related Diseases.* http://www.cdc.gov/nceh/ehhe/water/drinkingwatr.httm. Accessed 4 July 2009.

Cochran, G. A., and Cheney, R. A. 1982. *Mortality Trends in Philadelphia: Age- and Cause-Specific Death Rates, 1870–1930. Demography* 19, no. 1, 97–123.

Culp, G. L., and Culp, R. L. 1974. *New concepts in water purification.* Van Nostrand Reinhold Co., New York, NY.

Eaton, A. E., Clesceri, L. S., Rice, E. W., and Greenberg, A. E. 2005. *Standards Methods for the Examination of Water and Wastewater,* 21st ed. American Public Health Association. Washington, DC.

EPA. 2001. *Public Health and Environmental Radiation Protection Standards for Yucca Mountain, NV; Final Rule.* Code of Federal Regulations, Title 40, Part 197. Environmental Protection Agency, Washington, DC.

————. 2003. *Bacterial Water Quality Standards for Recreational Waters (Freshwater and Marine Waters).* Report EPA-823-R-03-008 (June). Environmental Protection Agency, Washington, DC.

————. 2009. *Safe Drinking Water Act.* www.epa.gov/safewater. Accessed 1 July 2009.

Fleming, Hu. 2002. "Out of the Dark: Ultraviolet Disinfection Lights Up Municipal Drinking Water Treatment Systems." *Environmental Protection,* 13, no. 3 (March), 46–53.

Frontius, Julius. A. D. 97. *The Water Supply of the City of Rome.* 1973. Clemens, H., trans. New England Water Works Association. Boston, MA.

Glacier, Aviva. 2006. *Threatened Waters—Turning the Tide on Pesticide Contamination.* (February 2006). http:// www.beyondpesticides.org. Accessed 30 August 2009.

Greenberg, Michael R. 2009. *Why is Water an Issue? American Journal of Public Health* 99, no. 11 (November), 1927.

Keach, William, ed. 1997. *The Complete Poems/Samuel Taylor Coleridge.* 498. Penguin Books. City of Westminster. London, England.

Larmer, Brook. 2010. "The Big Melt." National Geographic 217, no. 4 (April), 60–79.

Laughlin, James, ed. 2001. "Alum Replacement Gains Popularity at Municipal Plants." *Water World* (November), 22–23.

Lee, Sherline H., Levy, Deborah A., Craun, Gunther F., Beach, Michael J., and Calderon, Rebecca L. 2002. "Surveillance for Waterborne-Disease Outbreaks—United States, 1999–2000." *Morbidity and Mortality Report* 51, no. SS-8 (22 November), 1–47.

Leland, D. E., and Damewood III, M. 1990. "Slow Sand Filtration in Small Systems in Oregon." *AWWA Journal* 82, no. 6 (June), 50–59.

Montaigne, Fen, 2002. "Water Pressure." *National Geographic* 202, no. 3 (September), 2–33.

Nationalatlas.gov. 2009. "Water Use in the United States." http://nationalatlas.gov/articles/water/a_wateruse.html. Accessed 9 September 2009.

Nordstrom, D. Kirk. 2002. "Worldwide Occurrences of Arsenic in Ground Water." *Science,* 296, issue 5576 (21 June), 2143, 2145.

Oates, Wallace E. 2002. "The Arsenic Rule: A Case for Decentralized Standard Setting?" *Resources,* issue 147 (Spring), 16–18. Resources for the Future. Washington, DC.

Rogers, Peter. 2008. "Facing the Fresh Water Crisis." *Scientific American,* 299, no. 2 (August), 46–53.

Rosenberg, Tina. 2010. "The Burden of Thirst." National Geographic 217, no. 4 (April), 96–115.

Royte, Elizabeth. 2010. "The Last Drop." *National Geographic* 217, no. 4 (April), 172–176.

Simon, Paul. 2001. "Thirsty World." *Environmental Protection* 12, no. 11 (November), 12–13.

Staub, Emily. 2008. "Guinea Worm Cases Hit All-Time Low: Carter Center, WHO, Gates Foundation, and U.K. Government Commit $55 Million Toward Ultimate Eradication Goal." News Release (5 December). The Carter Center. Atlanta, GA.

Surawicz, Christina. 2009. *Acute and Infectious Diarrhea.* GI Board Review Course (30 September). Washington, DC.

USGS. 2009. "Where is Earth's Water Located?" Science for a Changing World. U.S. Department of the Interior, U.S. Geological Survey. Washington, DC. http://ga .water.usgs.gov/edu/earthwherewater.html. Accessed 3 July 2009.

Wikipedia. 2009. "Water Fluoridation." http://en.wikipedia.org/wiki/Water_fluoridation. Accessed 17 August 2009.

8. Liquid Wastes

Ai, H., Wang, Q., Fan, X., Xie, W., and Shinohara, R. 2005. "Effect of Hydraulic Retention Time on the Efficiency of Vertical Tubular Anaerobic Sludge Digester Treating Waste Activated Sludge." *Environmental Technology* 26, no. 7 (July), 725–732.

Bass, Carolle. 2009. "Tiny Troubles: How Nanoparticles are Changing Everything From our Sunscreen to our Supplements." *E Magazine* XX, no. 3 (July/August), 21–24, 26, 28–29.

Book, Sue. 2009. "Neuse River Kill May be Slowing." *Sun Journal,* New Bern, NC (18 September), 7.

CEQ. 1993. *23rd Annual Report, Council on Environmental Quality,* unnumbered figure, 5. The 1993 Report of the Council on Environmental Quality, Executive Office of the President. Washington, DC.

———. 1998. *Environmental Quality—The World Wide Web.* The 1997 Report of the Council on Environmental Quality, Executive Office of the President. Washington, DC.

Circuit Court. 2005. *Second Circuit Court Decision on CAFOs: Waterkeeper Alliance v EPA.* (399 F 3d 486) (28 February).

Clivus Multrum, Inc. 2011. Homepage. http://www.clivusmultrum.com. Accessed 11 January 2011.

Dix, Stephen P. 2001. "Onsite Wastewater Treatment: A Technological and Management Revolution: Part 1." *Water Engineering & Management* 148, no. 9 (September), 24–26, 28.

Eaton, A. E., Clesceri, L. S., Rice, E. W., and Greenberg, A. E. 2005. *Standard Methods for the Examination of Water and Wastewater.* 21st ed. American Public Health Association. Washington, DC.

Edwards, Peter. 1992. *Reuse of Human Wastes—in Aquaculture—A Technical Review. Water and Sanitation Report no. 2.* UNDP-World Bank Water and Sanitation Program, Washington, DC.

EPA. 1998. *Concentrated Animal Feeding Operations, Final Rulemaking.* CFR 40, Parts 122.21, and 122.26(c)(1)(i). Environmental Protection Agency. Washington, DC.

———. 2006. *U.S. Census Data on Small Community Housing and Wastewater Practices.* Report 832-F-99-060. Originally based on 1990 Census, but updated in 2006. Office of Water Resources. Environmental Protection Agency. Washington, DC.

———. 2009a. *Quality of our Nation's Waters: A Summary of National Water Quality Inventory: 1998.* Environmental Protection Agency. Washington, DC.

———. 2009b. *National Pollution Discharge Elimination System: Stormwater Program.* Office of Wastewater Management. Environmental Protection Agency. Washington, DC. http://cfpub.epa.gov/npdes/home.cfm?program_id=6. Accessed 3 February 2009.

Fernando, J. 2008. *Determination of Coefficient of Permeability from Soil Percolation Test.* 12th International Conference of International Association for Computer Methods and Advances in Geomechanics. Goa, India (1–6 October).

Freeman, A. Myrick, III, 2000. "Water Pollution Policy." In Portney, Paul R., and Stavins, Robert N., eds., *Public Policies for Environmental Protection,* 2nd ed., 169–213. Resources for the Future, Washington, DC.

Furukawa, David H. 1999. "The Key to Future Water Supplies: Desalination." *World of Water 2000: The Past, Present, and Future.* Supplement to PennWell Magazine, 148–150. Published by WaterWorld and Water & Wastewater International, Tulsa, OK.

Gloyna, Ernest. 1971. *Waste Stabilization Ponds.* Monograph Series no. 60. World Health Organization, Geneva, Switzerland.

Gray, Albert C. 1999. "Wastewater—Watershed Management Is Key to Next Millennium." *World of Water 2000: The Past, Present, and Future.* Supplement to *PennWell Magazine,* 114–118. Published by WaterWorld and Water & Wastewater International, Tulsa, OK.

Griffiths, Charles. 2003. *MultiCriteria Cost Effectiveness Using Water Quality Index.* EPA NCEE Symposium on Cost Effective Analysis for Multiple Benefits (9 September). Environmental Protection Agency and North Carolina Ecosystem Enhancement Program. Willard Hotel, Washington, DC.

Guy, Brenda, and Catanzaro, Mike. 2002. "Proper Operation, Maintenance, and Servicing of Aerobic Wastewater Treatment Systems," *Journal of Environmental Health* 64, no. 8 (April), 23–24.

Hetrick, Scott. 2001. *Bio-Kinetic Wastewater Treatment Systems,* Norweco, Norwalk, OH.

Holden, Constance. 2002. "Random Samples: Dead Zone Grows." *Science* 297, issue 5584 (16 August), 1119.

ISO. 2007. *Environmental Management Systems.* ISO 14001. International Organization for Standards, Washington, DC.

Logan, Terry J. 1999. "Biosolids—Challenges and Options in the Next Millennium." *World of Water 2000: The Past, Present, and Future.* Supplement to *PennWell Magazine,* 130, 132–134. Published by WaterWorld and Water & Wastewater International, Tulsa, OK.

Lovley, Derek R. 2001. "Anaerobes to the Rescue." *Science* 293, issue 5534 (24 August), 1444–1446.

NRC. 2008. *Urban Stormwater Management in the United States.* Prepublication copy (15 October). National Research Council. National Academies Press, Washington, DC.

Pitois, S., Jackson, M. H., and Wood. J. B. 2001. "Sources of Eutrophication Problems Associated with Toxic Algae: An Overview." *Journal of Environmental Health* 64, no. 5 (December), 25–32.

Rabalais, Nancy. 2010. "Gulf of Mexico Dead Zone Smaller Than Expected, But Severe." Science Daily. http://www.sciencedaily.com/ Accessed 2 February 2010.

Ruiz, Gregory M., Rawlings, Tonya R. Dobbs, Fred C., Drake, Lisa A., Mullady, Timothy, Huq, Anwarul, and Colwell, Rita R. 2000. "Global Spread of Microorganisms by Ships" (letter to editor). *Nature* 408, no. 6808 (2 November), 49.

Sakamoto, Gail. 2000. "UV Disinfection of Reclaimed Wastewater: The North American Experience." *Environmental Protection* 11, no. 10 (October), 20–25.

Saravanane, R., and Tamijevendane, S. 2009. "Antibiotic Liquid Waste Disposal—A Potential Threat and Environmental Compatibility." *Current Science* 96, no. 10 (25 May).

Satchell, Michael. 1996. "Hog Factories: Cheap Meat, Costly Problems." *U.S. News and World Report,* 58–59 (22 January).

Sims, Danny, and Bentley, Dennis E. 2001. "Relieving Wastewater Treatment Capacity Constraints through Improved Processing of Aerobically Digested Solids," *EM* (November), 32–36.

Smith, Daniel P. 2009. "Modular Nitrogen Removal in Distributed Sanitation Water Treatment Systems." *Environmental Engineer,* 8 (Spring).

Toyota Motor Corporation. 2007. *Environmental Update, 45th Issue* (August). Torrance, CA.

U.S. Congress. 1972. *Federal Water Pollution Control Act Amendments.* Public Law 92-500. Washington, DC.

———. 1976. *Resource Conservation and Recovery Act.* Public Law 84-580, 40 USC 6901 et seq. Washington, DC.

———. 1977. *Clean Water Act.* Public Law 95-217. Washington, DC.

———. 1990. *Pollution Prevention Act of 1990.* Public Law 101-508. Washington, DC.

WaterWorld. 2009. "Addressing Combined, Sanitary Sewer Overflows in the Urban Environment." *WaterWorld* 25, no. 2 (February), 1, 11.

Weinhold, Bob. 2002. "Water Pollution: Up a Chemical Creek." *Environmental Health Perspectives* 110, no. 7 (July), A 390.

Wikipedia. 2009. "Sewage Treatment or Domestic Wastewater Treatment." http://en.wikipedia.org/wiki/html. Accessed 20 December 2009.

Wolfe, Pamela. 1999. "History of Wastewater." *World of Water 2000: The Past, Present, and Future.* Supplement to *PennWell Magazine,* 24–36. Published by WaterWorld and Water & Wastewater International, Tulsa, OK.

9. Solid Wastes

Black, Harvey. 2002. "The Hottest Thing in Remediation." *Environmental Health Perspectives* 110, no. 3 (March), A 146–A 148.

CEQ. 1997. *Environmental Quality: The Twenty-Fifth Anniversary Report of the Council on Environmental Quality.* Executive Office of the President. Washington, DC.

Chen, William. 2002. "Greyfields: Revitalizing Communities through Mixed-Use Redevelopment." *EM* (October), 29–30.

Chu, Steven. 2009. "Appointment of Blue Ribbon Panel." News Release (29 January). U.S. Department of Energy. Washington, DC.

Debnam, Betty. 2009. "Sorting Through Recycling." *Sun Journal,* New Bern, NC (1 September), 8.

Editorial Staff. 2009. "The Landfill Lady: Debra Reinhart, Ph.D., PE." *Environmental Engineer* 45, no. 4 (Fall), 10–17.

———. 2010. "DOE Withdraws Repository License Application." *Nuclear News* 53, no. 4 (April), 63.

EPA. 1986a. *RCRA Orientation Manual.* Report EPA/530-SW-86-001. Environmental Protection Agency. Washington, DC.

———. 1986b. *Solving the Hazardous Waste Problem: EPA's RCRA Program.* Report EPA/530-SW-86-037. Environmental Protection Agency. Washington, DC.

———. 1993a. *Safer Disposal for Solid Waste: The Federal Regulations for Landfills.* Report EPA/530-SW-91-092. Environmental Protection Agency. Washington, DC.

———. 1993b. "Guidance to Hazardous Waste Generators on the Elements of a Waste Minimization Program." Environmental Protection Agency. *Federal Register* 58, no. 102 (28 May), 31114–31120. Environmental Protection Agency. Washington, DC.

———. 2008. "Public Health and Environmental Radiation Standards for Yucca Mountain, Nevada: Final Rule." *Federal Register* 73, no. 200 (15 October), 61256–61289. Environmental Protection Agency. Washington, DC.

———. 2009. Report on the Environment. "Quantity of Municipal Solid Waste Generated and Managed." http://cfpub.epa.gov/eroe/index.cfm?fuseaction=detail. viewInd&lv=list.listByAlpha&r=216598&subtop=228. Accessed 25 December 2009.

Fox, Robert D. 1996. "Physical/Chemical Treatment of Organically Contaminated Soils and Sediments," *EM* (May), 28–34.

Golaine, Andrea. 1991. *Superfund: Money Squandered in the Name of Public Health. Priorities,* 31–34 (Fall). American Council on Science and Health. New York.

Greenpeace. 2007. "Greenpeace Investigates Levels of Toxic Materials Found in Laptop Computers." http://74.125.113.132/search?q=cache:NHr38xI7IoJ:www .associatedcontent.com/article/426952/greenpeace_investigations_levels_of_ toxic.html. Accessed 31 August 2009.

Harris, Mark. 2008. "E-mail from America: Buy-Back Gadgets." *Sunday Times.* Seattle, WA. http://technology.timesonline.co.uk/tol/news/tech_and_web/personal_ tech/article4538181.ece. Accessed 10 March 2009.

Hayden, Thomas. 2002. "Science and Technology—Trashing the Oceans." *U.S. News and World Report* 133, no. 17 (4 November), 58–60.

Holden, Arthur, Park, Chu, June-Soo, Kim, Vivian, Choi, Michelle, Shi, Grace, Chin, Yating, Chin, Tiffany, Chun, Christina, Linthicum, Janet, Walton, Brian J., and

Hooper, Kim. 2009. "Unusual Hepta- and Octabrominated Diphyl Ethers and Nonabrominated Diphenyl Ether Profile in California, USA, Peregrine Falcons (Falco Peregrinus): More evidence for Brominated Diphenyl Ether-209 Debromination." *Environmental Toxicology and Chemistry* 28, no. 9, 1906–1911.

Hughes, Joseph B. 1996. "Biological Treatment of Hazardous Waste." *Frontiers in Engineering*, 37–39. National Academy Press. Washington, DC.

Huo, Xia, Peng, Lin, Xu, Xijin, Zheng, Liangkai, Qiu, Bo, Qi, Zongli, Zhang, Bao, Han, Dai, and Piao, Zhongxian. 2007. "Elevated Blood Lead Levels of Children in Guiyu, an Electronic Wasteland Recycling Town in China." *Environmental Health Perspectives* 115, no. 7, 1113–1117.

Isler, Margaret, and Lee, Martin R. 2002. *Environmental Protection Issues in the 107th Congress* (Updated July 3, 2002). Congressional Research Service. Library of Congress. Washington, DC.

Jurdi, Mey. 2002a. "Transboundary Movement of Hazardous Wastes into Lebanon, Part 1: The Silent Trade." *Journal of Environmental Health* 64, no. 1 (January–February), 9–14.

———. 2002b. "Transboundary Movement of Hazardous Wastes into Lebanon, Part 2: Environmental Impacts and the Need for Remedial Actions." *Journal of Environmental Health* 64, no. 1 (January–February), 15–19.

Lavelle, Marianne. 2002. "Arsenic and Barbeque." *U.S. News and World Report 133*, no. 10 (16 September), 58–59.

Le, Phuong, Associated Press. 2009. "Researcher's Track 3,000 Pieces of Seattle Trash." *Sun Journal*, New Bern, NC (14 September), 13.

Link-Wills, Kimberly. 2002. "Revolutionary Technology." *Georgia Tech Alumni Magazine* 79, no. 1 (Summer), 45–47.

Moeller, Dade W. 2009. "Lauriston S. Taylor Lecture: Yucca Mountain Radiation Standards, Dose/Risk Assessments, Thinking Outside the Box, and Recommendations." *Health Physics* 97, no. 5 (November), 376–391.

———. 2010. "The Tragedy of Yucca Mountain." *Radwaste Solutions* 17, no. 5 (September/October), 52–57.

———. 2011. "The Tragedy of Yucca Mountain. Part II." *Radwaste Solutions* 18, no. 1 (January/April), 25–30.

Moody, Dave, and Sharif, Farok. 2009. "WIPP @ 10: Securing the Past—Piloting the Future." *Radwaste Solutions* 16, no. 3 (May/June), 16–19.

Morgan, Russell. 2006. "Tips and Tricks for Recycling Old Computers." SmartBiz. http://www.smartbiz.com/article/articleprint/1525.-1/58. Accessed 17 March 2006.

NCRP. 2003. *Management Techniques for Laboratories and Other Small Institutional Generators to Minimize Off-Site Disposal of Low-Level Radioactive Waste.* Report no. 143. National Council on Radiation Protection and Measurements. Bethesda, MD.

O'Connell, Kim A. 2003. "Poison Planks." *Waste Age* 34, no. 10 (October), 40–44.

Ogunseitan, Oladele, Schoenung, Julie M., Saphores, Jean-Daniel M., and Shapiro, Andrew W. 2009. "The Electronics Revolution: From E-Wonderland to E-Wasteland." *Science* 326, issue 5953 (30 October), 670–671.

Padgett, C. B. 2001. "Oregon Survey Trumpets State's Efforts." *Waste Age* 32, no. 12 (December), 6–7.

Portney, P. R, and Stavins, R. N. 2000. "Introduction." In Portney, P. R, and Stavins, R. N., eds., *Public Policies for Environmental Protection*, 2nd ed., 1–10. Resources for the Future. Washington, DC.

Shea, Cynthia. 1988. "Plastic Waste Proliferates." *World-Watch* I, no. 2 (March–April), 7–8.

Sperber, JoAnn. 2002. "Sweden, Finland Pursue Deep Geologic Repository." *Nuclear Energy Insight* (August), 2. Nuclear Energy Institute. Washington, DC.

Tom, Patricia Anne. 2001. "Good Wood Gone Bad." *Waste Age* 32, no. 8 (August), 36–51.

U.S. Congress. 1976. *Resource Conservation and Recovery Act*. Public Law 84-580, 40 USC 6901 et seq. Washington, DC.

———. 1980. *Comprehensive Environmental Response and Liability Act*. Public Law 107-296. Washington, DC.

———. 1987. *Nuclear Waste Policy Act*, Public Law 100-203. Washington, DC.

———. 1990. *Pollution Prevention Act*, Public Law 101-508. Washington, DC.

———. 1992. *Federal Facility Compliance Act*. Public Law 102-386, 42 USC 6901 et seq. Washington, DC.

USDOE. 1999. *Implementation Guide for Use with DOE M 435.1*. Publication DOE G 435.1. U.S. Department of Energy, Washington, DC.

USNRC. 2009. *2009–2010 Information Digest*. Report NUREG-1350 21, 74–76. U.S. Nuclear Regulatory Commission. Washington, DC.

Verbit, S. R. 2001. "New Law May Unlock Potential of Brownfields." *Environmental Protection* 12, no. 8 (August), 31–33.

Wolpin, Bill. 2002. "A Moveable Beast" (editorial). *Waste Age* 33, no. 3 (March), 4.

10. Animals, Insects, and Related Pests

Associated Press. 2010. "Without a Wing, There's Not Much of a Prayer for Female Mosquitoes." *The Washington Post* (9 March), E3.

ATSDR. 2002a. *Toxicological Profile for Aldrin/Dieldrin (Update)*. U.S. Department of Health & Human Services, Agency for Toxic Substances and Disease Registry. Atlanta, GA.

———. 2002b. *Toxicological Profile for DDT/DDD/DDE (Update)*. U.S. Department of Health & Human Services, Agency for Toxic Substances and Disease Registry. Atlanta, GA.

Bill & Melinda Gates Foundation. 2010. "Bill and Melinda Gates Foundation Commits $10 Billion for Vaccine Research and Development." http://www.gates foundation.org/vaccines/Pages/default.aspx. Accessed 3 February 2010.

Breman J., Alilio, M., and Mills, A. 2004. "Conquering the Intolerable Burden of Malaria: What's New, What's Needed: A Summary." *American Journal of Tropical Medicine & Hygiene* 71, 1–15.

Breman J. G., and Holloway, C. N. 2007. "Malaria Surveillance Counts." *American Journal of Tropical Medicine & Hygiene* 77 (Supplement 6), 36–47.

Brown, Peter. 2009. "Night Stalker." *Scientific American* 301, no. 2 (August), 16, 18.

Campbell, Carlos C. 2009. "Mosquitoes Against Malaria?" (editorial). *New England Journal of Medicine* 364, no. 5 (30 July), 522–525.

Canby, Thomas Y. 1977. "The Rat –Lapdog of the Devil." *Reader's Digest* 11 (July), 60–87.

Carter Center. 2009. "The Carter Center River Blindness (Onchocerciasis) Program." http://www.cartercenter.org/health/river_blindness/index.htm. Accessed 8 September 2009.

CDD. 2008. "Surveillance for Lyme Disease—United States, 1992–2006." *Morbidity and Mortality Weekly Report* 57, no. SS10 (3 October), 1–9.

———. 2009. "Dengue Fever." http://www.cdc.gov/ncidod/dvbid/dengue/. Accessed 8 September 2009.

Conniff, Richard. 1977. "The Malevolent Mosquito." *Reader's Digest* 111, no. 66 (August), 153–157.

Diawara, Lamine, Traore, Mamadou O., Badji, Alioune, Bissan, Yiriba, Doumbia, Konimbad, Goita, Soula F. Konate, Lassana, Mounkoro, Kalifa, Sarr, Moussa D., Seek, Amadou F., Seck[1], Amadou F., Toé, Laurent, Tourée, Seyni, and Remme, Jan H. F. 2009. "Feasibility of Onchocerciasis Elimination with Ivermectin Treatment in Endemic Foci in Africa: First Evidence from Studies in Mali and Senegal." *PLoS Neglected Tropical Diseases* 3, no. 7 (July), 497.

Dougherty, Elizabeth. 2009. "New Target Against Flu Viruses May Extend Vaccine Potency." *Focus* (6 March), 1 and 6. News from Harvard Medical, Dental, and Public Health Schools. Boston, MA.

Duplaix, Nicole. 1988. "Fleas, the Lethal Leapers." *National Geographic* 173, no. 5 (May), 672–694.

Editorial Staff. 2009. "Briefings, New Clues in Bee Deaths." *Time* 174, no. 9 (7 September), 13.

Enserink, Martin. 2000. "Malaysian Researchers Trace Nipah Virus Outbreak to Bats." *Science* 289, issue 547 (28 July) 9, 518–519.

Galvin, Thomas J., and Wyss, John H. 1996. "Screwworm Eradication Program in Central America." *Annals American Academy of Science* 791, 233–240.

Graham, L. C. 2009. "Phorid Fly Update." Alabama Cooperative Extension System, Auburn, AL. http://www.aces.edu/dept/extcomm/newspaper/feb22a02.html. Accessed 7 October 2009.

Hayashi, Alden M. 1999. "Attack of the Fire Ants." *Scientific American* 280, no. 2 (February), 26, 28.

Holden, Constance. 2008. "Random Samples: Place Your Bats." *Science 320*, issue 5877 (9 May), 725.

IAEA. 1999. *Thematic Plan for IAEA Technical Cooperative Sterile Insect Technical Cooperation: Sterile Insect Technique for Old and New World Screwworm*. Report TP-NA-D4-01 1998. International Atomic Energy Agency, Vienna, Austria.

Korb, Judith, Weil, Tobias, Hoffmann, Katharina, Foster, Kevin R., and Rehli, Michael. 2009. "A Gene Necessary for Reproductive Suppression of Termites." *Science*, 324, issue 5928 (8 May), 758.

Kristiansen, Kirsten, and Skovmand, Ole. 1985. "A Method for the Study of Population Size and Survival Rate of Houseflies." *Entiomologia Experimentalis et Appocata* 38 (August), 145–150. Springer, Netherlands.

Lecrubier, Aude. 2002. "Pest Control: How to Fool a Fly." *Popular Science* 260, issue no. 1 (January), 42.

Marques, A. 2008. "Chronic Lyme Disease: A Review." *Infectious Disease Clinics of North America*, 22, issue 2, 341–360.

Matuschka, Franz-Rainer, and Spielman, Andrew. 1993. "Risk of Infection from and Treatment of Tick Bite." *Lancet* 342, no. 8870 (28 August), 520–530.

Matuschka, Franz-Rainer, Endepols, Stefan, Richter, Dania, and Spielman, Andrew. 1997. "Competence of Urban Rats as Reservoir Hosts for Lyme Disease Spirochetes." *Journal of Medical Entomology* 34, no. 4 (July), 489–493.

Medical News TODAY. 2009. "A Milestone Toward Ending River Blindness in the Western Hemisphere by 2012: Escuintla, Guatemala Biggest Endemic Area Yet to Stop Transmission." http://www.medicalnewstoday.com/articles/144278. php. Accessed 8 September 2009.

Molyneux P., Hotez, J., and Fenwick, A. 2005. "Rapid-impact Interventions: How a Policy of Integrated Control for Africa's Neglected Tropical Diseases Could Benefit the Poor." *PLoS Medicine* 2, no. 11, e336.

Normile, Dennis. 2009. "Scientists Puzzle Over Ebola-Reston Virus in Pigs." *Science* 323, issue 5913 (23 January), 451.

Richardson, R. H., Ellison, J. R., and Averhoff, W. W. 1982. "Autocidal Control of Screwworms in North America." *Science* 215, issue 4531 (22 January), 361–370.

Roses, Mirta. 2009. "PAHO/WHO Makes an Appeal to Fight Dengue." *Disasters: Preparedness and Mitigation in the Americas*, no. 111 (April), 7.

Russell, Colin A., Jones, Terry C., Barr, Ian G., Cox, Nancy J., Garten, Rebecca, J., Gregory, Vicky, Gust, Ian D., Hampson, Alan W., Hay, Alan J., Hurt, Aeron, C., de Jong, Jan C., Kelso, Anne, Klimov, Alexander I., Kageyama, Tsutomu, Komadina, Naomi, Lapedes, Alan S., Lin, Yi P., Musterin, Ana, Obuchi, Masatsugu, Obuchi, Odagiri, Takato, Osterhaus, Albert, D. M. E., Rimmelzwaan, Guus F., Shaw, Michael W., Skepner, Eugene, Stohr, Klaus, Tashiro, Masato, Fouchier, Ron, A. M., and Smith, Derek J. 2008. "The Global Circulation of Seasonal Influenza A (H3N2) Viruses." *Science* 320, issue 5874 (18 April), 340–346.

Satchell, Michael. 2000. "Rocks and Hard Places—DDT: Dangerous Scourge or Last Resort?" *U.S. News & World Report* 129, no. 23 (December), 64–65.

Shaw, Jonathan. 2001. "The Landscape Infections." *Harvard Magazine* 104, no. 2 (November/December), 42–47.

Spielman, Andrew, and Kimsey, Robert B. 1997. "Zoonosis." In *Encyclopedia of Human Biology*, 2nd ed., 803–812. Academic Press, New York, NY.

Spielman, Andrew, and D'Antonio, Michael. 2001. *Mosquito: A Natural History of Our Most Persistent and Deadly Foe.* Hyperion. New York, NY.

Telford, Sam R. III, Pollack, Richard J., and Spielman, Andrew. 1991. "Emerging Vector-Borne Infections." *Infections Disease Clinics in North America* 5, no. 1, 7–17.

WHO. 2009. *Fact Sheet No. 117,* 1211 (March). World Health Organization. Geneva, Switzerland.

Williams-Guillen, Kimberly, Perfecto, Ivette, and Vandermeer, John. 2008. "Bats Limit Insects in a Neotropical Agroforest System." *Science* 320, issue 5872 (4 April), 70.

Wilson, E. O. 2002. "Hot Spots—Preserving Pieces of a Fragile Biosphere." *National Geographic* 201, no. 1 (January), 86–89.

Wolfe, Nathan. 2009. "Preventing the Next Pandemic." *Scientific American* 300, no. 4 (April), 76–81.

Zimmerman, Robert. 2009. "Biological Struggle to Solve Bat Deaths." *Science* 324, issue 5931 (29 May), 1134–1135.

11. Injury Prevention and Control

Associated Press. 2010. "Highway Deaths at Lowest Levels Since 1950s." *Sun Journal,* New Bern, NC (12 March), 9.

Berg, Katherine, Hines, Marilyn, and Allen, Susan. 2002. "Wheelchair Users at Home: Few Home Modifications and Many Injurious Falls." *American Journal of Public Health* 2, no. 1 (January), 48.

CBS News. 2009. *60 Minutes* (2 August).

CDC. 1997. *Childhood Pedestrian Deaths during Halloween—United States, 1975–1996, Morbidity and Mortality Weekly Report* 46, no. 42, 987–990 (24 October). Centers for Disease Control and Prevention. Atlanta, GA.

———. 1998. "Deaths Resulting from Residential Fires and Prevalence of Smoke Alarms—United States 1991–1995." *Morbidity and Mortality Weekly Report* 47, no. 38 (2 October), 803–806. Centers for Disease Control and Prevention. Atlanta, GA.

———. 1999a. "Playground Safety-United States,1991–1995." *Morbidity and Mortality Weekly Report* 47, no. 39 (2 October), 803–806. Centers for Disease Control and Prevention. Atlanta, GA.

———. 1999b. "Playground Safety-United States, 1998–1999." *Morbidity and Mortality Weekly Report* 48, no. 16 (30 April), 329–332. Centers for Disease Control and Prevention. Atlanta, GA.

———. 2000. "Unpowered Scooter Related Injuries—United States, 1998–2000." *Morbidity and Mortality Weekly Report* 49, no. 19 (17 May), 1109–1110. Centers for Disease Control and Prevention. Atlanta, GA.

———. 2001. "Achievements in Public Health, 1900–1999: Motor-Vehicle Safety: A 20th-Century Public Health Achievement." *Morbidity and Mortality Weekly Report* 48, no. 18 (14 May), 36–374. Centers for Disease Control and Prevention. Atlanta, GA.

Davis. David E., Jr. 2009. "Pitiless Advice for This Rattner." *Car and Driver* 155, no. 2 (August), 22.

De Haven, H. 1942. "Mechanical Analysis of Survival in Falls from Heights of Fifty to One Hundred and Fifty Feet." *War Medicine* 2, 586–596.

Dingus, T. A., Neale, T. A., Sudweeks, V. L., and Ramsey, R. R. 2006. *The 100-Car Naturalistic Driving Study, Phase II—Results of the 100-Car Field Experiment.* Report DOT HS 810 53, Virginia Tech Transportation Institute. Blacksburg, VA.

Gibson, J. J. 1961. "Contribution of Experimental Psychology to the Formulation of the Problem of Safety: A Brief for Basic Research." In Herbert H. Jacobs, ed., *Behavioral Approaches to Accident Research,* Association for the Aid of Crippled Children, New York, NY.

Haddon, William, Jr. 1970. "On the Escape of Tigers: An Ecologic Note." *American Journal of Public Health* 60, no. 12 (December), 2229–2234.

Harvard School of Public Health. 2009. Program of the National Violent Firearms Injury Statistical System. http://www.hsph.harvard.edu/hicrc/nviss/. Accessed 12 October 2009.

Homer, Jenny, and French, Michael. 2008. "Motorcycle Helmet Laws in the United States from 1990 to 2005: Politics and Public Health." *American Journal of Public Health*, no. 3, 415–423.

ICRP. 1984. *Protection of the Public in the Event of Major Radiation Accidents: Principles for Planning*. Publication 40. International Commission on Radiological Protection. Pergamon Press. Elmsford, NY.

IIHS. 1994. "Best and Worst 1982–92 Passenger Vehicles with Lowest and Highest Drive Death Rates during 1989–93." *IIHS Status Report* 2, no. 11 (8 October), 2. Insurance Institute for Highway Safety, Arlington, VA.

———. 1999. *The Year's Work, 1999*. Insurance Institute for Highway Safety, Arlington, VA.

Lee, John D. 2009. "Can Technology Get Your Eyes Back on the Road?" *Science*, 324, issue 525 (17 April), 344–346.

MADD. 2008. *Ignition Interlocks*. Brochure (August). Mothers Against Drunk Driving. Irving, TX.

———. 2009. *Repeat Offenders—Recommendations for More Severe Restrictions on Drunk Drivers*. Brochure (November). Mothers Against Drunk Driving. Irving, TX.

Mandelblit, Bruce D. 2001. "Fighting Back—How to Prevent Workplace Violence." *Industrial Safety and Hygiene News* 35 (September), 70–71.

Miller, Matthew. 2002. "Mortal Allies: Guns and Suicide." *Harvard Public Health Review* 60 (Summer). Harvard School of Public Health. Boston, MA.

Miller, Matthew, Azrael, Deborah, and Hemenway, David. 2002. "Firearm Availability and Unintentional Firearm Deaths, Suicide, and Homicide among 5–14-Year-Olds." *Journal of Trauma* 52, no. 2 (February), 267–275.

Moeller, Dade W. 2003. "Thinking Outside the Box: Benefits of Reductions in Vehicular Accidents in Avoiding Radiation Doses." *Health Physics News* XXXI, no. 2 (February), 2, 4.

Nader, Ralph. 1965. *Unsafe at Any Speed*. Grossman Publishers. New York, NY.

Neeley, Grant W., and Richardson, Lilliard E. 2009. "The Effect of State Regulations on Truck-Crash Fatalities." *American Journal of Public Health 99*, no. 3 (March), 408–415.

NHTSA and IIHS. 2009. *Study: More Drunks on the Road at Night*, National Highway Traffic Administration and Insurance Institute for Highway Safety. Washington, DC. *New Bern Sun Journal*, New Bern, NC, 24 (26 July).

NIOSH. 2001. *Building Safer Highway Work Zones: Measures to Prevent Worker Injuries from Vehicles and Equipment*. DHHS/NIOSH Publication no. 2001–128. National Institute for Occupational Safety and Health. Cincinnati, OH.

———. 2003. *Work-Related Roadway Crashes*. DHHS (NIOSH) Publication 2003–11. National Institute for Occupational Safety and Health. Cincinnati, OH.

NSC. 1990. *Accident Facts, 1990 Edition*. National Safety Council. Itasca, IL.

———. 2001. *Injury Facts, 2001 Edition*. National Safety Council. Itasca, IL.

———. 2002. *Injury Facts, 2002 Edition*. National Safety Council. Itasca, IL.

———. 2008. *Injury Facts, 2008 Edition*. National Safety Council. Itasca, IL.

———. 2009. *Injury Facts, 2009 Edition*. National Safety Council. Itasca, IL.

NSKC. 2004. *Rural Injury Fact Sheet.* National Safe Kids Campaign. Washington, DC.

Phillips, William G., ed. 2001. "Automotive Technology-New Car Option: A Crystal Ball." *Popular Science* 25, no. 6 (December), 37.

Waller, Julian A. 1994. "Reflections on a Half Century of Injury Control." *American Journal of Public Health* 84, no. 4 (April), 664–670.

Walsh, Bryan. 2009. "How Green Is He?" *Time* 173, no. 21 (1 June), 34.

Wilson, Fernando A., and Stimpson, Jim P. 2010. "Trends in Fatalities from Distracted Driving in the United States, 1999 to 2008." *American Journal of Public Health* 100, no. 11 (November), 2213–2218.

Wilson J. A., Stimpson, J. P., and Hilsenrath, P. E. 2009. "Gasoline Prices and Their Relationship to Rising Motorcycle Fatalities, 1990–2007." *American Journal of Public Health,* vol. 99 no. 10 (October), 1753–1758.

Wilson, Kevin A. 2009. "Can Computers Teach Us to Drive?" *Autoweek* 58, no. 18 (7 September), 20–22, 24.

12. Electromagnetic Radiation

ACGIH. 2008. *TLVs and BEls-Based on the Documentation of the Threshold Limit Values for Chemical Substances and Physical Agents and Biological Exposure Indices.* American Conference of Industrial Hygienists. Cincinnati, OH.

ACS. 2009. *Cancer Facts & Figures: 2009.* American Cancer Society. Atlanta, GA.

Agnew, J., K. Grainger, I. Clark, and Driscoll, C. 1998. *Protection from UVR by Clothing.* Radiological Protection Bulletin, no. 200, 14–17 (April). National Radiological Protection Board. Chilton, Didcot, Oxon, United Kingdom.

Alverez, Robert. 2010. "Flying and Excess Radiation." *East Texas Review.* http://www.easttexasreview.com/newspaper.htm?ArticleID=1096. Accessed 2 April 2010.

Bair, William J. 1997. "Radionuclides in the Body: Meeting the Challenge (Lauriston S. Taylor Lecture)." *Health Physics* 73, no. 3 (September), 423–432.

Clarke, Roger H. 2008. "The Risks from Exposure to Ionizing Radiation." In Till, John E., and Grogan, Helen A., eds., *Radiological Risk Assessment and Environmental Analysis,* chapter 12, 531–550, figure 12.7, p. 548. Oxford University Press, Oxford, NY.

Darby, S., Hill, D., Auvinen A., Barros, J. M., Dios, P., Baysson, H., Bachicchio, F., and Falk, R. 2005. "Radon in Homes and Risk of Lung Cancer: Collaborative Analysis of Individual Data from 13 European Case-control Studies." *British Medical Journal* 330, 223–227.

Eisenbud, Merril. 1973. *Environmental Radioactivity,* 2nd ed. Academic Press. New York, NY.

EPA. 2003. *Assessment of Risks from Radon in Homes.* Report EPA 402-R-03-003. Office of Health and Environmental Assessment. Environmental Protection Agency. Washington, DC.

Evans, R.D. 1974. "Radium in Man." *Health Physics* 27 (November), 497–510.

Goans, Ronald E., and Christiansen, Doran M. 2009. "Evaluation of Acute Radiation Injury—Part II: Thermography for Fluoroscopy Injuries." *Health Physics News* XXXVII, no. 11 (November), 13–16.

Herberman, Ronald . 2008. "Cell Phone Safety Precautions." Memorandum (20 August). University of Pittsburgh Cancer Center. Pittsburgh, PA.

ICNIRP. 2009a. "ICNIRP Statement on the "Guidelines for Limiting Exposure to Time-Varying Electric, Magnetic, and Electromagnetic Fields (Up to 300 GHz)." *Health Physics* 97, no. 3 (September), 257–258. International Commission on Non-Ionizing Radiation Protection. Oberschleissheim, Germany.

———. 2009b. "Amendment to the ICNIRP "Statement on Medical Magnetic Resonance (MR) Procedures: Protection of Patients." *Health Physics* 97, no. 3 (September), 259–261. International Commission on Non-Ionizing Radiation Protection. Oberschleissheim, Germany.

ICRP. 1991. *1990 Recommendations of the International Commission on Radiological Protection*. Publication 60. Annals of the ICRP 21, nos. 1–3. Pergamon Press. Elmsford, NY.

———. 1994. *Age-Dependent Doses to Members of the Public from Intake of Radionuclides, Part 3*. Publication 67. Annals of the ICRP 23, no. 3/4. Elsevier Science Ltd. Tarrytown, NY.

———. 1996. *Age-Dependent Doses to Members of the Public from Intake of Radionuclides, Part 3. Ingestion Dose Coefficients*. Publication 69. Annals of the ICRP 25, no. 1. Elsevier Science Ltd. Tarrytown, NY.

———. 1999. *Protection of the Public in Situations of Prolonged Exposure*. Publication 82. Annals of the ICRP 29, nos. 1–2. Elsevier Science Ltd. Tarrytown, NY.

———. 2000. *Prevention of Accidental Exposures to Patients Undergoing Radiation Therapy*. Publication 86, Annals of the ICRP 30, no. 3. Elsevier Science Inc. Tarrytown, NY.

———. 2004. *Managing Patient Dose in Computed Tomography*. Publication 93. Annals of the ICRP 34, no. 1. Elsevier Inc. New York, NY.

———. 2007a. *Managing Patient Dose in Multi-Detector Computed Tomography (MDCT)*. Publication 102. Annals of the ICRP, 37, no. 1. Elsevier. Orlando, FL.

———. 2007b. *The 2007 Recommendations of the International Commission on Radiological Protection*. Publication 103. Annals of the ICRP, 37, nos. 2–4. Elsevier. Orlando, FL.

———. 2007c. *Scope of Radiation Protection Control Measures*. Publication 104. Annals of the ICRP 37, no. 5. Elsevier. Orlando, FL.

———. 2007d. *Radiological Protection in Medicine*. Publication 105. Annals of the ICRP 37, no. 6. Elsevier. Orlando, FL,

———. 2008. *Radiation Dose to Patients from Radiopharmaceuticals. A third amendment to ICRP Publication 53. Also includes Radiation Exposure of Hands in Radiopharmacies*. Publication 106. Annals of the ICRP 38, nos. 1–2. Elsevier. St. Louis, MO.

Kirschner, Susan Kantra. 2002. "FDA, It's Time to Study Cellphone Radiation." *Popular Science* 261, no. 3 (September), 16.

Lapp, Ralph E., and Andrews, Howard L. 1948. *Nuclear Radiation Physics*. Prentice-Hall. New York, NY.

Little J. B., Radford E. P., Jr., McCombs H. I., Hunt V. R., and Nelson, C. 1965. "Distribution of Polonium-210 in Pulmonary Tissues of Cigarette Smokers." *New England Journal of Medicine* 273, 1343–1351.

Little, John B. 1993. "Biologic Effects of Low-Level Radiation Exposure." In Taveras, J. M., Juan, M., and Ferruci, J. T., eds., *Radiologic Physics and Pulmonary Radiology*: 1, chapter 13. J. B. Lippincott. Philadelphia, PA.

Mettler, Fred A., Thomadsen, Bruce R., Bhargavan, Mythreyi, Gilley, Debbie B., Gray, Joel E., Lipoti, Jill A., McCrohan, John, Yoshizumi, Terry T., and Mahesh, Mahadevappa. 2008. "Medical Radiation Exposure in the U.S. in 2006: Preliminary Results." *Health Physics* 95, no. 5, 502–507 (November).

Moeller, Dade, and Sun, Casper. 2009. "Thinking Outside the Box: Collective Dose from Brazil Nuts: Providing Perspective." *Health Physics News* XXXVII, no. 3 (March), 13–14.

NCRP. 1987a. *Ionizing Radiation Exposure of the Population of the United States.* Report no. 93. National Council on Radiation Protection and Measurements. Bethesda, MD.

———. 1987b. *Radiation Exposure of the U.S. Population from Consumer Products and Miscellaneous Sources.* Report no. 95. National Council on Radiation Protection and Measurements. Bethesda, MD.

———. 2000. *Radiation Protection Guidance for Activities in Low-Earth Orbit.* Report no. 132. National Council on Radiation Protection and Measurements. Bethesda, MD.

———. 2001. *Evaluation of the Linear-Nonthreshold Dose-Response Model for Ionizing Radiation.* Report no. 136. National Council on Radiation Protection and Measurements. Bethesda, MD.

———. 2003. *Presidential Report on Radiation Protection Advice: Screening of Humans for Security Purposes Using Ionizing Radiation Scanning Systems.* Commentary No. 16. National Council on Radiation Protection and Measurements. Bethesda, MD.

———. 2009. *Ionizing Radiation Exposure of the Population of the United States.* Report no. 160. National Council on Radiation Protection and Measurements. Bethesda, MD.

NIEHS. 2009. *Electric and Magnetic Fields.* National Institute of Environmental Health Sciences. http://www.niehs.nih.gov/health/topics/agents/emf/. Accessed 28 October 2009.

NRC. 1972. *The Effects on Populations of Exposure to Low Levels of Ionizing Radiation. BEIR I Report.* Committee on the Biological Effects of Ionizing Radiation. National Research Council, National Academy of Sciences. Washington, DC.

———. 1980. *The Effects on Populations of Exposure to Low Levels of Ionizing Radiation: 1980, BEIR III Report.* Committee on the Biological Effects of Ionizing Radiations, National Research Council. National Academy Press. Washington, DC.

———. 1990. *Health Effects of Exposure to Low Levels of Ionizing Radiation, BEIR V Report.* Committee on the Biological Effects of Ionizing Radiations, National Research Council. National Academy Press. Washington, DC.

———. 2006. *Health Risks From Exposure to Low Levels of Ionizing Radiation, BEIR VII Phase 2 Report.* Committee to Assess Health Risks from Exposure to Low Levels of Ionizing Radiation. National Research Council. National Academies Press. Washington, DC.

O'Hagan, John, and Hill, Robert. 1998. "Laser Pointers." *Radiological Protection Bulletin,* no. 199 (March), 15–20. National Radiological Protection Board. Chilton, Didcot, Oxon, United Kingdom.

Roentgen, Wilhem. 1895. "On a New Kind of Ray." Paper presented before the Würzburg Physical and Medical Society, Germany, 1895. English translation. Stanton, Arthur. 1896. *Nature* 53, 274.

Ropeik, David, and Gray, George. 2002. *Risk: A Practical Guide for Deciding What's Really Safe and What's Dangerous in the World around Us.* Houghton Mifflin. Boston, MA.

USNRC. 2009. *2009–2010 Information Digest.* Report NUREG-1350 21, 31. U.S. Nuclear Regulatory Commission. Washington, DC.

Valberg, Peter A. 2001. "Do Power-Line Electric and Magnetic Fields (EMF) Affect Health?" *Trends in Risk and Remediation* (September), 4–5. Gradient Corporation. Cambridge, MA.

WHO. 2010. *Electromagnetic Fields and Public Health: Mobile Phones.* Fact sheet number 193 (May). World Health Organization. Geneva, Switzerland.

13. Occupational, Population, and Environmental Standards

ACGIH. 2008. *TLVs and BEIs—Based on the Documentation of the Threshold Limit Values for Chemical Substances and Physical Agents & Biological Exposure Indices.* American Conference of Governmental Industrial Hygienists. Cincinnati, OH.

Cohen, Bernard L. 1995. "Test of Linear-No Threshold Theory of Radiation Carcinogenesis for Inhaled Radon Decay Products." *Health Physics* 68, no. 2 (February), 157–174.

EPA. 2001. "Public Health and Environmental Radiation Protection Standards for Yucca Mountain, NV; Final Rule." Environmental Protection Agency, Code of Federal Regulations, 40 CFR, Part 197, *Federal Register* 66, no. 114 (12 June), 32074–32135.

Evans, Robley D. 1974. "Radium in Man." *Health Physics* 27, no. 5, 497–510.

FRC. 1960. *Background Material for the Development of Radiation Protection Standards.* Federal Radiation Council, Report no. 1. U.S. Department of Health, Education, and Welfare. Washington, DC.

ICRP. 1975. *Reference Man: Anatomical, Physiological and Metabolic Characteristics.* Publication 23. International Commission on Radiological Protection. Pergamon Press. Oxford, England.

———. 1977a. *Recommendations of the International Commission on Radiological Protection.* Annals of the ICRP 1, no. 3. Publication 26. International Commission on Radiological Protection. Pergamon Press. New York, NY.

———. 1977b. *Problems Involved in Developing an Index of Harm.* Publication 27. Annals of the ICRP 1, no, 4. International Commission on Radiological Protection. Pergamon Press. New York, NY.

———. 1985. *Quantitative Basis for Developing a Unified Index of Harm.* Publication 45. Annals of the ICRP 15, no, 3. International Commission on Radiological Protection. Pergamon Press. New York, NY.

———. 1989. *Age-dependent Doses to Members of the Public from Intake of Radionuclides: Part 1.* Publication 56. Annals of the ICRP 20, no. 2. International Commission on Radiological Protection. Pergamon Press. Elmsford, NY.

———. 1991. *1990 Recommendations of the International Commission on Radiological Protection.* Publication 60. Annals of the ICRP 21, nos, 1–3. International Commission on Radiological Protection. Pergamon Press. New York, NY.

————. 1993. *Age-dependent Doses to Members of the Public from Intake of Radionuclides: Part 2—Ingestion Dose Coefficients.* Publication 67. Annals of the ICRP 23, nos. 3/4. International Commission on Radiological Protection. Elsevier Science Ltd. Tarrytown, NY.

————. 1995. *Age-dependent Doses to Members of the Public from Intake of Radionuclides: Part 3—Ingestion Dose Coefficients.* Publication 69. Annals of the ICRP 25, no. 1. International Commission on Radiological Protection. Elsevier Science Ltd. Tarrytown, NY.

————. 2003. *A Framework for Assessing the Impact of Ionizing Radiation on Non-Human Species.* Publication 91. Annals of the ICRP 33, no. 3. International Commission on Radiological Protection. Elsevier Ltd. Kidlington, Oxford, UK.

————. 2006. *Assessing Dose of the Representative Person for Purposes of Radiation Protection of the Public and the Optimization of Radiological Protection of the Public, and the Optimization of Radiological Protection: Broadening the Process.* Publication 101. Annals of the ICRP 36, no. 3. International Commission on Radiological Protection. Publication 101. Elsevier. Orlando, FL.

————. 2007a. *The 2007 Recommendations of the International Commission on Radiological Protection.* Publication 103. Annals of the ICRP 37, nos. 2–4. International Commission on Radiological Protection. Elsevier. Orlando, FL.

————. 2007b. *Scope of Radiological Protection Control Measures.* Publication 104. Annals of the ICRP 37, no. 5. International Commission on Radiological Protection. Elsevier. Orlando, FL.

————. 2008. *Environmental Protection: the Concept and Use of Reference Animals and Plants.* Publication 108. Annals of the ICRP 38, nos. 4–6. International Commission on Radiological Protection. Elsevier. St. Louis, MO.

Los Alamos Scientific Laboratory. 1995. "Radium—The Benchmark for Alpha Emitters." *Los Alamos Science* 23, 224–233. Los Alamos National Laboratory, Los Alamos, NM.

Muller, Hermann J. 1927. "Artificial Transmutation of the Gene." *Science* 66, issue 1699 (11 July), 84–87.

NCRP. 1991. *Effects of Ionizing Radiation on Aquatic Organisms.* Report no. 109. National Council on Radiation Protection and Measurements. Bethesda, MD.

————. 1993. *Limitation of Exposure to Ionizing Radiation.* Report no. 116. National Council on Radiation Protection and Measurements. Bethesda, MD.

————. 2002. *Operational Safety Program for Astronauts in Low-Earth Orbit. A Basic Framework.* Report 142, table 2.4, p. 21. National Council on Radiation Protection and Measurements. Bethesda, MD.

————. 2006. *Information Needed to Make Radiation Protection Recommendations for Space Missions Beyond Low-Earth Orbit.* Report no. 153. National Council on Radiation Protection and Measurements. Bethesda, MD.

NRC. 1972. *The Effects on Populations of Exposure to Low Levels of Ionizing Radiation.* BEIR I Report. Advisory Committee on the Biological Effects of Ionizing Radiation. National Research Council. National Academy of Sciences, National Academy Press. Washington, DC.

Osborne, Richard V. 2009. "The ICRP Reflects the Views of Health Physicists in its New Recommendations." *Health Physics News* XXXVII, no. 6 (June), 10–11.

Parsons, Peter A. 2002. "Radiation Hormesis: Challenging LNT Theory via Ecological and Evolutionary Considerations." *Health Physics* 82, no. 4 (April), 513–516.

Roentgen, Wilhem. 1895. "On a New Kind of Ray." Paper presented before the Würzburg Physical and Medical Society, Germany, 1895. English translation. Stanton, Arthur. 1896. *Nature* 53, 274.

USNRC. 2005. *Standards for Protection against Radiation.* CFR, Title 10, part 20. U.S. Nuclear Regulatory Commission. Washington, DC.

14. Effluent Monitoring and Environmental Surveillance

Achard, F., Eva, H.D., Stibig, H-J., Mayaux, J., Gallego, P., Richards, T., and Malingreau, J.-P. 2002. "Determination of Deforestation Rates of the World's Humid Tropical Forests." *Science* 297, issue 5583 (9 August), 999–1002.

Aubrecht, C., Elvidge, C. D., Longcore, T., Rich, C., Safran, J., Strong, A. F., et al. 2008. "A Global Inventory of Coral Reef Stressors Based on Satellite Observed Nighttime Lights." *Geocarto International* 23, no. 6 (6 December), 467–479.

Beaugelin-Seiller, K., Boyer, P., Garnier-Laplace, J., and Adam, C. 2002. "CASTE-AUR: A Simple Tool to Assess the Transfer of Radionuclides in Waterways." *Health Physics* 83, no. 4 (October), 539–542.

Brown, Kathryn. 2002. *Water Scarcity: Forecasting the Future with Spotty Data. Science* 297, issue 5583 (9 August), 926–927.

Eaton, A. E., Clesceri, L. S., Rice, E. W., and Greenberg, A. E. 2005. *Standard Methods for the Examination of Water and Wastewater,* 21st ed. American Public Health Association. Washington, DC.

Ebersold, Peter J., and Nicholas Barker. 2003. "Having a Field Day." *Environmental Protection* 14, no. 3 (April), 45–49.

EPA. 1988. *Clean Air Act Assessment Package-1988 (CAP-88).* National Computer Center. Research Triangle Park, NC. //www.epa.gov/rpdweb00/assessment/CAP88/aboutcap88.html. Accessed 14 January 2010.

——. 1993. *R-EMAP: Regional Environmental Monitoring and Assessment Program.* Report EPA/625/R-93/012. Washington, DC.

——. 2001a. "Environmental Protection Agency: Public Health and Environmental Radiation Protection Standards for Yucca Mountain, NV; Final Rule." *Federal Register* 6, no. 114 (13 June), 32074–32137.

——. 2001b. *Measuring Pollution with Infrared Rays.* National Risk Management Research Laboratory. *NRMRL News* (September), 1. Washington, DC.

EPA, et al. 2001. *Multi-agency Radiological Laboratory Analytical Protocols Manual (MARLAP).* Reports EPA 402-B-01-2001; NIST PB2001-106745; and NUREG-1576. Environmental Protection Agency, U.S. Departments of Defense and Energy; the National Institute of Standards and Technology, U.S. Department of Commerce; and the U.S. Nuclear Regulatory Commission. Washington, DC.

Fucillo, Karen. 2009. *Environmental Surveillance and Monitoring Program.* New Jersey Department of Environmental Protection, Trenton, NJ (September).

Griffith, Jerry A., and Carolyn T. Hunsaker. 1994. *Ecosystem Monitoring and Ecological Indicators: An Annotated Bibliography.* Report EPA/620/R-94/021. Environmental Protection Agency. Washington, DC.

Guerney, Kevin R., and Raymond, Leigh. 2008. "Targeting Deforestation Rates on Climate Change Policy: A "Preservation Pathway" Approach." *Carbon Balance and Management Journal* 3 (3 March).

Howard, H. T., Pacific, K. H., and Pacific, J. A. 2002. "The Evolution of Remote Sensing." *Environmental Protection* 13, no. 4 (April), 28, 30–34.

HPS. 1999. *Sampling and Monitoring Releases of Airborne Radioactive Substances from the Stacks and Ducts of Nuclear Facilities.* American National Standard ANSI/HPS N13.1-1999. Health Physics Society, McLean, Virginia.

Hunsaker, C. T., Goodchild, M. F., Friedl, M. A., and Case, T. J. 2001. *Spatial Uncertainty in Ecology: Implications for Remote Sensing and GIS Applications.* Springer-Verlag. New York, NY.

ICRP. 2006. *Assessing Dose of the Representative Person for the Purpose of Radiation Protection of the Public,* and *The Optimization of Radiological Protection: Broadening the Process.* Publication 101. Annals of the ICRP 36, no. 3. Elsevier. Oxford, UK.

———. 2008. *Environmental Protection: the Concept and Use of Reference Animals and Plants.* Publication 108. Annals of the ICRP 38, nos 4–6. Elsevier. St. Louis, MO.

ISO. 2009. *International Standards on Environmental Management.* ISO-14001. International Organization for Standards, Washington, DC.

Lambert, Kathy F., and Van Bowersox. 2002. "Environmental Monitoring and National Security: Is There a Connection?" *EM* (August), 17–22.

Lodge, James P., ed. 1988. *Methods of Air Sampling and Analysis. Intersociety Committee Report.* CRC Press. Boca Raton, FL.

Malakoff, David. 2002. "Microbiologists on the Trail of Polluting Bacteria." *Science* 295, issue 5564 (29 March), 2352–2353.

Moeller, Dade W. 2005. "Environmental Health Physics." *Health Physics* 88, no. 6 (July), 676–696.

NCRP. 1991, *Effects of Ionizing Radiation on Aquatic Organisms.* Report 109. National Council on Radiation Protection and Measurements. Bethesda, MD.

———. 1996. *Screening Models for Releases of Radionuclides to Atmosphere, Surface, Water, and Ground.* Reports 123 I and 123 II. National Council on Radiation Protection and Measurements. Bethesda, MD.

Newman, Alan. 1995. "Major U.S. Human Exposure Assessment Survey Gets under Way." *Environmental Science and Technology* 29, no. 9 (September), 398A–399A.

NRC. 2000. *Ecological Indicators for the Nation.* Board on Environmental Studies and Toxicology, and Water Science and Technology Board. National Research Council. National Academy Press. Washington, DC.

PNNL. 2008. *Hanford Site Environmental Report for Calendar Year 2008. Report PNNL-2008-18427.* Pacific Northwest National Laboratory. Richland, WA. Available at http://hanford-site.pnl.gov/envreport/2008/index.htm.

USDOE. 1991. *Environmental Regulatory Guide for Radiological Effluent Monitoring and Environmental Surveillance.* Report DOE/EH-0173T (January). Washington, DC.

———. 2002. *A Graded Approach for Evaluating Radiation Doses to Aquatic and Terrestrial Biota.* Technical Standard DOE-STD-1153–2002. Washington, DC. Available at http://www.hss.doe.gov/nuclearsafety/ns/techstds/standard/standard.html.

USNRC. 2005. *Standards for Protection Against Radiation.* Code of Federal Regulations, Title 10, Part 20. U.S. Nuclear Regulatory Commission. Washington, DC.

Weinhold, Bob. 2002. "Air Pollution: U.S. Air Only Fair." *Environmental Health Perspectives* 110, no. 8 (August), A452.

Wikipedia. 2009. "Environmental Protection Agency." http://en.wikipedia.org/wiki/United_States. Accessed 14 January 2009.

Wilkins, Bernard. 1999. "Pigeons—A Novel Form of Airborne Radionuclides." *Radiological Protection Bulletin,* no. 215 (September), 7–11. National Radiological Protection Board. Chilton, DidCot, Oxfordshire, UK.

Wing, S., S. Freedman, and L. Band. 2002. "The Potential Impact of Flooding on Confined Animal Feeding Operations in Eastern North Carolina." *Environmental Health Perspectives* 110, no. 4 (April), 387–391.

15. Risk Assessment and Management

Associated Press. 2009. "Cancer Death Rate Drops Again in 2006." *Sun Journal,* New Bern, NC (27 May), A5.

ASTDR. 2001. *Guidance Manual for the Assessment of Joint Toxic Action of Chemical Mixtures.* Draft for Public Comment. Agency for Toxic Substances and Disease Registry. Atlanta, GA.

———. 2003. *FY202 Profile and Annual Report.* Report ATSDR-PE-RP-2003-0001. Agency for Toxic Substances and Disease Registry. Atlanta, GA.

———. 2006. *Toxicological Profile for Hydrogen Sulfide,* Agency for Toxic Substances and Disease Registry. Atlanta, GA.

———. 2007. *Toxicological Profile for Benzene.* Agency for Toxic Substances and Disease Registry. Atlanta, GA.

Clarke, Roger H. 2008. "The Risks from Exposure to Ionizing Radiation." In Till, John E., and Grogan, Helen A., eds., *Radiological Risk Assessment and Environmental Analysis,* chapter 12, 531–550. Oxford University Press, Oxford, NY.

Cooke, Roger M. 2009. "A Brief History of Quantitative Risk Assessment." *Resources* no. 172 (Summer), 8–9. Resources for the Future.

EPA. 1998. *Guidelines for Ecological Risk Assessment.* Report EPA/630/R095.002F. National Center for Environmental Assessment. Environmental Protection Agency. Washington, DC.

———. 2001. "Public Health and Environmental Radiation Protection Standards for Yucca Mountain, NV: Final Rule." 40 CFR, Part 197. *Federal Register* 66, no. 114 (12 June), 32074–32135 12.

———. 2009. *The Inside Story: A Guide to Indoor Air Quality.* http://www.epa.gov/iaq/pubs/insidest.html. Accessed 4 November 2009.

Espey D. K., Wu X-C., Swan, J., Wiggins, C., Jim, M. A., Ward, E., Wingo, P.A, Holly, L. H., Howe, H. L., Reis, L. A. G., Miller, B. A., Jemal, A., Ahmed, F., Cobb, N., Kaur, J. S., Edwards, B. K. 2007. "Annual Report to the Nation on the Status of Cancer, 1975–2004, Featuring Cancer in American Indians and Alaska Natives." *Cancer* 110, no. 10, 2119–2152.

Gage, Stephen D. 2001. "Harmonizing Controls for Chemicals and Radionuclides." *Health Physics* 80, no. 4 (April), 338–389.

Graham, John D. 1993. "The Legacy of One in a Million." *Risk in Perspective* 1, no. 1 (March), 1–2. Harvard Center for Risk Analysis, Boston, MA.

Graham, John D. 1999. "Making Sense of the Precautionary Principle." *Risk in Perspective* 7, issue 6 (September), 1–6. Harvard Center for Risk Analysis, Boston, MA.

Healy, B. 2008. "Unlocking the Secrets of Cancer." *U.S. News and World Report* 145, no. 10, 46–47.

ICRP. 2006. *Assessing Dose of the Representative Person for the Purpose of Radiation Protection of the Public,* and *The Optimisation of Radiological Protection: Broadening the Process.* Publication 101. Annals of the ICRP 36, no. 3. International Commission on Radiological Protection. Elsevier. Oxford, UK.

———. 2008. *Environmental Protection: The Concept and Use of Reference Animals and Plants.* Publication 108. Annals of the ICRP 38, nos. 4–6. International Commission on Radiological Protection. Elsevier. St. Louis, MO.

Johnson, B. L. 1992. "Principles of Chemical Risk Assessment: The ATSDR Perspective." In *Proceedings of a Conference on Chemical Risk Assessment in the DoD: Science Policy and Practices,* 1–35. American Conference of Governmental Industrial Hygienists. Cincinnati, OH.

Kaplan, Stanley, and B. John Garrick. 1981. "On the Quantitative Definition of Risk." *Risk Analysis* 1, no. 1, 11–27.

Karam, Andy, 2008. "Thinking About Radiation Safety in a Post-Cancer World." *Health Physics News* XXXVII, no. 6 (June), 7.

Knacker T., Duis K., and Coors, A. 2007. *Environmental Risk Assessment of Pharmaceuticals.* Project no. SSPI-CT-2003-511135. ERA Pharm. http://www.epa.gov/iaq/pubs/insidest.html tp://www.eraphar.org/. Accessed 7 October 2009.

Moeller, John A. 2005. *From Malabar to the Moon: America's Early Space Days.* Tennessee Valley Publishing. Knoxville, TN.

Moeller, Dade W. 2009. "Lauriston S. Taylor Lecture: Yucca Mountain Radiation Standards, Dose/Risk Assessments, Public Interactions: Thinking Outside the Box, Evaluations, and Recommendations." *Health Physics* 97, no. 5, 376–391 (November).

———. 2010. "The Tragedy of Yucca Mountain." *RadWaste Solutions* 17, no. 5 (September/October), 52–57.

Moeller, Dade W., and Sun, Casper. 2010. "Carcinogens in Cigarettes, the Role of ^{210}Po and ^{210}Pb, and Associated Health Impacts." *Health Physics* 98, no. 5 (May), 674–679.

NCRP. 1991. *Effects of Ionizing Radiation on Aquatic Organisms.* Report no. 109. National Council on Radiation Protection and Measurements. Bethesda, MD.

Rhomberg, Lorenz R. 2001. "Parsing the Precautionary Principle." *Trends in Risk and Remediation,* 4 (Spring). Gradient Corporation, Cambridge, MA.

Schmidt, Charles W. 1999. "A Closer Look at Chemical Exposures in Children." *Environmental Science and Technology* 4, no. 2 (1 February), 72A–75A.

Till, John E. 2008. "Radiological Assessment Process." In Till, John E., and Helen A. Grogan, eds., *Radiological Risk Assessment and Environmental Analysis,* 1–30. Oxford University Press. New York, NY.

United Nations. 1992. *Rio Declaration on Environment and Development.* http://www.unep.org/Documents.Multilingual/Default.asp?documentid=78&articleid=1163. Accessed 10 January 2011.

USNRC. 1962. *Reactor Site Criteria.* CFR, Title 10, part 100. Appendix A. U.S. Nuclear Regulatory Commission. Washington, DC.

————. 1975. *Reactor Safety Study: An Assessment of Accident Risks in U.S. Commercial Nuclear Power Plants. Report WASH-1400,* NUREG 75/014. U.S. Nuclear Regulatory Commission. Washington, DC.

————. 1991. *Severe Accident Risks: An Assessment for Five U.S. Nuclear Power Plants.* Report NUREG-1150. U.S. Nuclear Regulatory Commission. Washington, DC.

————. 2009. *2009–2010 Information Digest.*" Report NUREG-1350 21; 74–76. U.S. Nuclear Regulatory Commission. Washington, DC.

Wiener, Jonathan Baert, and Graham, John D. 1995. "Resolving Risk Tradeoffs." In Graham, John D., and Wiener, Jonathan Baert, eds., *Risk vs. Risk: Tradeoffs in Protecting Health and the Environment,* 226–271. Harvard University Press, Cambridge, MA.

Wilson, Richard, and Crouch, Edmund A. C. 2001. *Risk-Benefit Analysis,* 2nd ed. Harvard University Press. Cambridge, MA.

16. Energy Resources and Conservation

Adams, Rod. 1995. "Light Water Reactors: Adapting a Proven System." *Atomic Insights,* 1, no 7. Adams Atomic Engines, Inc. Annapolis, MD.

American Wind Energy Association. 2008. "Wind Energy Industry Creates Jobs, Shines as Growing Bright Spot in the Midst of a Faltering Economy." http://www.awea.org/newsroom/releases/Wind_Industry_Creates Jobs_10_Oct.html. Accessed 27 April 2010.

Amin, S. Massoud. 2010. "Securing the Electricity Grid." *The Bridge* 40, no. 1 (Spring), 13–20. National Academy of Engineering, Washington, DC.

Associated Press. 2010. "Town Prays for Missing Miners." *Sun Journal,* New Bern, NC (7 April), 10.

BBC NEWS. 2010. "£70m Tidal Power Scheme Goes on Display in Anglesey." (http://www.bbc.co.uk/news/uk-wales-north-west-wales-11037069. Accessed 4 December 2010.

Bosshard, Peter. 2009. *China's Three Gorges Dam: A Model of the Past.* International Rivers. Berkeley, CA.

Brink, Elizabeth, and Dvorak, Wil. 2009. *River Revival.* International Rivers. Berkeley, CA.

Carroll, Chris. 2009. "Can Solar Save Us?" *National Geographic* 216, no. 3 (September), 52–53.

Charles, Dan, 2009. "Corn-Based Ethanol Flunks Key Test." *Science* 324, issue 5927 (1 May), 587.

Clery, Daniel. 2009a. "Fusion's Great Bright Hope." *Science* 324, issue 5925 (17 April), 326–330.

————. 2009b. "ITER Gets Nod for Slower, Step-by-Step Approach. Great Bright Hope." *Science* 324, issue 5935 (26 June), 1627.

————. 2010. "Test Shots Show Laser-Fusion Experiment is on Target." *Science* 327, issue 5965 (29 January), 514.

Consumer Reports. 2009. "Cash for Clunkers. Top 10 Most Popular New Cars and Trade-Ins." *Cars Blog.* http://74.125.113.132/search?q=cache:FT8EHeyPPg8J:blogs.con-

sumerreports.org/cars/2009/08/cash-for-clunkers-top-10-most-popular-new-cars
-and-trade-ins.html+consumer+reports,+cash+for+clunkers,+August+26,+2009&
cd=1&hl=en&ct=clnk&gl=us. Accessed 10 February 2010.

Curry, Andrew. 2009. "Deadly Flights." *Science* 325, issue 5939 (24 July), 386–387.

Dewan, Shailla. 2008. "Tennessee Ash Flood Larger Than Initial Estimate." *New York Times* (26 December).

———. 2009. "TVA to Pay $43 Million on Projects in Spill Area." *New York Times* (15 September).

Electric Power Research Institute. 2006. "Exploring Tidal Power in the San Francisco Bay." http://www.pge.com/about/environment/pge/features/tidalpower.shtml. Accessed 4 September 2009.

Evans, Robley D. 1969. "Engineer's Guide to the Elementary Behavior of Radon Daughters." *Health Physics* 17, no. 2, 229–252.

Galbraith, Kate. 2008. "Power from the Restless Sea Stirs the Imagination." *New York Times* (22 September).

Hutchinson, Alex. 2008. "Solar's New Dawn." *Popular Mechanics* 185, no. 11; 62–67.

Johnson, George. 2009. "Plugging Into the Sun." *National Geographic* 216, no. 3 (September), 28–51.

Kannberg, J. D., Chassin, D. P., DeSteese, J. G., Hauser, S. G., Kintner-Meyer, M. C., Pratt, R. G., Schienbein, L. A., and Warwick, W. M. 2003. *GridWise™: The Benefits of a Transformed Energy System.* Pacific Northwest National Laboratory. U.S. Department of Energy.

Kerr, Richard A. 2009. "How Much Coal Remains?" *Science* 323, issue 5920 (13 March), 1420–1421.

Kofman, Jeffrey. 2009. "Bolivia Uyuni Salt Flats Hold Promise of Greener Future." Reporter's Notebook. *Nightline* (5 August), ABC News.

Kunzig, Robert. 2009. "The Canadian Oil Boom: Scraping Bottom." *National Geographic* 215, no. 3 (March), 34–59.

Mattson, Kevin. 2009. *What the Heck Are You Up to, Mr. President?* Bloombury Publishing, New York, NY.

McElroy, Michael B., Lu, Xi, Chris, P., Nielsen, Chris, and Wang, Yuxuan. 2009. "Potential for Wind-Generated Electricity in China." *Science* 325, issue 5946 (11 September), 1378–1380.

National Highway Transportation Safety Administration. 2009. *Joint Rulemaking to Establish CAFÉ and GHG Emissions Standards for Model Years 2012–2016: Final Rule.* Washington, DC.

NEI. 2009a. "How to Strengthen Climate Change, Energy Policy with New Nuclear Plants." *Nuclear Energy Insight* (November), 1. Nuclear Energy Institute. Washington, DC.

———. 2009b. "Nature Conservancy: Nuclear Plants Protect Against Sprawl." *Nuclear Energy Insight* (November), 8. Nuclear Energy Institute. Washington, DC.

Normile, Dennis, 2009. "New Facility Propels Korea to the Fusion Forefront." *Science* 323, issue 5917 (20 February), 1003–1004.

OGJ Newsletter. 2008. "Worldwide Look at Energy Resources and Production." *Oil & Gas Journal* 106, no. 48 (22 December), 22–23.

Pacella, Rena Marie. 2010. "The Next-Gen Wind Turbine." *Popular Science* 275 (April), 42–43.

Richter, Daniel deB., Jr., Jenkins, Dylan H., Karakash, John T., Knight, Josiah, McCreery, Lew R., and Nemestothy, Kasimir P. 2009. "Wood Energy in America." *Science* 323, issue 5920 (13 March), 1432–1433.

Sagoff, Jared. 2009. "Charging Ahead." *Argonne Now* 04, no. 02 (Fall), 16–18. Argonne National Laboratory. Argonne, IL.

Schauer, Ron, and Clayton, Mark. 2010. "U.S. Urged New Safety Standards Days Before Middletown Explosion." *Christian Science Monitor* (8 February).

Schuler, Richard E. 2010. "The Smart Grid. A Bridge Between Emerging Technologies, Society, and the Environment." *The Bridge* 40, no. 1 (Spring), 42–49. National Academy of Engineering, Washington, DC.

Scruggs, Jeff, and Jacob, Paul. 2009. "Harvesting Ocean Wave Energy." *Science* 323, issue 5918 (27 February), 1176–1178.

Sharp, Phil. 2009. "Reflections on Three Decades of Energy Policy." *Resources,* no. 171 (Winter–Spring), 3–4. Resources for the Future, Washington, DC.

Sowder, Andrew, Machiels, Albert, Hamel, Jeffrey, Wenzinger, Ed, and Dixon, Tim. 2009. "Readiness of the U.S. Reactor Fleet for MOX Fuel Utilization." Advances in Nuclear Fuel Management, Meeting IV (12–15 April), American Nuclear Society. Hilton Head, SC.

Staedter, Tracy. 2002. "Wave Power: How the Rolling Sea Generates Electricity." *Technology Review* 105, no. 1 (January/February), 86–87. Massachusetts Institute of Technology, Cambridge, MA.

Swanson, Richard M. 2009. *Photovoltaics Power Up. Science* 324, issue 5929 (15 May), 891–892.

Themelis, Nickolas J., and Priscilla A. Ulloa. 2007. "Methane Generation in Landfills." *Renewable Energy* 37, no. 7 (June), 1243–1257.

Titus, Brian D., Douglas G. Maynord, Caren C. Dymond, Graham Stinson, and Werner, Kurz. 2009. "Wood Energy: Protect Local Ecosystems." *Science* 324, issue 5933 (12 June), 1389–1390.

Truly, Richard H. 2002. "New Energy Frontiers." *The Bridge* 32, no. 2 (Summer), 5–10. National Academy of Engineering, Washington, DC.

U.S. Department of Energy. 2009a. "Solar Energy Technologies Program." http://www.1.eere.energy.gov/solar/program_areas.html. Accessed 15 June 2009.

———. 2009b. "Geothermal Heat Pumps (GHPs) Use Shallow Land Energy to Heat and Cool Buildings." http://www1.eree.energy.gov/geothermal/geothermal_basics.htm. Accessed 30 August 2009.

———. 2009c. *Energy Saving Estimates of Light Emitting Diodes in Niche Lighting Applications.* Building Energy Technologies Program, Office of Energy Efficiency and Renewable Energy. Washington, DC.

Viscarolasaga, Efrain. 2008. "Tide is Slowly Rising in Interest in Ocean Power." *The Journal of New England Technology* (1 August).

Walker, G. 1980. *Stirling Engines.* Clarenden Press, Gloucestershire, UK.

Walsh, Bryan. 2009. "Thank God It's Thursday." *Time* 174, no. 9 (7 September), 58.

Wikipedia. 2009a. "Geothermal Power in Iceland." http://en.wikipedia.org/wiki/Geothermal_power_in_Iceland. Accessed 16 June 2009.

———. 2009b. "Solar Power." http://en.wikipedia.org.wiki/Solar_power. Accessed 4 September 2009.

———, 2009c. "Stirling Engine." http://en.wikipedia.org.wiki/Stirling_engine. Accessed 4 September 2009.

———. 2010. "World Energy Resources and Consumption." http://en.wikipedia .org/wiki/World_energy_resources_and_consumption#cite_ref-29. Accessed 8 April 2010.

Wilkes, Justin. 2010. "Wind in Power: 2009 European Statistics." Report (February). The European Wind Energy Association.

Worrell, Ernst, Phylipsen, Dian, Einstein, Dan, and Martin, Nathan. 2000. *Energy Use and Energy Intensity of the U.S. Chemical Industry.* Energy Analysis Department, Environmental Energy Technologies Division, Ernest Orlando Berkeley Lawrence National Laboratory. U.S. Department of Energy.

17. Disasters and Terrorism

Abramovitz, J. N. 2001. "Averting Unnatural Disasters." In *State of the World*, 123–142. The WorldWatch Institute. W.W. Norton & Company, New York, NY.

AMA. 2002. *Bioterrorism Awareness: A Definitive Resource for State-of-the-Art Medical and Clinical Information for Responding to Biological or Chemical Attack.* CD-ROM. American Medical Association. Chicago, IL.

Associated Press. 2009. "Australia: Authorities to Re-Examine Wildfire Survival Strategies." *The Wall Street Journal* (11 February), A10.

———. 2010a. "Volcano Closes Global Skies." *Sun Journal*, New Bern, NC (16 April), 11.

———. 2010b. *Tornadoes Kill 10 People in the Southeast. Sun Journal*, New Bern, NC (25 April), 10.

Cais, Ph., Reichstein, M., Viovy, N., Granier, A., Ogee, J., Allard, V., Aubinet, M., Buchmann, N., Bernhofer, Chr., Carrara, Aa., Chevallier, F., De Noblet, N., Friend, A. D., Friedlingstein, P., Grunwald, T., Heinesch, B., Keronen, P., Knohl, A., Krinner, G., Loustau, D., Manca, G., Matteucci, G., Miglietta, F., Ourcival, J. M., Papele, D., Pilegaard, K., Ramba, S., Seufert, G., Soussana, J. F., Sanz, M. J., Schulze, E. D., Vesala, T., and Valentini, R. 2003. "Earth-wide Reduction in Primary Productivity Caused by Heat and Drought in 2003." *Nature* 437 (22 September), 483–484.

Carmona, R.H., Darling, R.G., Knoben, J.E., and Michael, J.M., eds. 2010. *Public Health Emergency Preparedness & Responses: Principles and Practices.* Commissioned Officers Association, U.S. Public Health Service, Landover, MD.

CBS News. 2010. "Gulf Oil Disaster." *60 Minutes* (16 May).

CBS Evening News. 2010. "Report on Blizzard in the Northeastern Region of the United States" (12 February). New York, NY.

CDC. 1997a. "Tornado-Associated Fatalities—Arkansas," 1997. *Morbidity and Mortality Weekly Report*, 46, no. 19 (16 May), 412–416. Centers for Disease Control and Prevention. Atlanta, GA.

———. 1997b. "Tornado Disaster—Texas, May 1997." *Morbidity and Mortality Weekly Report*, 46, no. 45 (14 November), 1046–1073. Centers for Disease Control and Prevention. Atlanta, GA.

———. 2000. "Biological and Chemical Terrorism: Strategic Plan for Preparedness and Response, Recommendations of the CDC Strategic Planning Workgroup." *Morbidity and Mortality Weekly Report* 49, no. RR-4 (21 April), 1, 5–6. Centers for Disease Control and Prevention. Atlanta, GA.

Center for Arms Control and Non-Proliferation. 2009. http://armscontrolcenter. org/policy/iran/?gclid=CNXn9_fTs54CFZho5QodpVcFsA. Accessed 30 November 2009.

Chang, Stephenie E. 2009. "Infrastructure Resilience to Disasters." *Resources* 39, no. 4 (Winter), 36–41. Resources for the Future, Washington, DC.

Chapin, Douglas M., Cohen, Karl P., Davis, W. Kenneth, Kintner, Edwin E., Koch, Leonard J., Landis, John W., Levenson, Milton, I. Mandil, Harry, Pate, Zack T., Rockwell, Theodore, Schreisheim, Alan, Simpson, John W., Squire, Alexander, Starr, Chauncey, Stone, Henry E., Taylor, John J., Todreas, Neil E., Wolfe, Bertram, and Zebroski, Edwin L. 2002. "Nuclear Power Plants and Their Fuel as Terrorist Targets." *Science* 297, no. 5589 (20 September), 1997, 1999.

Cochran, Thomas B., and McKinzie, Matthew G. 2009. "Detecting Nuclear Smuggling." *Scientific American* 298, no. 4 (April), 98–102, 104.

Cunningham, Richard. 2009. "What We do When Storms Hit Hard." Special Energy Supplement, *The Lamp*, 3–6, The Hibbert Group, Trenton, NJ,

Fink, Sheri. 2007. "The Science of Doing Good," *Scientific American* 297, no. 5 (November), 98–104.

Fleming, Melissa. 2002. "International Symposium Focuses on Global Nuclear Terrorism." *Health Physics* 82, no. 1 (January), 120–121.

Gwin, Peter. 2009. "Crude Currents." *National Geographic* 215, no. 4 (April), 33.

Haring, Markus. 2010. "All Shook Up. A New Geothermal Energy Method Could Trigger a Side Effect, Earthquakes." *Popular Science* 276, no. 4 (April), 29.

Hymann, David L. 2008. *Control of Communicable Disease Manual*, 19th ed., American Public Health Association, Washington, DC.

ICRP. 2005. *Protecting People against Radiation Exposure in the Event of a Radiological Attack*. Publication 96. Annals of the ICRP 35, no. 1. International Commission on Radiological Protection. Elsevier, Orlando, FL.

Ketterlinus, Robert D., ed. 2008. *Youth Violence: Interventions for Health Care Providers*. American Public Health Association, Washington, DC.

Lehrer, Jim (Jim Lehrer News Hour). 2010, "Review of Katrina Hurricane on its 5-Year Anniversary." Public Broadcasting System (25–26 August).

McClatchy Newspapers. 2010. "Federal Laws Point to Criminal Charges in Gulf Oil Spill." *Sun Journal*, New Bern, NC (13 May), 14.

Mitchell, John. 1999. "In the Wake of the Spill—Ten Years After Exxon Valdez." *National Geographic* 195, no. 3 (March), 96–117.

Moeller, John A. 2005. *From Malabar to the Moon: America's Early Space Days*. Tennessee Valley Publishing. Knoxville, TN.

Morrison, Jim. 2010. *Hurricanes Over the Horizon. Miller-McCune* (May/June), 58–67. Santa Barbara, CA.

National Research Council. 2001. *Environmental Performance of Tanker Designs in Collision and Grounding: Method for Comparison*. Special Report 259. Transportation Research Board, National Academy Press, Washington, DC.

———. 2002. *Making the Nation Safer: The Role of Science and Technology in Countering Terrorism.* National Academy Press. Washington, DC.

———. 2006. *Health Risks from Exposure to Low Levels of Ionizing Radiation: BEIR VII Phase 2 Report.* The National Academies Press, Washington, DC.

National Safety Council. 2009. *Injury Facts—2009 Edition, 30.* Itasca, IL.

NCRP. 2001. *Management of Terrorist Events Involving Radioactive Material.* Report no. 138. National Council on Radiation Protection and Measurements. Bethesda, MD.

Neville, Angela. 2001. "Seeing Community Right-To-Know Laws in a New Light" (editorial). *Environmental Protection* 12, no. 12 (December), 6.

NIOSH. 2003. *Guidance for Filtration and Air-Cleaning Systems to Protect Building Environments from Airborne Chemical, Biological, or Radiological Attacks.* DHHS (NIOSH) Publication No. 2003-136, National Institute for Occupational Safety and Health, U.S. Department of Health and Human Welfare. Cincinnati, OH.

Okrent, David, and Moeller, Dade W. 1981. "Implications for Reactor Safety of the Accident at Three Mile Island, Unit 2." *Annual Review of Energy* 6. Annual Reviews, Inc. Palo Alto, CA.

PAHO. 2010. "Avoiding the Urbanization of Disaster Risk" [Chapter 1 of *World Disaster Report 2010: Focus on Urban Risk,* published by the International Federation of Red Cross and Red Crescent Societies]. *Disasters: Preparedness and Mitigation in the Americas* 114 (October), 9. Pan American Health Organization. Washington, DC.

Parfit, Michael, 1998. "Living with Natural Disasters." *National Geographic* 194, no. 1 (July), 2–39.

Qureshi, K., Gershom, R. M., Sherman, M. F., Straub, T., Gebbie, E., McCollum, M., Erwin, M. J., and Morse, S. S. 2005. "Health Care Workers' Ability and Willingness to Report to Duty During Catastrophic Disasters." *Journal of Urban Health, Bulletin of the Academy of Medicine* 82, 378–388.

Ring, Joseph P. 2004. "Radiation Risks and Dirty Bombs." *Operational Health Physics,* Supplement to *Health Physics* 86, no, 2 (February), S42–S47.

Robock, Alan. 2002. *The Climatic Aftermath. Science* 295, no. 5558 (15 February), 1242–1244.

Schwartz, Harold M., Flood, Ann Barry, Gougelet, Robert M., Rea, Michael E., Nicolalde, Roberto J., and Williams, Benjamin B. 2010. "A Critical Assessment of Biodosimetry Methods for Large-Scale Incidents." *Health Physics* 98, no. 2 (February), 95–108.

Servat, Kevin. 2010. "Navigating Through Crisis." *Water and Wastes Digest* 50, no. 4 (April), 10, 12.

Smith, Tanalee. 2009a. Associated Press. "108 Killed in Australian Wildfires." *Sun Journal,* New Bern, NC (9 February), A1, A3.

Smith, Tanalee. 2009b. Associated Press. "Australian Wildfire Deaths Rise." *Sun Journal,* New Bern, NC (10 February), A1, A3.

Swenson, Dan D, and Marshall, Bob. 2005. "Flash Flood: Hurricane Katrina's Inundation of New Orleans." *Times-Picayune.* http://www.nola.com/katrina/graphics/credits.swf. Accessed 22 December 2009.

Talmazan, Yuliya. 2010. "Haiti Earthquake Death Statistic Update." http://www.now public.com/world/haiti-earthquake-death-statistics-update-eu-says-200-000 -dead-2559557.html. Accessed 31 January 2010.

Trichopoulos D, X. Zavitsanos, C. Koutis, P. Drogari, C. Proukakis, E. Petridou. 1987. "The Victims of Chernobyl in Greece: Induced Abortions After the Accident." *British Medical Journal* 295, no. 6606 (21 October), 1100.

U.S. Census Bureau. 2008. *2006–2008 American Community Survey.* U.S. Department of Commerce. Washington, DC.

U.S. Congress. 1986. *The Superfund Amendments and Reauthorization Act.* 6 US Code, Public Law 107-296 (17 October). Washington, DC.

——. 2009. *Hazardous Material Transportation Act of 2009.* H.R. 4016, 111th Session (4 November). Washington, DC.

U.S. Mine Rescue Association. 2009. Homepage. http://www.usmra.com/. Accessed 30 November 2009.

USNRC. 2007. *Review of NUREG-9654, Supplement 3, Criteria for Protective Action Recommendations for Severe Accidents.* U.S. Nuclear Regulatory Commission. Washington, DC.

——. 2009a. *2009–2010 Information Digest.* Report NUREG-1350 21, 90–95. U.S. Nuclear Regulatory Commission. Washington, DC.

——. 2009b. *Review of NUREG-9654, Supplement 3, Criteria for Protective Action Recommendations for Severe Accidents.* U.S. Nuclear Regulatory Commission. Washington, DC. http://www.nrc.gov/reading-rm/doc-collections/nuregs/ contract/cr6953/vol2/. Accessed 30 November 2009.

Wikipedia. 2010a. *2010 Haiti Cholera Outbreak.* http://en.wikipedia.org/wiki/2010_ Haiti_Cholera_Outbreak. Accessed (4 December 2010).

——. 2010b. "Prestige Oil Spill." http://en.wikipedia.org/wiki/Prestige_oil_spill. Accessed 16 April 2010.

Wisner, B, Blaikie, P., T. Cannon, Davis, I. 2004. *At Risk: Natural Hazards, People's Vulnerability, and Disasters,* 2nd ed. Routledge. New York, NY.

WVU. 2004. *Pest Control Following Disasters.* West Virginia University Extension Service Disasters and Emergency Management Manual. Section 12.9.1. Morgantown, WV.

18. A Global View

Alberts, Bruce. 2010. "An Education That Inspires" (editorial). *Science,* 330, issue 6001 (2 October), 427.

Alsop, Peter. 2009. "Invasion of the Longhorns." *Smithsonian* 40, no. 8 (November), 42–49.

AMA. 1989. *Stewardship of the Environment.* Council on Scientific Affairs, Report G, 1–89. American Medical Association. Chicago, IL.

Associated Press. 2009a. "Beetles, Wildfire: Double Threat in Warming World." *Sun Journal,* New Bern, NC (24 August), 13.

——. 2009b. "U.S. to Place Limits on Power Plant Water Pollution." *Sun Journal,* New Bern, NC (16 September), 22.

——. 2010. "Delegates Told No Secret Deal at Climate Talks." *Sun Journal,* New Bern, NC (6 December), 10.

Balmford, A., Bruner, A.Z., Cooper, P., Costanza, R., Farber, S., Green, R. E., Jenkins, M., Jefferiss, P., Jessamy, V., Madden, J., Munro, K., Myers, N., Naeem, S., Paavola, J., Rayment, M., Rosendo, S., Roughgarden, J., Trumper, K., and Turner, R. K. 2002. "Economic Reasons for Conserving Wild Nature." *Science* 297, issue 5583 (9 August), 950–953.

Beaugrand, Gregory, Reid, Philip C., Ibanez, Frederic, Lindley, J. Alistair, and Edwards, Martin. 2002. "Reorganization of North Atlantic Marine Copepod Biodiversity and Climate." *Science* 296, issue 5573, 1692–1694.

Brown, Gordon. 2009. "Copenhagen or Bust: The Time is Now for an International Deal on Climate Change." *Newsweek,* 56 (28 September).

Bugliarello, G. 2001. "Rethinking Urbanization." *The Bridge* 31, no. 1 (Spring), 5–12. National Academy of Engineering. Washington, DC.

Carpenter, B. 2002. *Sound and Fury. U.S. News & World Report* 133, no. 24 (December 23), 50–92.

CEQ. 1997. *Environmental Quality—The Twenty-Fifth Anniversary Report of the Council on Environmental Quality.* Council on Environmental Quality. Executive Office of the President. Washington, DC.

Chapple, Steve. 2009. "Reefs at Risk." *Reader's Digest* (August), 136–145.

Chu, Steven. 2009. *In the Great Ship Titanic. Newsweek* CLIII, no. 16 (20 April), 28.

Darwin, Charles R. 1859. *On the Origin of Species by Means of Natural Selection.* John Murray, Albemarle Street, London, England.

———. 1985. *On the Origin of the Species by Means of Natural Selection* (Penguin Classic edition), Chapter 4; 185.

De´ath, Glenn, Lough, Janice M., and Fabricius, Katharina E. 2009. "Declining Coral Calcification on the Great Barrier Reef." *Science* 323, issue 5910 (2 January), 116–119.

Debnam, Betty. 2009. "The Mini Page: Vanishing Animals, Disappearing Frogs and Bats." *Sun Journal,* New Bern, NC (20 October), 8.

Encyclopedia Britannica Online. 2010. "Nuclear Test Ban Treaty." Accessed 13 November 2010.

EPA. 2009. *Quality of our Nation's Waters: A Summary of National Water Quality Inventory: 1998.* Environmental Protection Agency. Washington, DC.

———. 2010. *Chesapeake Bay Program.* http://www.epa.gov/Region3/chesapeake/ Chesapeake Program Office, Region 3 Office, Environmental Protection Agency. Annapolis, MD. Accessed 20 April 2010.

Fitter, A. H., and Fitter, R. S. R. 2002. "Rapid Changes in Flowering Time in British Plants." *Science* 296, issue 5573 (31 May), 1689–1691.

Greenberg, Michael. 2001. *Earth Day Plus 30 Years: Public Concern and Support for Environmental Health. American Journal of Public Health* 91, no. 4 (April), 559–562.

Greene, Katie. 2002. "Bigger Populations Needed For Sustainable Harvests." *Science* 296, issue 5571 (17 May), 1229–1230.

Harvell, C. Drew, Mitchell, Charles E., Ward, Jessica R., Altizer, Sonia, Dobson, Andrew P., Ostfeld, Richard S., and Samuel, Michael D. 2002. "Climate Warming and Disease Risks for Terrestrial and Marine Biota." *Science* 296, issue 5576 (21 June), 2158–2162.

Hayden, Thomas. 2001. "Deep Trouble—Overfishing Has Torn the Sea's Web of Life—Mending It Won't be Easy." *U.S. News & World Report* 131, no. 9 (10 September), 68–70.

Holland, Jennifer S. 2009. "The Vanishing Frogs." *National Geographic* 215, no. 4 (April), 138–153.

Huggins, D. R., and Reganold, J. P. 2008. "No-Till Farming: The Quiet Revolution." *Scientific American* 299, no. 1(July), 70–77.

Johnson, Jeff. 1995. "EPA Must Look to the Future, Says Science Advisory Board." *Environmental Science and Technology* 29, no. 3 (March), 112A–113A.

Johnston, Renee. 2009. *Climate Change: The Role of the U.S. Agriculture Sector and Congressional Action. Report for Congress* (31 July). Congressional Research Service. Washington, DC.

Kaiser, Jocelyn. 2001. "Bold Corridor Project Confronts Political Reality." *Science* 293, issue 5538 (21 September), 2196–2197, 2199.

Kearney, Bill. 2001. "A Wetland Gained for a Wetland Lost?" *INFOCUS* 1, no. 2 (Fall/Winter), 17–18. The National Academies of Sciences, Washington, DC.

———. 2002. "Bringing Back the Big Muddy." *INFOCUS* 2, no. 1 (Spring), 15–16. The National Academy of Sciences. Washington, DC.

Kennedy, Donald. 2002. "POTUS and the Fish." *Science* 297, issue 5581 (26 July), 477.

Kintisch, Eli. 2009. "Projections of Climate Change Go from Bad to Worse, Scientists Report." *Science* 323, issue 5921, 1546–1547.

Krouse, Eric. 2009. Personal communication (26 September). Manlius, NY.

Lehmann, Caroline E. R. 2010. "Savannas Need Protection" (letter to editor). *Science* 327, issue 5966 (5 February), 642–643.

Leslie, Mitch. 2002. "Trouble in Paradise." *Science* 295, issue 5558 (15 February), 1199.

McGinn, Daniel. 2009. "The Greenest Big Companies in America." *Newsweek* (28 September), 34–54.

Miller, Greg. 2010. "In Central California, Coho Salmon Are on the Brink." *Science* 327, issue 5965 (29 January), 512–513.

Montgomery, D. R. 2008. "Pay Dirt." *Scientific American*, 299, no. 1 (July), 76.

Myers, Norman. 2002. "A Convincing Call for Conservation." Review of *The Future of Life* by Edward O. Wilson (Knopf, New York, NY, 2002). *Science* 295, issue 5554 (18 January), 447–448.

NASA. 2010. "World's Lakes Getting Hotter, More Than Air." *Sun Journal*, New Bern, NC (25 November), 13.

National Research Council. 2008. *Progress Toward Restoring the Everglades: The Second Biennial Review, 2008.* Committee on Independent Scientific Review of Everglades Restoration Progress. National Academies Press. Washington, DC.

NRDC. 2010. "Lethal Sounds: The Use of Military Sonar Poses Deadly Threat to Whales and Other Marine Mammals." http://www.nrdc.org/wildlife/marine/sonar.asp National. Resources Defense Council. Washington, DC. Accessed 21 April 2010.

Penuelas, Josep, and Filella, Iolanda, 2001. "Responses to a Warming World." *Science* 294, issue 5543 (26 October), 793, 795.

Pickrell, John. 2004. "U.S. Navy Sonar May Harm Killer Whales, Expert Says." *National Geographic News* (31 March). Washington, DC.

Powell, Alvin. 2002. "Agreeing on What to Argue About: Environmental Report Tries to Make Sense of Flood of Data." *Harvard University Gazette* XCVIII, no. 2 (25 September), 27–28.

Prugh, Thomas, and Assadourian, Erik. 2003. "What Is Sustainability, Anyway?" *World-Watch* 16, no. 2 (September/October), 10–21.

Reisner, Marc. 2000. "Unleash the Rivers." *Time* 155, no. 17 (April–May), 66–71.

Schmidt, Karen F. 2001. "A True-Blue Vision for the Danube." *Science* 294, issue 5546 (16 November), 1444–1445, 1447.

Sills, Jennifer, ed. 2010. "Climate Change and the Integrity of Science." *Science* 328, issue 5979 (7 May), 689–690.

Stegeman, John J., and Solow, Andrew R. 2002. "Environmental Health and the Coastal Zone." *Environmental Health Perspectives* 110, no. 11 (November), A660–A661.

Stokstad, Erik. 2009. "Debate Continues Over Rainforest Fate—With a Climate Twist." *Science* 323, no. 5913 (23 January), 448.

Thigpen, David E. 2002. "The Fight over Big Muddy's Flow." *Time* 160, no. 2 (8 July), 72.

Tibbetts, John. 2002. "Coastal Cities—Living on the Edge." *Environmental Health Perspectives* 110, no. 11 (November), A674–A681.

United Nations. 2002. *World Summit on Sustainable Development.* Johannesburg, South Africa.

U.S. Census Bureau. 2009. World Population Clock. http://www.census.gov/main/www/popclock.html. Accessed 20 October 2009.

U.S. Congress. 1977. *Clean Water Act and Amendments.* Public Law 95-217. Washington, DC.

———. 1987. *Water Quality Act.* Public Law 100-4. Washington, DC.

———. 1990. *Clean Air Act Amendments of 1990.* Public Law 101-549. Washington, DC.

van Mantgem, Phillip J., Stephenson, Nathan L., Byrne, John C., Daniels, Lori D., Franklin, Jerry F., Fule, Peter Z., Harmon, Mark E., Larson, Andrew J., Smith, Jeremy M., Taylor, Alan H., and Veblen, and Thomas, T. 2009. "Widespread Increase in Tree Mortality Rates in the Western United States." *Science* 323, issue 5913 (23 January), 521–524.

Weis, Virginia M., and Denis Allemand. 2009. *What Determines Coral Health? Science* 324, issue 5931 (29 May), 1153–1155.

Weise, Elizabeth. 2009. "Frog-killing Fungus Seeps Deep into Skin." *USA Today* (23 October), 2A.

Wilson, Edward O. 2000. "Biodiversity—Vanishing Before Our Eyes." *Time* 155, no. 17 (April–May), 29–34.

———. 2002. "Hot Spots—Preserving Pieces of a Fragile Biosphere." *National Geographic* 201, no. 1 (January), 86–89.

Withgott, Jay. 2002. "California Tries to Rub Out the Monster of the Lagoon." *Science* 295, issue 5563 (22 March), 2201–2202.

WMO. 2007. *Scientific Assessment of Ozone Depletion: 2006.* Global Ozone Depletion
    Research and Monitoring Project. Report no. 50. World Meteorological Orga-
    nization. Geneva, Switzerland.
Wolinsky-Nahmias, Yael, and Kim, So Young. 2008. "International Public Opinion
    on the Environment: Responses to Inequality and Globalization." Paper pre-
    sented at the annual meeting of the 31st ISPP Annual Scientific Meeting (23
    May), Sciences PO. Paris, France.

TABLE CREDITS

Table 1.1 Prepared by author.

Table 1.2 Based on *Harvard Report on Cancer Prevention,* Center for Cancer Prevention (Boston, MA: Harvard University, 1996).

Table 1.3 Based on U.S. Department of Health and Human Services, *Healthy People 2010: Understanding and Improving Health,* 2nd ed., (Washington, DC: Government Printing Office, November 2000).

Table 1.4 CDC. 1999. "Ten Great Public Health Achievements—United States, 1900–1999." *Morbidity and Mortality Weekly Report* 48, no. 12 (2 April), 241–243. Centers for Disease Control and Prevention. Atlanta, GA.

Table 2.1 C. D. Klaassen, "Principles of Toxicology," in C. D. Klaassen, Mary O. Amdur, and J. Doull, *Casarett and Doull's Toxicology: The Basic Science of Poisons,* 3rd ed, (New York: Macmillan Publishing Company, 1986), table 2-1, p. 12. Adapted with permission of The McGraw-Hill Companies.

Table 2.2 C. D. Klaassen, "Principles of Toxicology," in C. D. Klaassen, Mary O. Amdur, and J. Doull, *Casarett and Doull's Toxicology: The Basic Science of Poisons,* 3rd ed, (New York: Macmillan Publishing Company, 1986), table 2-2, p. 13. Adapted with permission of The McGraw-Hill Companies.

Table 2.3 Charnley, Gail, 2001, "Lecture Outline on Principles of Toxicology." Presentation in course entitled "Risk," Harvard School of Public Health, Boston, MA.

Table 2.4 Charnley, Gail, 2001, "Lecture Outline on Principles of Toxicology." Presentation in course entitled "Risk," Harvard School of Public Health, Boston, MA.

Table 3.1 Based on National Research Council, *Environmental Epidemiology: Public Health and Hazardous Wastes* (Washington, DC: National Academy Press, 1991), table 3-4, p. 120.

Table 3.2 National Research Council, *Radiation Dose Reconstruction for Epidemiological Uses* (Washington, DC: National Academy Press, 1995), table 7-2, p. 73.

Table 4.1 Prepared by author.

Table 4.2   Centers for Disease Control and Prevention, "Prevention of Leading Work-Related Diseases and Injuries," *Morbidity and Mortality Weekly Report* 32, no. 2 (21 January 1983), table 1, 25.

Table 4.3   Based on information in National Safety Council, *Injury Facts—2009 Edition* (Itasca, NY: NRC, 2009) p. 48.

Table 4.4   John Tibbetts, "Under Construction: Building a Safer Industry," *Environmental Health Perspectives* 110, no. 3 (March 2002), A134–A141.

Table 4.5   American Conference of Governmental Industrial Hygienists, *2008 TLVs and BEIs—Based on the Documentation of the Threshold Limit Values for Chemical Substances and Physical Agents, & Biological Exposure Indices* (Cincinnati, OH: ACGIH, 2008).

Table 4.6.   American Conference of Governmental Industrial Hygienists, *2008 TLVs and BEIs—Based on the Documentation of the Threshold Limit Values for Chemical Substances and Physical Agents, & Biological Exposure Indices* (Cincinnati, OH: ACGIH, 2008).

Table 4.7   Based on American Conference of Governmental Industrial Hygienists, *2008 TLVs and BEIs—Based on the Documentation of the Threshold Limit Values for Chemical Substances and Physical Agents, & Biological Exposure Indices.* (Cincinnati, OH: ACGIH, 2008).

Table 5.1   U.S. Environmental Protection Agency, *National Ambient Air Quality Standards (NAAQS),* http://www.epa.gov/air/criteria.html. Accessed 2 August 2009.

Table 5.2   U.S. Environmental Protection Agency, *Air Quality and Emissions Trends,* http://www.epa.gov/airtrends/aqtrends.html. Accessed 8 January 2010.

Table 5.3   Based on Suzan M. Zummo, and Meryl H. Karol, "Indoor Air Pollution: Acute Adverse Health Effects and Host Susceptibility," *Journal of Environmental Health* 58, no. 6, table 3 (January/February 1996), 27–29.

Table 6.1   Based on Centers for Disease Control and Prevention, "Diagnosis and Management of Foodborne Illnesses—A Primer for Physicians," *Morbidity and Mortality Weekly Report* 50, no. RR-2, 1–60 (26 January 2001), and *Wikipedia,* "Bovine Spongiform Encephalopathy," http://wikipedia.org.wiki/Bovine_spongiform_encephalopathy. Accessed 7 July 2010.

Table 6.2   Based on Sandra A. Hoffmann, "Attributing U.S. Foodborne Illness to Food Consumption," *Resources* 172 (Summer 2009), pp. 14–18.

Table 6.3   Based on Abraham S. Berenson, ed., *Control of Communicable Diseases Manual,* 16th ed., (Washington, DC: American Public Health Association, 1995), p. 194.

Table 6.4   Based on information from Institute of Food Technologists, "*Government Regulations of Food Safety: Interaction of Scientific and Societal Forces,*" *Food Technology* 46, no. 1 (January 1992).

Table 6.5   Based on Institute of Food Technologists, "Government Regulation of Food Safety: Interaction of Scientific and Societal Factors," *Food Technology* 46, no. 1 (January 1992), 73–80.

Table 7.1   Steven Babin, Howard S. Burkom, Zaruhi R. Mnatsakamyan, Liane C. Ramac-Thomas, Michael W. Thompson, Richard A. Wojcik, Sheri Happel Lewis, and Cynthia Yund, "Drinking Water Security and Public Health Disease Outbreak

Surveillance," *The Johns Hopkins University APL Technical Digest* 27, no. 4 (2008), 403–411, table 1, p. 404,

Table 7.2 Prepared by author, with updates, based on Sandra Postel, *Increasing Water Efficiency,* in Lester R. Brown et al., eds., *State of the World,* Worldwatch Institute Report (New York: W.W. Norton, 1986), table 3-6, p. 55.

Table 7.3 Prepared by author.

Table 7.4 Russell Derickson, *Primary Drinking Water Standards* (Brookings, SD: Cooperative Extension Service, College of Agriculture & Biological Sciences, South Dakota State University, 1994), http://agbiopubs.sdstate.edu/articles/ExEx1025.pdf. Accessed 18 August 2009.

Table 7.5 Lee Harms, Cindy L. Wallis-Lage, and Ron Ratnayaka, "What the Future Holds for Water and Wastewater: Disinfection," in *World of Water 2000: The Past, Present, and Future,* Supplement to *PennWell Magazine* (Tulsa, OK: WaterWorld and Water & Wastewater International, 1999), 152–156.

Table 7.6 Prepared by author.

Table 8.1 Prepared by author.

Table 8.2 Prepared by author based on multiple references, including the National Research Council, *Managing Wastewater in Coastal Urban Areas* Committee on Wastewater Management for Coastal Urban Areas (Washington, DC: National Academy Press, 1993), table ES.1, p. 5.

Table 8.3 Based on U.S. Environmental Protection Agency, *Quality of our Nation's Waters: A Summary of National Water Quality Inventory: 1998* (Washington, DC: EPA, 2000), supplemented by information from the National Safety Council, *Injury Facts, 2001 edition* (Itasca, IL: NSC 2001), p. 147, and subsequent updates.

Table 9.1 Based on information at Zero Waste America, "Waste & Recycling, Data, Maps, & Graphs," http://www.zerowasteamerica.org/Statistics/htm. Accessed 22 August 2010.

Table 9.2 U.S. Environmental Protection Agency, *RCRA Orientation Manual,* Report EPA/530-SW-86-001, II-9, III-4, and IV-3 (Washington, DC: EPA, 1986).

Table 9.3 U.S. Environmental Protection Agency, *Solving the Hazardous Waste Problem: EPA's RCRA Program,* Report EPA/530-SW-86-037 (Washington, DC: EPA, 1986), p. 8.

Table 9.4 Based on U.S. Nuclear Regulatory Commission, *Regulation of the Disposal of Low-Level Radioactive Waste: A Guide to the Nuclear Regulatory Commission's 10 CFR Part 6* (Washington, DC: Office of Nuclear Material Safety and Safeguards, 1989) figure 1, p. 2a; and idem. *Information Digest: 1995 Edition.* Report NUREG-1350, vol. 7 (Washington, DC: U.S. Government Printing Office, 1989).

Table 9.5 Prepared by author based on references cited in footnotes in table and Greenpeace, *Greenpeace Investigates Levels of Toxic Materials Found in Laptop Computers,* http://74.125.113.132/search?q=cache:NHr38xI7IoJ:www.associatedcon tent.com/article/426952/greenpeace_investigations_levels_of_toxic.html. Accessed 31 August 2007.

Table 9.6 U.S. Environmental Protection Agency, *Low-Level Mixed Waste: A RCRA Perspective for NRC Licensees* Report EPA/530-SW-90-057 (Washington, DC: EPA, 1990) p. 23.

Table 9.7 Prepared by author.

Table 9.8   Prepared by author.

Table 9.9   Based on information in U.S. Nuclear Regulatory Commission, *2009–2010 Information Digest,* Report NUREG-1350 (Washington, DC: USNRC, 2009), pp. 21, 74–76.

Table 10.1   Based on information in Nathan Wolfe, "Preventing the Next Pandemic," *Scientific American* 300, no. 4 (April 2009), 76–81.

Table 10.2   Based on information in Nathan Wolfe, "Preventing the Next Pandemic," *Scientific American* 300, no. 4 (April 2009), 76–81.

Table 10.3   Prepared by author.

Table 10.4   Based on Peter Wehrwein, "Pharmaco-philanthropy," *Harvard Public Health Review* (Summer 1999), 32–39.

Table 11.1   Based on National Safety Council, *Injury Facts: 2009 Edition* (Itasca, IL: NSC, 2009), p. 2.

Table 11.2   Based on National Safety Council, *Injury Facts: 2009 Edition* (Itasca, IL: NSC, 2009), p. 132.

Table 12.1   Prepared by author.

Table 12.2   International Commission on Radiological Protection, *Protecting People Against Radiation Exposure in the Event of a Radiological Attack,* Annals of the ICRP 35, no. 1 (2005), p. 5, table 1.

Table 12.3   International Commission on Non-Ionizing Radiation Protection, "General Approach to Protection Against Non-Ionizing Radiation," *Health Physics* 82, no. 4 (April 2002), 540–550, table 1.

Table 12.4   International Commission on Non-Ionizing Radiation Protection, "General Approach to Protection Against Non-Ionizing Radiation," *Health Physics* 82, no. 4 (April 2002), 540–550, table 2.

Table 12.5   Compiled by author. Based primarily on information provided by Wallace Friedberg, Kyle Copeland, Frances E. Duke, Keran O'Brian III, and Edgar B. Darden Jr., "Radiation Exposure during Air Travel: Guidance Provided by the Federal Aviation Administration for Air Carrier Crews," *Health Physics* 79, no. 5 (November 2000), 591–595.

Table 12.6   International Commission on Radiological Protection, *Managing Patient Dose in Multi-Detector Computed Tomography (MDCT),* Annals of the ICRP 37, no. 1 (Orlando, FL: Elsevier, 2007).

Table 12.7   Prepared by author.

Table 13.1   Prepared by author.

Table 13.2   Prepared by author based primarily on International Commission on Radiological Protection, *The 2007 Recommendations of the International Commission on Radiological Protection,* Annals of the ICRP 27, nos. 2–4 (Orlando, FL: Elsevier, 2007), table A.4.3, p. 182.

Table 13.3   Based on International Commission on Radiological Protection, *The 2007 Recommendations of the International Commission on Radiological Protection,* Annals of the ICRP 27, nos. 2–4 (Orlando, FL: Elsevier, 2007), table 6, p. 99.

Table 13.4   Based on National Council on Radiation Protection and Measurements, *Operational Safety Program for Astronauts in Low-Earth Orbit: A Basic Framework,* Report no. 142 (Bethesda, MD: NCRP 2002), table 2.4, p. 21. Report no. 142.

Table 14.1   Prepared by author.

Table 14.2   Based on data in J. P. Corley, D. H. Denham, R. E. Jaquish, D. E Michels, A. R. Oleson, and D. A. Waite, *A Guide for Environmental Radiological Surveillance at ERDA Installations,* Report ERDA-77-24 (Springfield, VA: U.S. Department of Energy, National Technical Information Service, 1977); and Dade W. Moeller, Jack M. Selby, David A. Waite, and John P. Corley, "Environmental Surveillance for Nuclear Facilities," *Nuclear Safety* 19, no. 1, 66–70 (January–February 1978), table 3, p. 73.

Table 14.3   Based on Jerry A. Griffith and Carolyn T. Hunsaker, "Ecosystem Monitoring and Ecological Indicators: An Annotated Bibliography," EPA/620/R-94/021 (Athens, GA: EPA, 1994).

Table 14.4   Based on Jerry A. Griffith and Carolyn T. Hunsaker, "Ecosystem Monitoring and Ecological Indicators: An Annotated Bibliography," EPA/620/R-94/021 (Athens, GA: EPA, 1994), table 1.

Table 15.1   Based on Michael Gochfeld and Joanna Burger, *Environmental and Ecological Assessment,* in Robert Wallace, ed., *Maxcy-Rosenau-Last Public Health and Preventive Medicine* (Stamford, CT: Appleton and Lange, 1998), pp. 435–441.

Table 15.2   Based on papers listed in the footnotes to the table.

Table 15.3a   Agency for Toxic Disease Registry and Toxic Substances, *The Agency for Toxic Substances and Disease Registry: FY 2002 Profile and Annual Report,* Report ATSDR-PE-RP-2003-0001 (Atlanta, GA: ATSDR , 2003), chapter 2, table 1, p. 31.

Table 15.3b   Agency for Toxic Disease Registry and Toxic Substances, *The Agency for Toxic Substances and Disease Registry: FY 2002 Profile and Annual Report,* Report ATSDR-PE-RP-2003-0001 (Atlanta, GA: ATSDR , 2003), chapter 3, p. 43.

Table 15.4   Based on John E. Till, *The Radiological Assessment Process,* chapter 1, pp. 1–30, in John E. Till and Helen A. Grogran, *Radiological Risk Assessment and Environmental Analysis* (New York, NY: Oxford University Press, 2008).

Table 15.5   Based on information in International Commission on Radiological Protection, *Recommendations of the International Commission on Radiological Protection,* Publication 60, Annals of the ICRP 21, no. 1–3 (Elmsford, NY: Pergamon Press, 1991); and International Commission on Radiological Protection, *The 2007 Recommendations of the International Commission on Radiological Protection,* Publication 103, Annals of the ICRP, 37 (2–4) (Orlando, FL: Elsevier, 2007), table A.4.5.

Table 15.6   Based on John D. Graham, *Annual Report* (Boston, MA: Center for Risk Analysis, Harvard School of Public Health, 1993), inside front cover.

Table 15.7   Based on Roger H. Clarke, *The Risks from Exposures to Ionizing Radiation,* in John E. Till and Helen A. Grogran, *Radiological Risk Assessment and Environmental Analysis* (New York, NY: Oxford University Press), table 12.4, p. 541.

Table 15.8   Based on D. J. Paustenbach, *The Practice of Exposure Assessment: A State-of-the-Art Review,* in A. Wallace Hayes, ed., *Principles and Methods of Toxicology,* 4th ed. (Philadelphia, PA: Taylor and Francis, 2001); and John E. Till and Helen A. Grogran, *Radiological Risk Assessment and Environmental Analysis* (New York, NY: Oxford University Press), chapter 1.

Table 16.1   Based on D. J. Paustenbach, *The Practice of Exposure Assessment: A State-of-the-Art Review,* in A. Wallace Hayes, ed., *Principles and Methods of Toxicology,* 4th ed. (Philadelphia, PA: Taylor and Francis, 2001); and John E. Till and Helen A.

Grogran, *Radiological Risk Assessment and Environmental Analysis* (New York, NY: Oxford University Press), chapter 1.

Table 16.2   Based on data from Paul J. Meler, *Life-Cycle Assessment of Electricity Generation Systems and Applications to Climate Change Policy Analysis*, Ph.D. dissertation, University of Wisconsin, Madison (August 2002). Emissions avoided were calculated on the basis of regional and national fossil-fuel emission rates from the U.S. Environmental Protection Agency and the Energy Information Administration, U.S. Department of Energy (2009).

Table 16.3   Based on Science Concepts, Inc., "The Impacts of Nuclear Energy on Utility Fuel Use and Utility Atmospheric Emissions, 1973–1990" (Chevy Chase, MD, Science Concepts, Inc, December 1991), table 3, 14.

Table 16.4   Based on National Council on Radiation Protection and Measurements, *Carbon-14 in the Environment*, Report no. 81 (Bethesda, MD: NCRB, 1985); and J. Tichler, K. Norden, and J. Congemi, *Radioactive Materials Released from Nuclear Power Plants: Annual Report, 1987*, Report NUREG/CR-2907 vol. 8 (Washington DC: U.S. Nuclear Regulatory Commission, 1989).

Table 16.5   Based on data in National Council on Radiation Protection and Measurements, *Carbon-14 in the Environment*, Report no. 81 (Bethesda, MD: NCRB, 1985); and J. Tichler, K. Norden, and J. Congemi, *Radioactive Materials Released from Nuclear Power Plants: Annual Report, 1987*, Report NUREG/CR-2907 vol. 8 (Washington DC: U.S. Nuclear Regulatory Commission, 1989; and Ralph Larsen, Senior Scientist, Environmental Protection Agency, Research Triangle Park, NC, Personal communication (May 2003).

Table 17.1   Prepared by author.

Table 17.2   Based on chart provided by PAHO, *Humanitarian Assistance in Disaster Situations—A Guide for Effective Donations* (Washington, DC: Pan American Health Association, 2002).

Table 17.3   Based on Stephanie Chang, "Infrastructure Resilience to Disasters," *The Bridge* 39, no. 4 (Winter 2009), 36–43, table 1, 37, National Academies of Engineering, Washington, DC,

Table 17.4   Based on CDC, *Biological and Chemical Terrorism: Strategic Plan for Preparedness and Response: Recommendations for the CDC Strategic Planning Workshop. Morbidity and Mortality Weekly Report* 49, no. RR-4 (21 April 2000), pp. 7–8, box 5.

Table 17.5   Based on CDC, "Biological and Chemical Terrorism: Strategic Plan for Preparedness and Response: Recommendations for the CDC Strategic Planning Workshop," *Morbidity and Mortality Weekly Report* 49, no. RR-4 (21 April 2000), box 3, pp. 5–6.

Table 17.6   Prepared by author.

Table 17.7   Prepared by author.

Table 17.8   Based on PAHO, *Humanitarian Assistance in Disaster Situations—A Guide for Effective Donations* (Washington, DC: Pan American Health Association, 2002).

Table 18.1   Based on CEQ, *Environmental Quality: The Twenty-fifth Anniversary of the Council on Environmental Quality* (Washington, DC: Council on Environmental Quality, Executive Office of the President, 1997), chapter 7, 129–131

Table 18.2   Based on Alvin Powell, "Agreeing on What to Argue About: Environmental Report Tries to Make Sense of Flood Data," *Harvard University Gazette* 98, no. 2 (25 September 2000), 27–28.

Table 18.3   Based on Alvin Powell, "Agreeing on What to Argue About: Environmental Report Tries to Make Sense of Flood Data," *Harvard University Gazette* 98, no. 2 (25 September 2000).

Table 18.4   Based on NRC, *Grand Challenges in Environmental Sciences* (Washington, DC: National Research Council, National Academy Press, 2001).

FIGURE CREDITS

Figure 1.1   Prepared by the author.

Figure 1.2   Wayne R. Ott, 1990, "Total Human Exposure: Basic Concepts, EPA, Field Studies, and Future Research," *Journal of the Air and Waste Management Association* 40, no. 7 (July 1990), figure 1, p. 968.

Figure 1.3   Based on UNIS, *World Population Growth Will Occur in Urban Areas* (Vienna, Austria: United Nations, 2008).

Figure 1.4   Based on *Wikipedia*, "World Population," http://en.wikipedia.org/wiki/World_population. Accessed 12 August 2009.

Figure 2.1   T. A. Loomis, *Essentials of Toxicology* (Philadelphia, PA: Lea & Febiger), figure 2.1, p. 15.

Figure 2.2   T. A. Loomis, *Essentials of Toxicology* (Philadelphia, PA: Lea & Febiger), figure 2.2, p. 16.

Figure 2.3   Edward J. Calabrese, Linda A. Baldwin, and Charles D. Holland, "Hormesis: A Highly Generalizable and Reproducible Phenomenon with Important Implications for Risk Assessment," *Risk Analysis* 19, no. 2 (1999), 261–281, figure 1.

Figure 3.1   National Research Council, *Environmental Epidemiology: Public Health and Hazardous Wastes* (Washington, DC: National Academy Press, 1991), figure 3-5, p. 122. Copyright 1990, by the American Chemical Society.

Figure 3.2   National Research Council, *Environmental Epidemiology: Public Health and Hazardous Wastes* (Washington, DC: National Academy Press, 1991), figure 7-1, p. 221.

Figure 3.3   Roger H. Clarke, "The Risks from Exposure to Ionizing Radiation," in John E. Till and Helen A. Grogan, eds., *Radiological Risk Assessment and Environmental Analysis* (Oxford, NY: Oxford University Press, 2008), chapter 12, figure 12.7, p. 548.

Figure 4.1   Adapted from Barry S. Levy and David W. Wegman, eds., *Occupational Health: Recognizing and Preventing Work-Related Disease*, 2nd ed. (Boston, MA: Little, Brown, 1988), figure 22.6, p. 21.

Figure 4.3   Prepared by author based on American Conference of Governmental Industrial Hygienists, *Industrial Ventilation: A Manual of Recommended Practices*, 20th ed. (Lansing, MI: Committee on Industrial Ventilation, 1988), figure VS-202, pp. 5–21.

Figure 4.4   Based on Karen Springer, "The Dangerous Desk," *Newsweek* 137, no. 13 (March), unnumbered figure, p. 67.

Figure 5.1   Richard Wilson and Edmund A. C. Crouch, *Risk-Benefit Analysis,* 2nd ed., (Cambridge, MA: Harvard Center for Risk Analysis, 2001), figure 2-2, p. 31 (cited as being taken from Beaver, 1953).

Figure 5.2   National Council on Radiation Protection and Measurements, *Deposition, Retention, and Dosimetry of Inhaled Radioactive Substances,* Report No. 125 (Bethesda, MD: NCRP, 1997), figure 99.1, p. 55.

Figure 6.1   U.S. Department of Agriculture, "Food Pyramid," http://www.mypyr amid.gov/. Accessed 6 October 2009.

Figure 6.2   Based on F. L. Bryan, *Foodborne Illnesses and Their Control* (Atlanta, GA: Centers for Disease Control and Prevention and Control, 1980), figure 4-1, p. 2.

Figure 7.1   G. A. Condran and R. A. Cheney, "Mortality Trends in Philadelphia: Age- and Cause-Specific Death Rates, 1870–1930," *Demography* 19, no. 1, 97–123.

Figure 7.2   Council on Environmental Quality, *Environmental Quality 23rd Annual Report* (Washington, DC: Council on Environmental Quality, Executive Office of the President, 1993), unnumbered figure, p. 226.

Figure 7.3   USGS, "Where is Earth's Water Located?" Science for a Changing World website (Washington, DC: U.S. Geological Service. U.S. Department of the Interior, 2009), figure titled "One estimate of global water distribution," http://ga.water.usgs.gov:80/edu/earthwherewater.html. Accessed 23 December 2009.

Figure 7.4   Roger M. Waller, *Ground Water and the Rural Home Owner* (Washington, DC: U.S. Geological Survey, 1989), unnumbered figure, p. 13.

Figure 7.5   Based on EPA, *Water Sense: Indoor Water Use in the United States* (Washington, DC: EPA, 2009), http://www.epa.gov/watersense/pubs/indoor.htm. Accessed 30 July 2009.

Figure 7.7   American Association of Vocational Instructional Materials, *Planning for an Individual Water System* (Athens, GA: American Association of Vocational Instructional Materials, 1973), figure 48, p. 58.

Figure 8.4   Based on CEQ, *23rd Annual Report of the Council on Environmental Quality* (Washington, DC: Executive Office of the President, 1993).

Figure 9.1   Washington State Department of Ecology, *Solid Waste Landfill Design Manual* (Olympia, WA: Washington State Department of Ecology, 1993).

Figure 9.2   Based on EPA, *Report on the Environment,* "Quantity of Municipal Solid Waste Generated and Managed," exhibit 4-12, http://cfpub.epa.gov/eroe/index.cfm ?fuseaction=detail.viewInd&lv=list.listByAlpha&r=216598&subtop=228. Accessed 25 December 2009.

Figure 9.3   Based on material provided by the U.S. Environmental Protection Agency, Washington, DC.

Figure 9.4   Adapted from U.S. Nuclear Regulatory Commission, *Recommendations to the NRC for Review Criteria for Alternative Methods for Low-Level Radioactive Waste Disposal,* NUREG/CR-5041, Washington, DC: U.S. Nuclear Regulatory Commission, 1987), p. 1, figures 2.8.1 and 2.8.2.

Figure 9.5   Based on information in U.S. Nuclear Regulatory Commission, *Information Digest: 2001 edition,* Report NUREG-1350 13 (Washington, DC: USNRC, 2001), figure 36, p. 76; and U.S. Nuclear Regulatory Commission, *Digest 2009–2010,* Report 1350 21 (Washington, DC: USNRC, 2010), p. 75.

Figure 9.6   Adapted from U.S. Nuclear Regulatory Commission, *Information Digest 2009–2010,* Report NUREG-1350 21(Washington, DC: USNRC, 2009), figure 43, p. 82.

Figure 10.1   *Publications for Rat Control and Prevention Programs, Public Health Reports* 80, no. 1 (January 1970), p. 40.

Figure 10.2   *The Life Cycle of the Mosquito, EnchantedLearning.com,* http://www.enchantedlearning.com/subjects/insects/mosquito/lifecycle_bw.GIF. Accessed 24 April 2010.

Figure 10.3   *House Fly Life Cycle* (Raleigh, NC: State of North Carolina Department of Environmental and Natural Resources Environmental Health Public Health, 2006).

Figure 11.1   Based on information in National Safety Council, *Injury Facts, 2009* (Washington, DC: NSC, 2009).

Figure 11.2   Based on information in National Safety Council, *Injury Facts, 2009* (Washington, DC: NSC, 2009).

Figure 11.3   Based on Insurance Institute for Highway Safety, *Twenty Years of Accomplishment by the Insurance Institute for Highway Safety* 2 (Arlington, VA: IIHS, 1989).

Figure 11.4   Prepared by author based on Insurance Institute for Highway Safety, *IIHS Status Report 28,* no. 9 (24 July 1993), p. 3.

Figure 11.5   Based on National Safety Council, *Injury Facts* (Itasca, IL: NSC, 2009).

Figure 11.6   Based on Courtney Humphries, "Injury Control: Child Firearms Deaths Tied to Gun Availability," *Focus: News of the Harvard Medical, Dental, and Public Health Schools,* Boston, MA (8 March), 1, 3, and 5

Figure 11.7   Based on National Safety Council, *Injury Facts* (Itasca, IL: NSC, 2009).

Figure 12.3   National Council on Radiation Protection and Measurements, *Screening of Humans for Security Purposes Using Ionizing Radiation Scanning Systems,* NCRP Commentary no. 16 (Bethesda, MD: NCRP, 2003), fig. 5.1, p. 18.

Figure 12.4   Based on data in Donald T. Oakley, *Natural Radiation Exposure in the United States,* Report no. ORP/SID-72-1, U.S. Environmental Protection Agency (Washington, DC: U.S. Government Printing Office, 1972).

Figure 12.5   Based on Fred A. Mettler, "Medical Radiation Exposure: The Largest (and Growing) Radiation Source in the U.S.," Herbert M. Parker Lecture, Herbert M. Parker Foundation, Richland, WA (7 October 2008).

Figure 12.6   U.S. Nuclear Regulatory Commission, *Information Digest: 2009–2010,* Report NUREG-1350, vol. 21 (Washington, DC: USNRC, 2009), figure 18, p. 33.

Figure 12.7   Based on significant modifications of figure 1.1 of National Council on Radiation Protection and Measurements, *NCRP Report 160* (Bethesda, MD: NCRP, 2009), page 11; and USNRC, *2000–2010. Information Digest,* Report NUREG-1350, vol. 21, 31 (Washington, DC: U.S. Nuclear Regulatory Commission, 2010), unnumbered figure, page 149.

Figure 13.1   Based on Robley D. Evans, "Radium in Man," *Health Physics* 27, no. 5 (1974), 497–510, figure 3, p. 504.

Figure 13.3   Based on data in D. C. Kocher, "Review of Radiation Protection and Environmental Radiation Standards for the Public," *Nuclear Safety* 29, no. 4 (October–December 1988), 463–475.

Figure 14.1   Dade W. Moeller, Jack M.Selby, David A. Waite, and John P. Corley, "Environmental Surveillance for Nuclear Facilities," *Nuclear Safety* 19, no. 1 (January–February), 66–70.

Figure 14.2   ICRP International Commission on Radiological Protection, *Principles of Monitoring for the Radiation Protection of the Public,* Publication 43, Annals of the ICRP 15, no. 1 (New York: Pergamon Press, 1985), figure 3, p. 16.

Figure 14.3   Based on World Health Organization, "Guidelines on Studies in Environmental Epidemiology," *Environmental Health Criteria* 27, Geneva, Switzerland, figure 3.1, p. 109.

Figure 15.1   U.S. Environmental Protection Agency, *Risk Assessment for Toxic Air Pollutants—A Citizen's Guide* (Research Triangle Park, NC: EPA, 1991), unnumbered figure, p. 8.

Figure 15.2   Richard Wilson and Edmund A,C. Crouch, *Risk Benefit Analysis,* 2nd ed. (Cambridge, MA: Harvard Center for Risk Analysis, 2001), figure 6-1, p. 151.

Figure 16.1   Based on EIA, U.S. Department of Energy—International Outlook—2009, *World Energy Demand and Economic Outlook,* http://www.eia.doe.gov/oiaf/ieo/world.html. Accessed 13 August 2009.

Figure 16.3   Based on the Council on Energy Awareness, Washington, DC.

Figure 16.4   Prepared by author based on U.S. Nuclear Regulatory Commission, *2009–2010 Information Digest,* Report NUREG 1350 21 (Washington, DC: USNRC, 2009), figure 17, p. 26.

Figure 16.5   Based on U.S. Nuclear Regulatory Commission, *2009–2010 Information Digest,* Report NUREG 1350 21 (Washington, DC: USNRC, 2009).

Figure 16.6   Based on Dan Charles, *Leaping the Energy Gap. Science* 325, no. 5942 (14 August 2009), "Energy-pie," p. 806.

Figure 17.1   Based on Dade W. Moeller, "A Review of Countermeasures for Radionuclide Releases," *Radiation Protection Management* 3, no. 5 (October 1986), figure 1, 72.

Figure 17.2   Based on an update of a figure published in USAEC, *Radioactive Iodine in the Problem of Radiation Safety* (translated from a USSR report). The update was based on information in NCRP, *Management of Terrorist Events Involving Radioactive Materials,* Report no. 138 (Bethesda, MD: National Council on Radiation Protection and Measurements, 2001), section 4.4.4, p. 48.

Figure 17.3   Based on Federal Radiation Council, *Background Material for the Development of Radiation Protection Standards: Protective Action Guides for Strontium-89, Strontium-90, and Cesium-137,* Report no. 7 (Washington, DC: FRC, 1965), figure 2, p. 15.

Figure 18.1   Based on *Summary of Biodiversity,* http://www.qc.ec.gc.ca/faune/biodiv/en/images/combiod.gif. Accessed 14 February 2010.

Figure 18.3   Based on National Snow and Ice Data Center, *Vanishing Ice* (Washington, DC: National Aeronautics and Space Administration, 2010), http://earthobservatory.nasa.gov/Features/vanishing/. Accessed 24 February 2010.

Figure 18.4   Based on National Aeronautics and Space Administration, *Average Global Temperature* (Washington, DC: NASA, Goddard Institute for Space Studies, 2009).

INDEX

accident prevention. *See* injury/injury prevention

Accredited Standards Developer organizations, 317

acetone, 75

acidic deposition, 423–424

acid rain, 335, 358

activated carbon, 154

activated sludge process, 180

acute toxicity studies, 29–32

aflatoxins, 125

Agency for Toxic Substances and Disease Registry (ATSDR), 37, 228, 347–348

Agreement States, 195

agriculture: antibiotic use in, 116; irradiation in, 130–131; water consumption in, 145–146, 174; pollution from, 165; liquid waste from, 173, 183; farmland ecosystem in, 183, 430–431; solid waste from, 195; insects' impact on, 228; farm-related injuries in, 262; contamination of food in, 288; no-till farming in, 431. *See also* pesticides

AIDS (acquired immune deficiency syndrome), 9, 35, 220

air, 87–112; monitoring and assessment of, 19–21, 50, 77–79, 317–320, 330–331, 335–337; indoor quality of, 50, 68, 102–108, 323, 326; epidemiological studies of, 56–57; government regulation of, 57, 87, 94–99, 108–110; workplace quality of, 68–70, 76–78, 79–82; health effects of pollution in, 88–91; outdoor quality of, 91–97; management of quality of, 92–96, 101–102, 105–110; visibility in, 101; acidic deposition in, 423–424

air bags, 18, 249

airborne particles, 8, 99, 110; health effects of, 304; monitoring of, 320, 322, 330–331; from agriculture, 430–431

air cleaning, 79–80, 92–94

air quality index (AQI), 102

air sampling, 77–78

air travel, 292

alcohol use, 4, 14–15, 17; motor vehicle operation and, 250–251, 264; injuries associated with, 259

allergic reactions, 26

allowance trading, 106–108

alpha radiation, 269

alveoli. *See* lungs/respiratory system

ambient environment, 3–4; pollution in, 4–5, 421; assessment of, 19–21. *See also* air; water

American Conference of Governmental Industrial Hygiene (ACGIH), 295, 304, 313

American National Standards Institute (ANSI), 315–317

American Public Health Association, 330

American Water Works Association, 330

Ames test, 33–34

amphibians, 425–426

anaerobic decomposition, 176